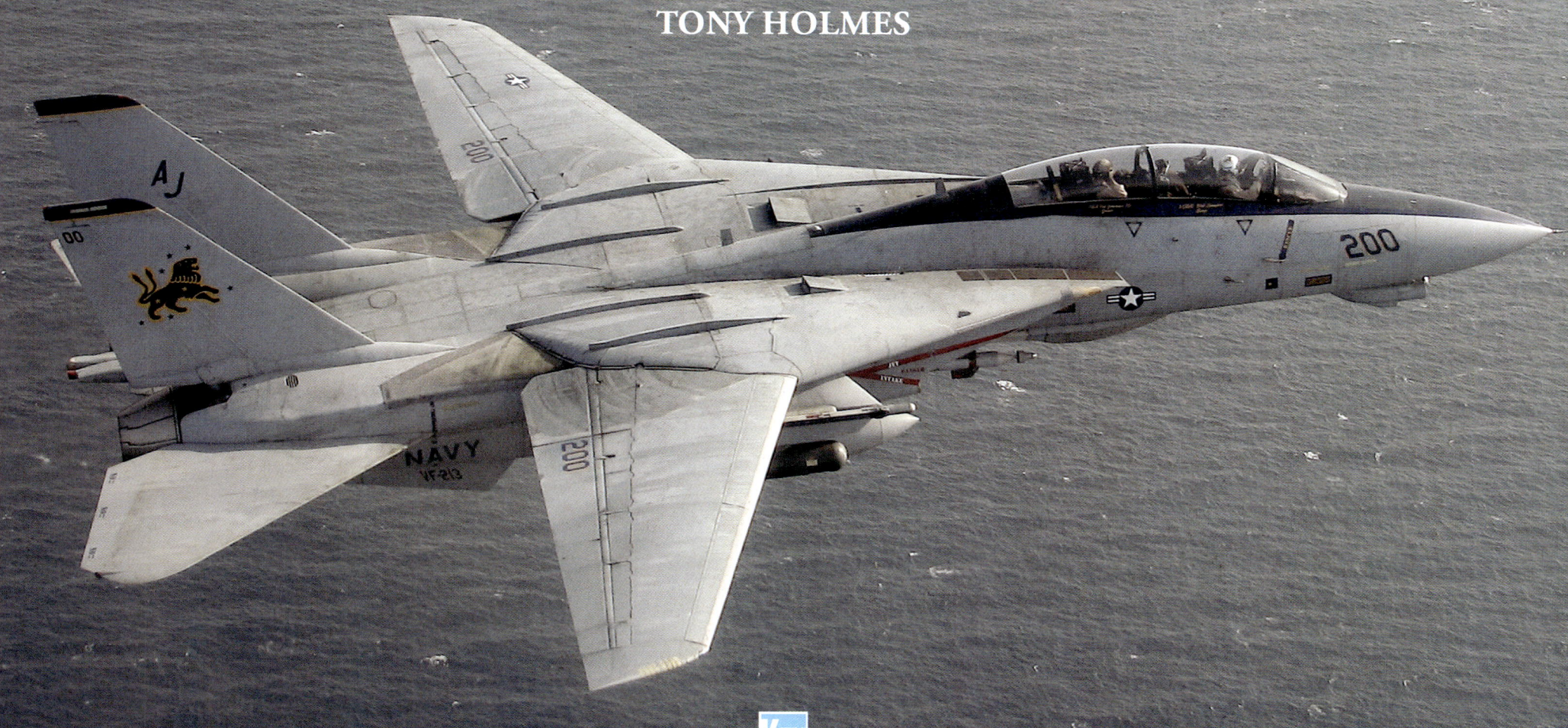

F-14 "BOMBCAT"
THE US NAVY'S ULTIMATE PRECISION BOMBER

TONY HOLMES

Front cover image: Capt Bill Sizemore, CAG of CVW-8, holds formation in F-14D "Black Lion 207" (BuNo 161166) behind a C-2A Greyhound during an OIF IV mission on February 2, 2006. (Richard Cooper)

Title page image: Carrying a GBU-12 LGB (just visible) and a GBU-38, VF-213's "Black Lion 200" (BuNo 164347) flies over the NAG in search of *Theodore Roosevelt* after completing an OIF IV mission in October 2005. This aircraft accrued 163.7 combat flying hours during operations between October 6, 2005 and February 7, 2006 – the least amount tallied by any of the 11 Tomcats embarked in *Theodore Roosevelt* by VF-213 for the unit's final operational deployment with the F-14. (Lt(jg) Scott Timmester)

Back cover image: Silhouetted against the setting sun, an F-14D from VF-31 patrols over Iraq, south of Baghdad, in November 2005 during an OIF IV mission. This photograph was taken by a RIO in a second Tomcat. (Lt(jg) Scott Timmester)

DEDICATION

I would like to dedicate this volume to Capts John "Bag" Hefti, Alex "Yogi" Hnarakis and Dale "Snort" Snodgrass, Cdr Neil "Waylon" Jennings and CWO5 Michael R Lavoie, all of whom, sadly, have passed away since the first edition of *F-14 BOMBCAT* was published in 2015.

Published by Key Books
An imprint of Key Publishing Ltd
PO Box 100
Stamford
Lincs PE9 1XQ

www.keypublishing.com

The right of Tony Holmes to be identified as the author of this book has been asserted in accordance with the Copyright, Designs and Patents Act 1988 Sections 77 and 78.

Copyright © Tony Holmes, 2025

ISBN 978 1 80282 560 2

All rights reserved. Reproduction in whole or in part in any form whatsoever or by any means is strictly prohibited without the prior permission of the Publisher.

Typeset by SJmagic DESIGN SERVICES, India.

CONTENTS

Acknowledgements 4
Foreword 5
Introduction 7

Chapter 1   "Bombcat" Pioneers 9
Chapter 2   "Bombcat" Blooded Over Bosnia 34
Chapter 3   LANTIRN Revolution 48
Chapter 4   LANTIRN Demo and Operational Debut 62
Chapter 5   Bombcat CO 77
Chapter 6   Air-to-Ground Training 102
Chapter 7   Operation *Southern Watch* 122
Chapter 8   Balkan Bombers 185
Chapter 9   Operation *Enduring Freedom* 203
Chapter 10 Strafing 244
Chapter 11 Operation *Iraqi Freedom I–III* 282
Chapter 12 Last Catfight 352

Glossary 381
Bibliography 384

ACKNOWLEDGEMENTS

Since 1987, I have written a handful of books and a considerable number of magazine articles on the F-14 Tomcat, the squadrons that have been equipped with the aircraft and the Naval Aviators and Naval Flight Officers (NFOs) that have flown it. I grew up with the jet, as US Navy carrier battle groups would routinely visit my hometown of Fremantle, Western Australia, for spells of "R&R" during their Indian Ocean and Pacific Ocean cruises in the 1980s.

In the late 1990s and early 2000s, after moving to England to further my career in aviation publishing, I was fortunate enough to visit a number of carriers operating off the coast of southern California and in the Northern Arabian Gulf in order to cover the aircraft's exploits as a bomber following its embracing of the strike fighter role. Several visits to Naval Air Station (NAS) Lemoore, NAS Fallon and NAS Oceana post Operations *Enduring Freedom* and *Iraqi Freedom* also gave me the opportunity to chronicle the jet's use as a "Bombcat" in both conflicts. I am, therefore, very familiar with the Tomcat story.

As always, when producing a volume such as this one, I have called upon my regular contacts for information and photographs to help illustrate the colorful career of a Naval Aviation icon. Such a request usually opens the floodgates, and once again it has been a job to pare down my selection from the hundreds of images supplied to the 350 seen in this revised edition of *F-14 BOMBCAT*. I would therefore like to thank Torsten Anft, Ian Black, Iwan Bögels, David F Brown, Ted Carlson,

The cover of the original *F-14 BOMBCAT* bookazine, published in June 2015, featured Richard Cooper's stunning shot of VF-213's F-14D 'Black Lion 207' (BuNo 161166) over the NAG in February 2006.

Marc Chiabaud, Joe Ciliberti, Cdr Pete Clayton, US Navy (Ret.), Richard Collens, Richard Cooper, Danny Coremans, Steve Czerviski, Neil Dunridge, Peter R Foster, Sergio Gava, Fiore Grossi, Michael Grove, Mark Hasara, Takashi Hashimoto, Erik Hildebrandt, Mike Kopack Snr, Gert Kromhout, Erik Lenten, Don Linn, Rick Llinares, Alfredo Maglione, Cdr Peter Mersky, US Navy (Ret.), Lt Cdr Rick Morgan, US Navy (Ret.), Mark Munzel, Troy Quigley, Angelo Romano, Dave Roof, Cdr Doug Siegfried, US Navy (Ret.), at the Tailhook Association, Erik Sleutelberg, Dan Stijovich, Warren Thompson, Simon Watson, Jinno Yukihisa and Martin Zijlstra for their efforts on my behalf. Patch guru Paul Newman also came up trumps, just as he has done on so many previous occasions.

I have had invaluable assistance with the text from the following individuals, all of whom have vast knowledge of the Tomcat from an operational standpoint – Admiral John Aquilino, US Navy (Ret.), Cdr Dave Baranek, US Navy (Ret.), Capt Dan Cave, US Navy (Ret.), Capt Pat Cleary, US Navy (Ret.), Capt Will Cooney, US Navy (Ret.), Cdr Doug Denneny, US Navy (Ret.), Capt Paul Filardi, US Navy (Ret.), Cdr James Flatley IV, US Navy (Ret.), Capt Brian Gawne, US Navy (Ret.), Capt Justin Halligan, US Navy, the late Capt Alex Hnarakis, US Navy (Ret.), Cdr Ken Hockycko, US Navy, the late Cdr Neil Jennings, US Navy (Ret.), Lt Cdr Mikal Kissick, US Navy (Ret.), Lt Cdr Van Kizer, US Navy (Ret.), Cdr Matthew Koop, US Navy, Lt Cdr Rick Morgan, US Navy (Ret.), Lt Cdr Jim Muse, US Navy (Ret.), Lt Cdr Dave Parsons, US Navy (Ret.), Cdr Michael Peterson, US Navy (Ret.), Lt Cdr Matt Pothier, US Navy (Ret.), Capt Rob Reed, US Navy (Ret.), Rear Admiral John Saccomando, US Navy, Cdr Larry Sidbury, US Navy (Ret.), Rear Admiral William Sizemore, US Navy (Ret.), Cdr Rob Soderholm, US Navy (Ret.), Capt Randy Stearns, US Navy (Ret.), Cdr Scott Timmester, US Navy (Ret.), Lt Tony Toma, US Navy (Ret.), Cdr Tom Twomey, US Navy (Ret.) and Admiral James Winnefeld, US Navy (Ret.).

Thank you also to Brad Elward and Robert K Wilcox for allowing me to quote text from their books, *TOPGUN – The Legacy* and *Black Aces High*, respectively.

I must also mention Cdr Jim "Puck" Howe, US Navy (Ret.), who has graciously written the foreword to this new edition of *F-14 BOMBCAT*. His longstanding connection with the aircraft lasted through to its retirement from US Navy service in September 2006, "Puck" serving as the final CO of VF-31.

Finally, I would like to thank Glenn Sands, formerly editor of *Airforces Monthly*, for commissioning the original *F-14 BOMBCAT* bookazine in 2015, and leading Naval Aviation historian Mike Crutch for his rigorous fact checking and proofreading skills. Gentlemen, I owe you both many beers.

FOREWORD

By Cdr Jim "Puck" Howe, US Navy (Ret.)

The F-14 Tomcat embodied a sense of tough love for all those associated with this magnificent aircraft. Whether building it, fixing it or flying it, nothing about the F-14 came easy. Yet, its challenges were uniquely rewarding, resonating with a precision that only exacting pursuits can achieve. From my initial encounter, cycling the controls as a student in VF-101, to my final flight in the last US Navy Tomcat to soar the skies, the grip of its allure never wavered. Moments of reflection on my time with the F-14 consistently evoke a smile, though admittedly there were a couple of less joyful memories, notably those TF30 engine stalls.

One defining attribute I hold dear about the Tomcat and its community was its "flexibility." During my early days in VF-101, the F-14 was only beginning to explore its air-to-ground capability, a facet present since its inception. Looking back, our lack of understanding and somewhat cavalier approach to this transformation was akin to entrusting a loaded gun to a child. Back then, the notion of the F-14 evolving into a "strike fighter" was met with mixed feelings among us purebred fighter pilots – aloof, arrogant and unapologetic. Yet, our leadership emphasized that a single-role aircraft wouldn't fare well in a budget-conscious US Navy. To secure the Tomcat's future, mastering bomb-dropping became imperative, ushering in the era of "flexibility."

In a mere five years, the F-14 transitioned from guarding against improbable Cold War scenarios to becoming

A veteran of fleet tours with VF-32 and VF-2, as well as a WTI and SFTI Instructor with SWATSLANT and VF-101, Cdr Jim "Puck" Howe was the US Navy's final Tomcat squadron CO in 2006 with VF-31. (Neil Dunridge)

the LANTIRN-equipped, self-escorting strike fighter of choice. With its two-seat cockpit, substantial payload and superior speed, range and recovery capabilities compared to the F/A-18, the F-14 transformed into the formidable "Bombcat." Much to the dismay of our Hornet counterparts, Tomcat squadrons were entrusted with the most challenging missions.

By the time Operation *Iraqi Freedom* [OIF] unfolded, the F-14 held a central role in carrier-borne strike packages. Equipped with the Joint Tactical Information Distribution System, Infrared Search and Track, Joint Direct Attack Munition [JDAM] and an exceptional Forward-Looking Infrared targeting system, the Tomcat, augmented with an extra set of eyes and a Phoenix missile, became virtually unstoppable in the strike role. Throughout the 30-day period of OIF in the spring of 2003, the Tomcat's prowess in strike capabilities reached its zenith.

Shortly before OIF, the F-14D was authorized to carry JDAM. In a testament to the Tomcat community's unparalleled efficiency, software engineers from Naval Air Weapons Station [NAWS] China Lake and Weapons School instructors from Naval Air Station [NAS] Oceana swiftly loaded the latest Operational Flight Program into all frontline F-14Ds. I vividly recall a late-night mission briefing with Lt Mark "Fun" Mhley, where he distilled the complexities of JDAM usage for me after I had told him, "Brother, I need the *Reader's Digest* version. Remember, I'm attention deficit, it's dark, and I have DCAG in my trunk. Make it fighter pilot-proof!" And, indeed, he did. Baghdad International Airport witnessed the collapse of three hangars as my JDAM hit dead-center. For that transformative 30-day period, Tomcat bombs reigned over fortified Iraqi positions, showcasing the F-14 at its most brilliant, and affirming that, in its distinguished history, we had truly saved the best for last.

Author's Note

Cdr Jim "Puck" Howe was one of the first Tomcat pilots selected to attend the US Marine Corps' Marine Aviation and Weapons Tactics Squadron One Weapons and Tactics Instructor (WTI) course. It is a six-week-long postgraduate level air-to-ground syllabus held at Marine Corps Air Station (MCAS) Yuma, Arizona, and for US Marine Corps pilots participating in that course, being a TOPGUN graduate was a prerequisite – Howe went through WTI first, then TOPGUN. The WTI course was where the US Navy got the idea for the Strike Fighter Tactics Instructor (SFTI) program, which it instigated in an effort to create more consistency among Naval Aviators and NFOs in all fleet squadrons.

"Puck" was a "strong advocate of the air-to-ground mission" for the Tomcat community, being one of the first F-14 Forward Air Control (Airborne) (FAC(A)) pilots. He was also an early convert to Night Vision Goggles, subsequently instructing fighter crews in their employment. The last Commanding Officer (CO) of VF-31, he was by then one of only three Naval Aviators qualified to create both FAC(A)s and FAC(A) instructors – a qualification he maintained while leading his squadron, despite onerous currency requirements. "Puck" logged 2,800 flying hours in the F-14 (he flew all three variants) and made 710 traps in the jet.

INTRODUCTION

Few aircraft of the Cold War era have attained such cult status and adoration among those that flew it, wanted to fly it or kept it flying as the mighty Grumman F-14 Tomcat. As big, bold and brash as the *Top Gun* film in which it played the starring role in 1986, the US Navy's ultimate fleet fighter epitomized what Naval Aviation was all about throughout its three decades of fleet service.

The Tomcat was the final creation of the Grumman Aircraft Corporation, which had been churning out fighters for the US Navy from its Bethpage plant in New York State since the 1930s. Wildcats, Hellcats, Tigercats, Panthers, Cougars and Tigers all graced the flightdecks of aircraft carriers from World War Two through to the Cold War. Like the various cats that these machines were named after, Grumman aircraft had a reputation for being tenacious fighters that proved their durability in combat. As the ultimate product of the Grumman "Iron Works," the F-14 would inherit all the qualities of its feline forebears. Like them, it would marry good performance with immense strength – a necessity if the aircraft was to withstand the violence of operations from a pitching carrier deck at sea. The Tomcat also possessed handling qualities that made the aircraft more than a match for many of its opponents.

The F-14 was a big fighter in every sense of the word, tipping the scales at 74,350lb (33,724kg) when fully loaded – Grumman's first fleet fighter, the biplane FF-1 of 1933, weighed in at just 4,828lb (2,190kg). Of the F-14's Cold War contemporaries, only the MiG-25 "Foxbat" was heavier, and pilots flying the Soviet interceptor had the luxury of operating the jet from vast

VF-2's final CAG aircraft, F-14D BuNo 163894, complete with OIF mission markings, flies in close formation with VFA-2's future 'Bullet 101', F/A-18F BuNo 165917, on May 31, 2003. The latter aircraft, flown by a VFA-122 crew from NAS Lemoore, rendezvoused over California specifically for this photo opportunity as VF-2 flew from *Constellation* to NAS Oceana after the unit's final deployment with the F-14. (US Navy)

runways in the USSR. When it comes to the unique environment of "blue water ops," the Tomcat remains the largest fighter to have been sent to sea.

The size of the F-14 meant that it could be adapted to perform other roles aside from that for which it was originally built – fleet defense against missile-equipped Soviet bombers. With the thawing of the Cold War in the late 1980s, the Tomcat community was faced with a stark choice. Turn the aircraft into a multi-role strike fighter platform such as its great rival, the F/A-18 Hornet, or stick steadfastly to the fighter mission and face almost certain extinction. The reality of this situation was brought home to Naval Aviators and NFOs flying the F-14 during Operation *Desert Storm* in 1991. While supporting the campaign to liberate Kuwait from Iraqi occupation, some 99 Tomcats embarked in five aircraft carriers logged more than 4,000 sorties as they performed combat air patrols, fighter escort missions for strike aircraft and aerial reconnaissance. However, the jet was usurped in its primary mission of air-to-air combat by the US Air Force's F-15C Eagle, which claimed 34 aerial victories to the F-14's one.

Lessons were quickly learned from *Desert Storm* to the extent that just weeks after the conflict had ended, Grumman's Field Service Department journal *Tomcat News* proclaimed, "We should take heed of the writing on the wall and continue to press forward into the world of mud-moving!" As a Naval Aviator or NFO in the fighter community at NAS Miramar or NAS Oceana in the 1970s and 1980s, expressing such sentiments would have been viewed as sacrilegious. Indeed, highly respected British aviation journalist Lindsay Peacock noted just that in his 1986 *Osprey Combat Aircraft Series* volume on the Tomcat:

> Although originally developed with air-to-ground applications in mind, the F-14 has never been seriously viewed as a "mud-mover," and any aspirations it has in this direction are extremely modest to say the least, weaponry which could be employed being confined to conventional "iron bombs." Nevertheless, it does possess the ability to take a respectable payload – 14,500lb (6,577kg) to be precise – and it is not beyond the realms of possibility that it could be pressed into use as a strike/close support fighter in a permissive environment. Since such situations are likely to be few and far between, there seems to be little risk of Tomcat fighter jockeys being asked to compromise their much vaunted superiority by engaging in such mundane activities as merely dropping bombs or strafing ground targets.

Yet, by 1992, the only way the Tomcat was going to keep its place on a carrier flightdeck was if it could drop bombs. Thanks to the aircraft's awesome load-carrying capacity, legendary long range and the advent of a bolt-on targeting sensor pod for precision bombing, the Tomcat evolved into the "Bombcat," and it went on to play a pivotal role in the US military's "War on Terror" from 2001 through to its retirement in 2006.

By then, the F-14 was far from being in the flush of youth. Nevertheless, its contribution in Operations *Enduring Freedom* (2001–02) and *Iraqi Freedom I–IV* (2003–06) meant that the aircraft approached phasing out with its claws well and truly bared. Such was the feeling that the jet evoked among those that flew it, Cdr Curt Seth, executive officer (XO) of the US Navy's last fighter squadron equipped with the Tomcat, made the following comment when asked how the F-14 compared with his new, considerably more modern, F/A-18E Super Hornet on the eve of his unit's transition to the latter type in the fall of 2006: "Although I am sure that I will love flying the Super Hornet, there was simply nothing bad about the Tomcat from a pilot's perspective. The F-14 was, and still is, simply a great aircraft."

Enough said!

Former A-6 Intruder pilot Cdr Curt Seth was XO of VF-31 when it transitioned from the Tomcat to the Super Hornet in the fall of 2006. (US Navy)

CHAPTER 1

"BOMBCAT" PIONEERS

One of the iconic Cold War fighters, the F-14 Tomcat came into its own as a precision strike aircraft in the twilight of its career when threatened with premature retirement.

The post-*Desert Storm* years were bleak ones for the US Navy's fighter community, with swingeing budget cuts seeing ten deployable Tomcat units and one fleet replacement squadron decommissioned due to the aircraft's astronomical maintenance costs and single-mission capability. However, just when it looked like the F-14's ocean-going days were numbered, a reprieve came thanks to the accelerated demise of another Grumman "Ironworks" product. The all-weather, long-range, A-6 Intruder bomber was chopped from service, again due to high maintenance costs and the supposed evaporation of its mission in the post-Cold War world.

With the Intruder on the verge of retirement, and the Tomcat seemingly following in its footsteps, the US Navy now found itself facing a shortage of tactical carrier aircraft to fulfill its global "policing" mission. When the original VFX specification that led to the F-14 was released in June 1968, it included an important secondary close air support (CAS) role, with a payload of up to 14,500lb (6,577kg). Grumman's publicity material for the winning design (303E) included artwork of the "Tomcat-to-be" toting heavy loads of air-to-ground ordnance. Clearly, the jet had the capability to drop bombs, although this mission requirement was ultimately abandoned by the US Navy. Nevertheless, Grumman, unlike McDonnell Douglas, which had directed its design team to keep the "rival" F-15 Eagle as light as possible in the company's quest to build the world's best fighter, accepted the need for air-to-ground capability and devoted time and resources to ensuring that the F-14 could, if required, operate in the fighter-bomber role.

F-14A BuNo 157990 was Full Scale Development (FSD) aircraft No 11, and as such, it was assigned to NAS Point Mugu in the early 1970s. Here, it conducted trials with a variety of ordnance, including this mixed mission load-out that saw the aircraft laden down with 14 Mk 82 500lb (226kg) dummy bombs fitted with Mk 15 Snakeye tail retarders, two AIM-9 Sidewinders and two AIM-7 Sparrows, plus external tanks. Stricken on December 31, 1985, this jet has been on display in the March Field Museum for almost 40 years. (*US Navy*)

Despite the US Navy having little to no interest in hanging dumb bombs beneath its then-new fleet interceptor, Air Test and Evaluation Squadron (VX) 5, whose job it was to develop and evaluate aircraft tactics and techniques for the delivery of airborne weapons, undertook an appraisal of the Tomcat as a bomber. According to Lt Cdr Lloyd Abel, who flew the F-14 in the memorable tower flyby scene in *TOP GUN*:

VX-5 [at NAS China Lake, California] wrote an extensive report on the F-14's ability to drop bombs in early 1975 – I was a Radar Intercept Officer [RIO] instructor at VF-124 then. The main problem it found with the aircraft in the fighter-bomber role centered on issues of over- and under-voltage from the Tomcat's generators due to the inability to regulate them (which was later rectified through the installation of a voltage monitor unit on the master test control panel), thereby affecting electrical fusing in carried conventional bombs – Mk 82s, for example. Another problem was bomb-to-bomb collisions in the "tunnel" [between the engine nacelles] because of curious aerodynamics generated above 275 knots [509km/h] in multiple bomb release. The same issue affected ground target accuracy for single bomb release. It would have been pretty risky to drop mechanically fused bombs conventionally in a hostile environment at 275 knots.

I am not sure where the assets came from to conduct these tests – probably from VX-4 or the Pacific Missile Test Center [PMTC], maybe even Grumman, but not VF-124. All assets that came to VF-124 from Grumman were dedicated to standing up squadrons.

The Mk 82 dummy bombs, attached to modified Phoenix missile pallets, were a tight fit in the "tunnel" between the engine nacelles. This photograph was one of a series of shots taken of BuNo 157990 at Point Mugu in 1973 when Grumman was keen to publicize the F-14's multi-mission capabilities. (US Navy)

Bombs were in fact fitted to at least one pre-production aircraft primarily in an effort to secure foreign sales. As early as 1973, and largely for publicity purposes, a Tomcat flew with 14 Mk 82 500lb (227kg) bombs attached to modified Phoenix missile pallets, as well as two AIM-7 Sparrow and two AIM-9 Sidewinder missiles on its wing glove pylons. The AN/AWG-15 weapons control panel (which had an air-to-ground attack function) was installed in the cockpit from the start, although this was viewed as little more than a jettison and air-to-air missile launch panel by Tomcat crews for many years.

The bomb separation issues noted by Lt Cdr Abel had first come to light when F-14 prototypes were flight-tested by the Naval Air Test Center (NATC) at NAS Patuxent River, in Maryland. During an air-to-ground evaluation flight, MiG killer Lt Cdr Curt Dosé, who was assigned as a test pilot to the Carrier Suitability Branch of the NATC, released ordnance from the centerline pallets of his F-14 but the bombs did not separate from the jet due to the slipstream that was generated within the "tunnel" between the nacelles on the underside of the aircraft. This left the bombs "flying in formation" beneath the Tomcat, even after he pulled G and slowed down.

Initially, BRU-21/22 racks were used during the bomb-drop trials at the NATC, these effectively being modified multiple/triple ejector racks. They were bolted to the centerline weapons rails, with four of the latter being fitted. Two BRU-21/22s (one forward and one aft)

F-14A BuNo 157984 was FSD No 5, and it also spent time at both Point Mugu and China Lake undertaking weapons trials with VX-5. Photographed on approach to Point Mugu in October 1975, the aircraft has been liberally adorned with calibration markings to assist engineers with analysis of footage shot during the trials' flights. The spotless Tomcat is carrying two Mk 82 500lb (226kg) dummy bombs fitted with Mk 15 Snakeye tail retarders on the forward stores' pallets. (Peter R Foster)

were fitted on each side of a single rail, allowing the latter to carry four Mk 82 bombs. The aft rail could only carry three Mk 82s per rack (two forward and one aft) because of a lack of space between the rack and the tailhook attachment point and adjoining fuel dump pipe. The BRU-21/22 racks failed to attain clearance for operational use due to the restricted release conditions caused by their weak ejection velocity, which, in certain conditions, caused the bombs to "float" in the "tunnel" between the nacelles and strike the bottom of the Tomcat – an alarming phenomenon experienced by Lt Cdr Dosé.

By the time the F-14 was again looked at as a potential strike fighter, these issues had been eradicated (modified BRU-10 racks were initially fitted, followed by purpose-built BRU-32s with AD-703 bomb racks from 1992) and the aircraft had more than a decade of fleet service behind it. Commenting on the early evaluation of the F-14 as a bomber, legendary Naval Aviator Rear Admiral Paul T Gillcrist noted in his 1994 volume *Tomcat! The Grumman F-14 Story*:

> The F-14A had a significant air-to-ground capability, but it was never made credible. The funds set aside to do the required air-to-ground testing were spent on other tests that were considered to be of higher priority. As a consequence, the Tomcat, by default, came to be considered a "single-mission" fighter. Of course, that is an over-simplification of a much more fundamental fact of military aviation; for as long as I can remember, fighter pilots have tended to look down their noses at the crude business of dropping bombs or, for that matter, anything from their airplanes. As a result, in the early 1980s, when the US Navy strike fighter community (Hornets) began to replace the light attack community (Corsair IIs), some far-sighted leaders in the fighter community foresaw the need to demonstrate the air-to-ground capability of the F-14. But too many fighter pilots dragged their feet and the inevitable happened – the F-14D program was cancelled in favor of the Super Hornet.

Bombs were again seen on a Tomcat in the mid 1980s when the F-14B prototype was loaded up with four Mk 83 "1,000 pounders," as well as AIM-7s and AIM-9s, during the F-14D development program. The D-model jet that evolved from the upgraded F-14B would, of course, become closely associated with the air-to-ground mission.

In defense of the fighter community, the US Navy had the A-6 Intruder as its primary bomber for the first two decades of the Tomcat's service, so there really was no requirement

to hang dumb bombs on the fleet's long-range interceptor. As one RIO from the 1980s remembered, "the A-6 crews were professionals at dropping bombs. We would have seriously invaded their turf. And there is no doubt that working air-to-ground detracts from your air-to-air proficiency." Nevertheless, some Naval Aviators and NFOs saw the writing on the wall for the F-14 as a single mission aircraft early on, one such individual being future VF-2 CO Cdr Doug Denneny;

> When I was in VF-14 in 1988, we picked up a new Gunner, CWO2 Chuck Maner, who came to us from NAS Cecil Field [in Florida] and the A-7 light strike community. Gunner and I talked a lot about the AN/AWG-15 panel and what could be done to turn our jets into "Bombcats" (non-precision only). It was a bit of heresy back then to talk about it, but we figured out we needed it to compete with the Hornets that were coming into our air wing someday. I submitted a TAC D&E [Tactical Development and Evaluation] proposal to Fighter Wing One, US Atlantic Fleet [FITWINGONE] to investigate the feasibility of dropping inert Mk 76 practice bombs from the F-14, and it was killed at that level. There was no interest by the powers that be to make us mud movers.

Away from the fleet squadrons, the Tomcat's air-to-ground capabilities were quietly being explored by Naval Air Systems Command (NAVAIR) following recommendations by the highly influential Naval Operational Advisory Group. Impressed by the success of the F/A-18, the latter recommended a shift away from single-role aircraft such as the A-6 and F-14. The group stated in 1987 that the Tomcat should be more flexibly employed, taking greater advantage of the aircraft's latent ground attack capability.

The NAVAIR trials culminated in the dropping of two inert Mk 84 2,000lb (907kg) iron bombs from an F-14A assigned to VX-4 on the NAS Point Mugu range in California on November 10, 1987. Operational Test and Evaluation Force flights followed in 1988, and frontline trials commenced in the summer of 1990, with Miramar-based F-14A+ units VF-24 and VF-211 of CVW-9 leading the way. VF-24's CO at that time was Cdr Timothy Prendergast, who subsequently recalled:

> The first two fleet squadrons to drop bombs were VF-24 and VF-211. True, developmental and operational test and evaluation squadrons had also done some of that before us. It was required,

FSD aircraft No 7, BuNo 157986 was used in the 1970s as the prototype for the proposed F-14B fitted with Pratt & Whitney F401-PW-P400 engines in place of the troublesome TF30-P-412s from the same manufacturer. This upgrade was quickly abandoned due to budget overruns associated with the Tomcat, and the jet was eventually fitted with General Electric F101-DFE turbofans in fall 1981 as part of the more successful Super Tomcat upgrade that led to the F-14A+ and F-14D. As this photograph clearly shows, BuNo 157986 also conducted flight trials loaded with four 2,000lb (907kg) Mk 84 dummy bombs, a pair of inert AGM-88 HARM and a solitary AIM-9L training round. This aircraft survives as an exhibit onboard the former USS *Intrepid* (CVS-11) in New York City. (via David F Brown)

NF-14A BuNo 159455, assigned to the NATC at NAS Patuxent River, undertook "Bombcat" air-to-ground capability trials at the Naval Weapons Center at China Lake between November 14 and December 20, 1990. During that period the aircraft performed 23 flights with a variety of air-to-ground ordnance. Seen here at "Pax River" in September 1990, this former fleet fighter (it served with VF-124, VF-1, VF-24, VF-143 and VF-302) was eventually scrapped at the naval air station in September 1997. (David F Brown)

for example, that every time a major new [mission] tape came out for the AN/AWG-9, the test and evaluation squadrons had to make sure that the F-14's air-to-ground capability hadn't been corrupted.

To quote from the FITRON TWO FOUR message I released that afternoon: "The fighter squadrons of CVW-9 today executed the first bombing mission by fleet squadrons in the history of the F-14 community. A composite division of four F-14A Plus aircraft [from VF-24] attacked the Cactus West target complex in R-2301W with [four] inert Mk 84 LDGP [Low Drag General Purpose]. Weapons systems functioned as advertised and reflected the accuracy previously demonstrated by the test community.

VF-211 followed suit exactly one week later, as this story in the winter 1990 issue of *The Hook* detailed: "The 'Fighting Checkmates' of VF-211 continue to set the pace in the Tomcat community by becoming the first [actually second] fleet F-14s to drop bombs. The historic day was August 15 when 'Checkmates' CDR Don McCort/LT Jim Mathews and LCDR

VF-24 (1990)

Charles Lawson/LT Don McClenney dropped Mk 83 inert 1,000lb bombs at Cactus West target range, 12 miles south-east of Yuma, Arizona."

Lt Jim Mathews, who was RIO for "Checkmate One," Cdr McCort, during the bomb drop mission, made history seven days later, as noted in the cover story of NAS Miramar's *Jet Journal* newspaper on September 14, 1990:

On August 22, 1990, Fighting Checkmate LT Jim "Nixon" Mathews became the first man in history to drop bombs from both an F-14A (Plus) Tomcat and an F-15E Strike Eagle.

LT Mathews then changed services temporarily as he was introduced to the air-to-ground mission Air Force-style. The Fighting Checkmates hosted the 550th Tactical Fighter Training Squadron [TFTS] from Luke Air Force Base from August 18 to 25. Checkmate aircrew were able to acquire some pertinent tactical knowledge about the air-to-mud job from the Strike Eagle. Needless to say, the 550th became very appreciative of the incredible air-to-air capabilities of the Checkmates and their Tomcat team.

All in all, it was a great detachment, and a great day for the now world-famous LT Jim "Mud" Mathews.

Above: The first of two inert Mk 84 bombs falls away from F-14A BuNo 159616 during trials over the Bloodsworth Island Range in Chesapeake Bay in 1990. This aircraft was also assigned to the NATC following more than a decade of fleet service. The diagonal slash and small chevrons beneath the cockpit denote the jet's previous assignment to VF-84. The Tomcat was stricken at Patuxent River in September 1994. (US Navy)

Left: "Armed" with two inert Mk 83 1,000lb (454kg) bombs, VF-211's F-14A+ BuNo 162913 heads for the Cactus West target range 12 miles south-east of Yuma on August 30, 1990 – just 15 days after the unit had dropped its first ordnance from a Tomcat. (Fiore Grossi)

Opposite: The pilot of "Nickel 104" gently breaks away from his section lead to reveal the pair of Mk 83s carried by the jet. VF-211, along with VF-24, were at the forefront of fleet trials with the "Bombcat" during the second half of 1990. (Fiore Grossi)

August 1990 also saw VX-4 publish the first edition of the *F-14 Warfare Tactics Guide* as part of the unit's remit for creating service doctrine for all of the US Navy and US Marine Corps frontline combat aircraft types. This document would form the cornerstone of training for fleet Tomcat units when it came to instructing fighter crews in the air-to-ground mission.

Strike U

VF-24 and VF-211 continued to explore the F-14's capabilities as a bomber when both units attended the Naval Strike Warfare Center (NSWC), better known as Strike U, at NAS Fallon, Nevada, as part of CVW-9's air wing work-ups in preparation for its next operational deployment onboard USS *Nimitz* (CVN-68). According to Cdr Timothy Prendergast, the units' early focus on turning the Tomcat into a bomber was backed by the CO of CVW-9, Capt Howard Petrea: "CAG-9 at the time, Capt 'Rusty' Petrea, deserves a 'shout-out' for his unwavering support of the 24/211 entry into the world of air-to-ground. Two months after the first fleet bombs had been dropped, CVW-9 went to Fallon, and CAG carved out enough weapons from the NCEA [Non-Combat Expenditure Allocation] for the air wing to allow us to fully participate in air wing strikes as 'Bombcat' strikers – another first."

Left: Lt Jim Mathews was one of the VF-211 RIOs involved in the unit's first ever bomb drop on August 15, 1990. (Fiore Grossi)

Below: VF-24 fully embraced the F-14's new-found bombing role when it attended Fallon for CVW-9's air wing work-ups in September–October 1990, with two of its aircraft arriving at the Nevada airfield wearing temporary, water-soluble camouflage. This is F-14A+ BuNo 163409, which also received a bomb tally and *Camel Smoker* nose art during the Fallon det. "Renegade 210" subsequently saw combat with VF-32 during Operation *Desert Fox* in December 1998. (Michael Grove via Mike Crutch)

"BOMBCAT" PIONEERS

The winter 1990 issue of *The Hook* also detailed VF-211's exploits with its Tomcats at Fallon: "This new-found air-to-ground capability, along with new fighter-to-fighter data link, helped the 'Fighting Checkmates' assist CVW-9 in its superb performance in an air wing det [detachment] to Fallon. For the first time in Strike U history, VF-211 'Bombcats' provided daily demonstrations of the multi-role F-14A+ strike fighter's capabilities in air superiority, air-to-ground delivery and tactical reconnaissance."

VF-24 summarized its operations at Fallon in the spring 1991 edition of *The Hook*:

The "Renegades" of Fighter Squadron 24 spent the summer months preparing to introduce the significant and long-awaited strike mission to the war-fighting capability of CVW-9. The squadron arrived at NAS Fallon on September 30 with a complement that included two tactically camouflaged strike-configured "Bombcats." Equipped with modified BRU-10 bomb racks to carry Mk 83 or Mk 84 general purpose [GP] bombs, the "Renegades" Tomcats quickly proved that "speed is life" no matter what the mission, and that the F-14A+ has lots of life left in it.

In the months of preparation for these initial strike missions, LCDR Phil Nelson developed the FRS [Fleet Replacement Squadron] training syllabus and mapped out the ground school courses and training requirements for the new Tomcat community air-to-ground mission. He also co-ordinated a complete revision of the operations capability and planned operational environment statement, which was endorsed by Commander, Fighter and Airborne Early Warning Wing, US Pacific Fleet [COMFITAEWWINGPAC] and forwarded to the Chief of Naval Operations [CNO] via Commander, Naval Air Force's US Pacific Fleet [COMNAVAIRPAC].

Along with all the paperwork, VF-24 aircrew completed extensive training, including Head-Up Display [HUD] symbology, low-altitude training, weapons system interface, target acquisition, target area tactics, weaponeering, strike planning and safety procedures. In the air, the squadron flew the first Tomcat no-drop bomb scoring flights at the Yuma TACTS [Tactical Aircrew Combat Training System] range.

During the Fallon det, "Renegade'" aircrews worked diligently to fully integrate the F-14A+ into all facets of strike planning and greatly expand the air wing's ability and flexibility to put hard-kill

VF-24 (1990)

"Renegade 212" (F-14A+ BuNo 163411) taxies at Fallon in October 1990, this aircraft being the second VF-24 Tomcat to receive a temporary desert-inspired camouflage scheme. Christened *Thief Of Baghdad* and also featuring a bomb tally, the aircraft was lost on March 15, 1993 when it broke up in-flight while serving with VF-101. (Michael Grove via Mike Crutch)

ordnance on a wide variety of targets. Even with 8,000lb of bombs aboard (four Mk 84s), the "Tomcat Pop" maneuver, made possible by the tremendous thrust provided by the twin General Electric F110 engines, allowed the aircraft to safely get into and out of the most heavily defended targets.

While the Miramar units were forging ahead with turning the F-14 into a "Bombcat," on the east coast Oceana-based Tomcat FRS VF-101 became the first FITWINGONE squadron to drop ordnance when an instructor crew in an F-14A+ expended two inert Mk 84s, as detailed in the winter 1990 edition of *The Hook*:

A new era in Oceana history was ushered in on September 12, 1990, when VF-101 CO Cdr C K Crandall and Lt D R Pine dropped a pair of Mk 84 2,000lb inert bombs from an F-14A+ Tomcat. The event was complemented when Capt R W Potter, Commander FITWINGONE, and Sqn Ldr A J Yule, Royal Air Force, pickled a single Mk 84 to a near perfect bullseye on the target range at Dare County, North Carolina.

In addition to filling the role as air-superiority fighter for the fleet, the multi-purpose Tomcat will fulfill an added mission: strike warfare. The "Grim Reapers" of VF-101 have been tasked as model manager for the Tomcat strike fighter program to develop an air-to-ground syllabus for the F-14,

The instructor cadre at VF-101 come together for a group photograph on the ramp at Oceana in 1990, just as the unit began to explore the air-to-ground role in the Tomcat. Skipper Cdr C K Crandall and his RIO Lt D R Pine were the first "Grim Reapers" to drop bombs (albeit two inert Mk 84s) on the Dare County range on September 12, 1990. RAF exchange officers Sqn Ldr Nick Randerson (pilot) and Flt Lt Nick Seward (navigator) can be seen at left in the front row. Fellow exchange pilot Sqn Ldr Alec Yule was involved in the September 12 bomb drop, flying the second F-14A+ to expend a single Mk 84 on that date. (US Navy)

as well as train fleet squadrons and replacement aircrew in strike warfare. VF-33, VF-102, VF-41 and VF-84 have completed the VF-101 course of instruction and FRS aircrew are expected to begin training in fiscal year 1992.

Carrier group commanders have long awaited the F-14 strike program for the additional flexibility it brings to the battle group. Tomcats can now carry larger payloads farther, while continuing to carry an offensive air-to-air load.

On the west coast, COMFITAEWWINGPAC FRS VF-124 established a similar syllabus at the same time, with VF-1 and VF-2 receiving instruction. The latter unit's command history for 1990 stated that it had completed eight days of "air-ground weapons training (BOMBCAT) at NAS Miramar and MCAS Yuma" in November of that year:

> VF-2 was the first sea-going fleet fighter squadron to successfully complete the new air-ground weapons delivery syllabus and deploy with a strike attack ("Bomb Cat") capability to support Operation *Desert Shield* '90. The unit had aggressively completed the CNO-approved Air-Ground Training program just one week prior to deployment. It achieved a 100 percent success rate with air-ground ordnance delivery, with a squadron CEP [Circular Error of Probability] of 75ft after expenditure of 12 [inert] Mk 82 GP bombs [on the Yuma range – this ordnance was the first to be dropped by F-14As assigned to a fleet unit].

However, according to a briefing report given by the Flag Panel at the Tailhook Association's Naval Aviation Symposium in September 1991, "CVW-2's Bombcats were up to speed and qualified for air-surface delivery when the air wing deployed onboard USS *Ranger* (CV-61) in December 1990, but they were unable to contribute to Naval Aviation's tonnage against Saddam owing to improperly manufactured bomb-rack components. A fix is still six months off."

Test and evaluation units continued to clear the F-14 for the carriage of various air-to-ground ordnance during the second half of 1990, with the NF-14As of the NATC being particularly busy. The work carried out by VX-4 and the NATC gained further importance on January 7,

Grumman aircraft dominate the flightdeck of *Ranger* as the carrier departs NAS North Island on December 8, 1990. Although the A-6Es of VA-145 and VA-155 provided the embarked CVW-2 with its air-to-ground firepower during its six-month-long war cruise that saw the vessel heavily involved in Operations *Desert Shield* and *Desert Storm*, VF-1 and VF-2 had completed their "Bombcat" training syllabus just a week prior to this photograph being taken. However, they were unable to contribute to "Naval Aviation's tonnage against Saddam owing to improperly manufactured bomb rack components." (US Navy)

1991, when the much-vaunted McDonnell Douglas/General Dynamics A-12 Avenger II was cancelled. Under development as an A-6 replacement, the stealthy Avenger was over budget and behind schedule when it was terminated. Now, the US Navy needed an all-weather day/night deep-strike interdiction platform to fill the role that would soon be vacated by the retiring Intruder. Many in the fighter community felt that the Tomcat was the jet for the job.

Operation *Desert Storm*

The aircraft's performance in Operation *Desert Storm* in January–February 1991 further underlined the fact that F-14 units urgently needed to broaden their offensive capabilities or face wholesale retirement. The ten Tomcat squadrons involved in the campaign would fly mainly strike escort and Tactical Airborne Reconnaissance Pod System (TARPS) missions during the war. The high hopes that the fighter community had had of adding to the jet's victory haul were stymied by the non-appearance of the Iraqi Air Force (IrAF) in its patrol sectors.

Up until the invasion of Kuwait, when the F-14 units were allocated a single Combat Air Patrol (CAP) station over Iraq, the Tomcat crews had been tasked with performing defensive CAPs for the carrier battle groups in the Red Sea and the Northern Arabian Gulf (NAG). When the jets did venture over enemy territory while escorting strike mission, the IrAF refused to engage them. Some crews felt that the Iraqi pilots chose to flee whenever they picked up emissions from the F-14's AN/AWG-9 radar.

There was also a feeling within the Tomcat community that USAF E-3 Airborne Early Warning and Control System (AWACS) controllers who ran the interceptions of IrAF aircraft favored the allocation of F-15Cs to deal with the enemy threat. "There was a lot of parochialism as to where the F-14 and F-15 fighter CAPs were placed," recalled Cdr Doug Denneny, who was then a lieutenant serving with VF-14. "The Eagles got the kills because it was the USAF's E-3 AWACS that were running the show up north. They would even call Navy guys off and then bring in Eagles for easy pickings. This could just be the ranting and raving of pissed-off Naval Aviators and NFOs, but from what I personally saw in OIF, there was probably a shred of truth in these stories."

In reality, Tomcats had been kept out of the aerial action because the US Navy had failed to develop the necessary systems – primarily up-to-date Identification Friend or Foe (IFF) equipment – and procedures required to integrate carrier air wings as part of a joint air component command. This meant that F-14 crews were unable to solve the strict Rules of Engagement (RoE) that would have allowed them to autonomously engage aerial targets using only their onboard sensors. Instead, they had to rely on controlling platforms such as USAF E-3s to give them their clearance to fire.

With the RoE criteria met, fighters with Beyond Visual Range air-to-air missiles such as the AIM-7 Sparrow and AIM-54 Phoenix could fire their ordnance at long range, safe in the knowledge that no friendly aircraft in the area would be shot down instead. USAF F-15C pilots could solve all the required RoE criteria for identifying enemy aircraft from within their own cockpits, prior to shooting them down. The F-14, conversely, lacked the IFF systems and software to meet all RoE criteria, which left its crew reliant on outside clearance to engage. The job of defeating the IrAF was therefore given to the Eagle pilots, who shot down 35 aircraft.

Armed with AIM-54C, AIM-7M and AIM-9L air-to-air missiles, and not a bomb in sight, two F-14As from VF-1 prepare to launch from *Ranger* in the NAG during *Desert Storm*. US Navy Tomcats would perform just six interceptions during the campaign, resulting in a solitary aerial kill. The latter took the form of a Mil Mi-8 helicopter that was shot down by VF-1 on February 6, 1991. At least three Tomcats also strafed targets. (Cdr Pete Clayton)

Ultimately, the F-14 performed just six intercepts in *Desert Storm*, resulting in a solitary aerial kill. The latter took the form of a Mil Mi-8 helicopter that was shot down with an AIM-9M by a crew from VF-1 on February 6, 1991. This victory went some way to evening up for the score for the Tomcat community, as VF-103 had lost a jet to an SA-2 surface-to-air missile (SAM) on January 21 – its pilot had been rescued and the RIO captured. This aircraft had been flying a TARPS mission, proving that these sorties were usually more action packed than the traditional fighter CAPs flown by the Tomcat in *Desert Storm*.

The importance of tactical reconnaissance grew as the conflict progressed, for the US Navy had quickly found that it could not rely on USAF tactical reconnaissance assets for up-to-date bomb damage assessment photography in the wake of air strikes by carrier-based attack aircraft. TARPS-equipped F-14s also helped in the daily hunt for Iraq's mobile "Scud" ballistic missiles.

Despite maintaining a mission-capable rate of 77 per cent, logging a total of 4,182 sorties and completing 14,248 flight hours (more than all other US Navy fixed-wing aircraft) during the 40-day air war, the F-14 had proven largely ineffective in its primary role as an interceptor. And at this point in the jet's history, being a fighter was still virtually the only role it could perform.

Live Ordnance

An important milestone in the Tomcat's evolution into a strike platform came shortly after *Desert Storm* had ended, when, in March 1991, VF-21 and VF-154 became the first fleet-assigned F-14A units to drop live air-to-ground ordnance during an air wing weapons detachment to Fallon with CVW-14. Two months later, VF-142 and VF-143 earned a similar distinction for the newly re-designated F-14B (which had previously been known as the F-14A+) during CVW-7's Fallon weapons detachment. As an early recipient of the B-model Tomcat, Oceana-based VF-142 had in fact been tasked by Commander, Naval Air Force, US Atlantic Fleet (COMNAVAIRLANT) with commencing development of an air-to-ground strike syllabus while deployed onboard USS *Dwight D. Eisenhower* (CVN-69) in 1990. By mid-cruise in June of that year, Tomcat crews had started simulating such attacks as part of a CVW-7 strike package.

With both VF-101 and VF-124 now having functioning training syllabi for the "air-to-mud" mission, more units checked out aircrew in the "Bombcat" during the course of 1991. One such squadron was CVW-3's VF-32, which completed the VF-101 syllabus in between an air-to-air gunnery exercise and supporting the *Dwight D. Eisenhower* battle group work-ups.

VF-154 (1991)

And it was from the latter vessel during the carrier's 1991–92 deployment with Sixth Fleet that the "Ghostriders" of VF-142 became the first F-14-equipped unit to fly from a carrier with live bombs. "Ambitions for the cruise are many," noted the squadron report in the spring 1992 issue of *The Hook*, "and one of the unit's primary goals is to exploit the F-14's air-to-ground capability while deployed. Led by LCDR Thomas Lansdale, the recent Flight Test Directorate F-14 air-to-ground lead RIO, VF-142 became the first fleet-deployed F-14 squadron to take off with and deliver ordnance at sea [against floating practice targets – earth ball-sized inflated red balloons, nicknamed 'killer tomatoes']."

Back at Miramar, VF-2 was also embracing the new multi-role Tomcat, as this brief report in the winter 1991 issue of *The Hook* noted:

On August 28, the "Bullets" opened a new chapter in the history of the F-14A. For the first time in its nearly 20-year naval career, the Tomcat did the big triple – fired a missile, dropped [Mk 83] bombs and strafed – all in the same flight. VF-2's XO, CDR Brian Flannery/LT(jg) Ken Umekubo, and LTs Jerry Goggin/Ruben Gavieres destroyed their target drone on the PMTC range with AIM-7 missiles. Following their successful firings, both aircraft then moved to the air-to-ground mission and dropped two sand-filled Mk 83 1,000lb [454kg] bombs from each jet on ground targets at San Clemente Island [off the coast of southern California]. To further demonstrate the Tomcat's capabilities, the "Bounty Hunters" then strafed the same targets with 20mm cannon. This display of the F-14's air-to-air and air-to-ground capabilities further verified the Tomcat's nearly limitless potential as a weapons system in adding firepower and flexibility to the Navy's carrier battle group.

By the fall of 1991, the Tomcat Advanced Strike Syllabus (TASS) had been established to replace the FRS bombing course that had initially introduced F-14 aircrew to the air-to-ground mission. CVW-17's VF-74 and VF-103, equipped with F-14Bs, were the first units to complete TASS, shortly after undertaking a more familiar Fleet Fighter Air Combat Maneuvering Readiness Program against A-4s, F-5s and F-16s of VF-43. "The next training evolution, however, was a radical new concept for the F-14 community," noted VF-103's report in the spring 1992 issue of *The Hook*:

In September, the "Sluggers" and sister-squadron VF-74 became the first F-14 squadrons in the Navy to go through TASS in order to introduce the Tomcat and its aircrew to the world of air-to-ground weapons. TASS enabled VF-103 to demonstrate the multi-mission capability of the F-14B while training aircrew in air-to-ground strike tactics.

From the outset during TASS, a total commitment from both aircrew and maintenance personnel enabled the "Sluggers" to put "bombs on target" from the start. Strike Weapons and Tactics School, US Atlantic Fleet [SWATSLANT, which devised and implemented TASS] specifically cited VF-103 for the professional manner in which it conducted operations. The unit proved that the F-14B has the capability to deliver heavy ordnance at long range and high speed [in order] to function as a force multiplier for a carrier battle group. While Phase One of the air-to-ground "Bombcat" program is limited to the Mk 80 series of iron bombs, Navy officials hope to expand the capability to include smart weapons as well. The 1990s promises to be the decade of the "Bombcat."

In October, following TASS, the "Sluggers" put the whole summer's training package together at NAS Fallon under the guidance of the NSWC. For the first time since *Desert Storm*, all CVW-17 assets were combined and employed to perform simulated strikes against a hostile, integrated air defense network. Not only did the "Sluggers" clear the sky of bandits, but for the first time it planned, briefed and led bombing missions against simulated targets.

According to leading Naval Aviation historian Mike Crutch, author of the peerless *CVW – US Navy Carrier Air Wing Aircraft 1975–2015* series of reference works:

With the Tomcat's awakened capability to deliver air-to-ground ordnance now spreading across the fleet, both VF-74 and VF-103 were the first to take things to a higher level in September 1991. TASS, delivered by instructors from SWATSLANT, developed the basic tactics already practiced to widen skills in attack profiles, weapons employment and strike planning to improve effectiveness of the non-"smart" weapons that were carried by Tomcats at the time. TASS was subsequently rolled out across the fleet, with SWATSLANT staff covering both coasts to train squadrons as they reached certain points of their turnaround cycle – this was usually just ahead of the air wing weapons detachment to Fallon, so that classroom teaching could be put into practice as early as possible. TASS was folded into the Advanced Attack Readiness Program (AARP) the following year.

The air-to-ground training undertaken by VF-74 and VF-103 was subsequently put to good use when, on deployment onboard USS *Saratoga* (CV-60) in 1992, CVW-17 became the first East Coast air wing to conduct "Bombcat" operations. The latter included dawn strikes by two four-aircraft divisions (one from each unit) against targets on the Wadi el Natrun range complex in Egypt, 700 miles (1,100km) from *Saratoga*, on September 14–15. Each F-14B dropped two Mk 83 bombs, before landing back at the Greek naval base at Souda Bay in Crete. Here, they re-armed and then attacked the Egyptian range for a second time, before landing back onboard *Saratoga* shortly before dusk. The aircraft had relied on US Navy and USAF aerial refueling to complete the strike missions.

With their wings locked in the 75-degree over-sweep position that was the norm for Tomcats chained down to the carrier flightdeck, F-14As from VF-21 and VF-154 crowd the stern of *Independence* as the carrier sails into Sydney Harbour on May 1, 1992. Immediately prior to this port call, both units had dropped ordnance on bombing ranges on the east coast of Australia. Four months after this photograph was taken, "Lance 206" (BuNo 160692) made history when it became the first bomb-armed Tomcat to participate in an OSW mission. (US Navy)

In October 1991, a month after VF-74 and VF-103 had completed TASS, VF-21 and VF-154 became the first COMFITAEWWINGPAC units to finish the syllabus, albeit while flying from NAS Cubi Point, in the Philippines. Both units had started the year assigned to CVW-14, and in March, as previously noted in this chapter, had been the first operational F-14 squadrons on the west coast to drop live ordnance from the new Tomcat bomb racks that were at last beginning to reach the frontline.

On August 24, VF-21 and VF-154 were transferred to CVW-5, forward deployed at Naval Air Facility Atsugi, in Japan. The veteran carrier USS *Midway* (CV-41) was being replaced by the larger USS *Independence* (CV-62), and this meant that CVW-5 could now include Tomcats within its makeup. Sailing to Japan embarked in "Indy" [*Independence*], the squadrons flew off to Atsugi in September and then headed south to Cubi Point in October for a three-week air wing work-up overseen by NSWC instructors. VF-21 and VF-154 completed the TASS syllabus as part of this training period.

The first units to undertake the TASS course at Miramar were CVW-15's VF-51 and VF-111, with the latter unit detailing its embracing of the strike fighter mission in the summer 1992 issue of *The Hook*:

> Seeking to assert the capability of its F-14A Tomcats as multi-role aircraft, VF-111 established a comprehensive air-to-ground training syllabus. Training lectures co-ordinated with dedicated strike missions that included low-level ingress to the target area gave squadron aircrews exposure to the strike mission. It did not take long for the accuracy of the F-14 as an attack aircraft and its capabilities as a strike platform to be evident.
>
> Prior to the CVW-15 Fallon detachment, VF-111 participated locally in the TASS taught by the SWATSLANT based at NAS Oceana. This was the first opportunity for Miramar-based fighters to receive dedicated air-to-ground training that will prove invaluable as "Sundowners" aircrew become experts in air-to-ground weapons delivery and the strike mission.
>
> The CVW-15 detachment to NAS Fallon was a great success. "Sundowners" flew the F-14 in the strike role and proved that the jet's combination of speed, long-range, accuracy in the target area and aerial lethality is unsurpassed. On April 14, 1992, the "Sundowners" scored a direct hit on target by dropping the squadron's 111th bomb during a strike mission on the Fallon range complex. To date, VF-111 has dropped more than 18 tons of air-to-ground ordnance.

Mid-April 1992 also saw VF-21 and VF-154 embark in *Independence* as part of CVW-5 for the air wing's first operational deployment following its acquisition of the Tomcat squadrons. An early highlight of the cruise was a visit to Sydney, in New South Wales, to mark the 50th anniversary of the Battle of the Coral Sea. As the carrier headed south, VF-21 and VF-154 exercised with the Royal Australian Air Force (RAAF) and took the opportunity to hone their air-to-ground skills on bombing ranges on the east coast of Australia. Among the VF-154 pilots to participate in the deployment was Lt Tim Byrne, who recalled: "We had personnel from VX-4 out on the boat with us for the first month of the cruise, and they oversaw our dropping of the Mk series bombs from different altitudes. We routinely performed high-level (starting at 30,000ft) 45-degree dive-bombing runs. The jets went supersonic almost immediately, and the [gunsight] pipper was buried off the bottom of the HUD – fun times."

Both Tomcat units made history later in the cruise when the *Independence* battle group transited west across the Indian Ocean to the Persian Gulf. In an article titled "Freelancers keep the peace" in the winter 1992 edition of *The Hook*, VF-21 highlighted its achievements whilst assigned to US Central Command (CENTCOM):

VF-154 (1992)

> Less than two years after their stand in the Arabian Gulf as part of Operation *Desert Shield*, the "Freelancers" of Fighter Squadron 21 returned to the area with their F-14As. The CVW-5 team was thrown into a frenzy of activity with President [George H W] Bush's announcement [on August 26] of a no-fly zone over Iraq south of the 32nd parallel. Exactly 24 hours later, at 1715hrs on August 27, the first division of Tomcats arrived over Iraq led by the skipper, Cdr Dan Cain. Operation *Southern Watch* [OSW] had begun. On the same mission that kicked off *Southern Watch*, LCDRs Jody "Jeffro" Richardson and Ted "Slapshot" Carter launched from *Indy* in "Freelancer 206" carrying two Mk 83 sand-filled blue bombs for the inaugural "Strikecat" flight, thereby flexing the muscles of the new F-14A role [in what was the US Navy's first operational "Bombcat" sortie].
>
> Through the efforts of the CVW-5 team, Operation *Southern Watch* was a success. The joint operation with the *Independence* battle group, USAF, British and French forces effectively strangled Iraqi flights south of the 32nd parallel. It was awesome to see *Indy*'s flightdeck configured for a launch of 14 Tomcats.
>
> While the "Freelancers" excelled in the air-superiority mission, they also expanded the Tomcat's strike role. On a subsequent training flight, "Jeffro" Richardson and LT Jay "Spock" Humphlett delivered bombs on a target in a display of "Strikecat" muscle to prove that no other aircraft can carry ordnance farther faster, drop it on target and still slug its way out of enemy airspace like the F-14.

Seven months after VF-21 had patrolled over southern Iraq in its F-14As "armed" with sand-filled blue bombs, CVW-9's VF-24 and VF-211, embarked in *Nimitz*, flew the first Tomcats committed to OSW loaded with live Mk 80 series ordnance. The "Bombcat" revolution was quickly gaining pace.

VF-101's Fallon Debut

With fleet F-14 units embracing the air-to-ground mission on both coasts, training squadrons also ramped up instruction in this area for replacement Naval Aviators and NFOs destined for service in Tomcats. A big step in this direction came in the fall of 1992 when VF-101 completed its first F-14 detachment to Fallon. During the two-week evolution, students dropped 232 Mk 76, BDU-45 and Mk 83 inert bombs. VF-101 also became the first F-14 squadron to expend live Mk 82 GP bombs from redesigned racks, destroying armored vehicle targets in 66 live drops on the Bravo 19 complex. According to a VF-101 report:

> The detachment achieved a 95 percent sortie completion rate in 144 sorties. Many of the events included opposed low levels that culminated in [the] first bomb on target within plus or minus ten seconds. On the det, 18 replacement pilots and RIOs completed strike-phase training in less than two weeks – a feat normally taking four. The average CEP for replacements was 198ft, and 79ft for instructors. Another first was F-14 Tomcat night bombing, with eight of 12 bombs impacting within 100ft of the target.

While VF-101 (and VF-124 at Miramar) were expanding their air-to-ground training, fleet units continued to make strides in the employment of the Tomcat as a bomber. As CVW-3 prepared for its 1992–93 Sixth Fleet cruise embarked in USS *John F. Kennedy* (CV-67), the air wing's VF-14 and VF-32 were well equipped in this respect, having aircraft, weaponry, bomb racks and, most importantly, Naval Aviators and NFOs who had been thoroughly trained to use the jet in the strike role. VF-14's preparation for the deployment was detailed in the squadron's command history for 1992:

> The "Tophatters" began the year based ashore at NAS Oceana undergoing training in preparation for a fall 1992 deployment. Since it first entered fleet service, the F-14 has had the capability to serve as a self-escorted strike fighter. This capability had lain dormant until very recently, and in 1991 dedicated programs such as Low Altitude Training [LAT] and TASS were introduced to hone these specific "mud-moving" skills. VF-14 commenced these programs in January 1992, with two solid months of LAT and TASS, as well as several schools including Joint Munitions Effectiveness Manual [JMEM], Strike Leader Attack Training Syllabus [SLATS] and a user course for AN/SYQ-21 Tactical Automated Mission Planning System [TAMPS].

VF-84's "Victory 202" (F-14A BuNo 160390) was loaded with four inert Mk 83 1,000lb (454kg) GP bombs and two AIM-9L and two AIM-7M drill rounds for a dedicated photo sortie from Oceana on November 1, 1992. This aircraft was transferred to Miramar-based VF-213 in late August 1993, and its pilot, Lt Kara Hultgreen, was killed when the jet crashed into the sea off the coast of San Diego while on approach to *Abraham Lincoln* on October 25, 1994. (US Navy)

Initially, aircrew were trained in the low altitude/strike environment, after which the "Tophatters" participated in TASS – a two-week exercise utilizing various low-level navigation routes in conjunction with bombing ranges along the East Coast to simulate both medium and long-range strikes. The "Tophatters" were able to use this training while hosting the first "Coastal Carnage" air-wing exercise of the year. By drawing participants from every CVW-3 squadron, and adversary "orange air" assets from VF-43, VF-14 led the air wing through a simulated opposed strike into the Dare County and Piney Island target complex in North Carolina. During this exercise the Tomcat showed its versatility as a swing-role fighter-bomber when LTs Vinny Zaccardi and Joe Merrell engaged and "destroyed" a bogey on the scripted low-level strike route and then completed their bombing mission, dropping Mk 76 bombs within 23ft of the bullseye. All that and they still had enough gas to provide strike escort on the way out.

March saw a continuation of the strike fighter program as the "Tophatters" accompanied the air wing to NAS Fallon for three weeks of overland strike training. During the unit's time here, it was instructed in how to perform the CAS mission in the air-to-ground capable F-14. The high speeds, low altitudes and exact timing required by CAS challenged every aircrew, and VF-14 responded superbly, scoring second among all air wing squadrons. Combat Search and Rescue [CSAR] in realistic scenarios also tested every aircrew. On every mission, VF-14 aircrews located and authenticated a crew simulated down in hostile territory. After locating the survivors, "Tophatter" Tomcats performed enemy ground suppression tactics to protect survivors and the rescue helicopter. The Fallon training culminated with a day-long simulated amphibious assault, during which all "Tophatter" aircrew were pushed to the limit while conducting both air superiority and flexible CAS missions.

Four F-14As from VF-84 conduct a formation attack on the Vieques Target Range during a squadron detachment to Naval Station (NAVSTA) Roosevelt Roads in early 1993. Each aircraft was armed with four 500lb (226kg) Mk 82 bombs. (US Navy via Cdr Peter B Mersky)

Perfectly illustrating the evolving ground attack role adopted by the F-14 community in the wake of *Desert Storm*, VF-14's CAG jet (F-14A BuNo 161855) is armed with two live AIM-7Ms and AIM-9Ls on the wing glove pylons and a pair of Mk 83 GP bombs fitted with BSU-85 high-drag fins on the centerline BRU-32 racks while flying over the Adriatic in March 1993. VF-14 and sister-squadron VF-32 completed numerous air-to-ground courses prior to joining the rest of CVW-3 for its 1992–93 Mediterranean cruise embarked in *John F. Kennedy*, the units dropping more than 200 iron bombs each during the ten-month work-up cycle. (US Navy via Cdr Peter B Mersky)

In May, VF-14 participated in the joint services exercise "Ocean Venture," contributing two days of strikes, CAS and air cover for Marine forces in the Cherry Point Target Complex in North Carolina. July and August saw CVW-3 deployed onboard JFK in the Caribbean Sea for Composite Training Unit Exercise [COMPTUEX], VF-14 dropping ordnance on the Vieques Target Range in Puerto Rico – LT Pete Ulrich and LCDR Phil Melfa became the first Tomcat aircrew to deliver air-to-ground ordnance from JFK with a precision strike on the range. The final training evolution prior to deployment was FLEETEX [Fleet Exercise], which took place over ten days in September off the coast of North Carolina. Operations consisted of strikes ashore and CAS for the attached amphibious group. With air-to-ground clearance in hand, VF-14 participated in every strike either as an air superiority Tomcat fighter or "Bombcat" strike fighter, delivering killer BDU-45 500lb [227kg] practice bombs on to targets at Cherry Point's BT-9/11 range. Upon successful completion, the JFK/CVW-3 team was ready for cruise. During the final weeks prior to deployment in October, VF-14 led the continually expanding F-14 air-to-ground effort. A series of live Mk 82 and Mk 83 bombs were dropped over five days, the first such live ordnance expended by Oceana Tomcats.

By then, both VF-14 and VF-32 had each dropped more than 200 iron bombs during the course of the ten-month work-up cycle.

As previously noted, with Tomcat units becoming more adept at performing the ground attack mission, so the TASS

VF-32 (1992–93)

VF-74 and sister-squadron VF-103 were early converts to the air-to-ground mission, being the first units to undertake the TASS course in the fall of 1991. By the time "Clubleaf 200" (F-14B BuNo 162921) was photographed at Roswell Industrial Air Center, New Mexico, on June 14, 1993, taxiing out "armed" with two inert Mk 83 bombs, VF-74 had become quite adept at the strike fighter mission. The unit was part of CVW-17's full strength commitment to Exercise *Roving Sands 93*, which, according to a USAF press release, was "the largest military exercise on US soil that allowed training in a joint environment to home command and control procedures, and integrate new systems in theater and air missile defense." Several air wings participated in the biennial exercise during the 1990s. (US Navy)

syllabus was upgraded and replaced by AARP during 1992 – the two-week course featured content virtually identical to that taught to F/A-18 pilots. Weapons were now being regularly expended by F-14 squadrons both at home during air wing work-ups and when deployed on ranges in the Mediterranean, North Africa and Kuwait. Yet despite the Tomcat community belatedly embracing the strike fighter role, the US Navy looked to an increasing number of multi-role F/A-18s on carrier flightdecks as the best way to provide post-Cold War mission flexibility. As a result, the fighter wings on both coasts saw their numbers decimated from April 1993 through to August 1996, when 14 fleet and reserve F-14 squadrons were disestablished.

Proof that the Tomcat was still not considered to be a truly frontline strike aircraft even after most fleet units had completed TASS and AARP evolutions in order to be classified as "Bombcat" capable had come as early as January 13, 1993. On that day, CVW-15, embarked in USS *Kitty Hawk* (CV-63) in the NAG, became the first air wing since *Desert Storm* to participate in bombing strikes on Iraqi targets when SAM sites and IrAF airfields were hit as part of OSW. Some 116 coalition strike aircraft were involved in the "one day war," with CVW-15's contribution consisting of three packages – two Hornet elements and an Intruder element. These aircraft employed laser-guided ordnance, unguided Mk 83 bombs and AGM-88 High-speed Anti-Radiation Missiles (HARM). The F-14As from VF-51 and VF-111 provided fighter escort for the strike aircraft, the Tomcats being armed with AIM-7s, AIM-9s and AIM-54s. None carried bombs,

VF-51 (1994)

The F-14A CAG jets from VF-14 (BuNo 161855) and VF-41 (BuNo 162703) replenish their tanks from VMGR-152 KC-130F BuNo 149816 over Arizona in October 1993 during MAWTS training at Yuma. Crews from both fighter units spent 26 days undertaking this intensive air-to-ground course. (US Navy)

however, which meant that both squadrons were little more than aerial spectators when the IrAF wisely chose not to oppose the strikes.

The fact that VF-51 and VF-111 had been left on the sidelines once again reflected the last-minute nature of the whole "Bombcat" upgrade for the F-14. As previously noted, there had initially been a shortage of bomb racks for the aircraft, which in turn meant that some air wings only had one of its two Tomcat units qualified to undertake bombing missions – the remaining squadron would specialize in TARPS instead. The biggest handicap facing Tomcat units at this point, however, was the restriction in the type of ordnance they could employ.

The basic "Bombcat" carried only unguided freefall air-to-surface weapons – the same capability first seen on the pre-production F-14s of the early 1970s. Bombs cleared for use by the jet included the 500lb (227kg) Mk 82 in its original low-drag configuration and in Snakeye or BSU-86 high-drag guise. The F-14 could also carry various types of the 1,000lb (454kg) Mk 83 bomb (including versions fitted with BSU-85 Air Inflatable Retard equipment) and the 2,000lb (907kg) Mk 84. Mk 7 dispenser-based weapons had also been cleared in the early 1990s, including the anti-armor Mk 20, Mk 99 and Mk 100 Rockeye and the CBU-78 Gator filled with 15 BLU-92/B anti-personnel and 45 BLU-91/B anti-tank mines. VF-21 had become the first fleet Tomcat unit to drop such munitions when it expended Mk 20 cluster bombs on June 5, 1993 during its AARP training phase. The Mk 7's limited operational employment is detailed elsewhere in this volume.

Although the combination of these weapons and the Tomcat's impressive endurance made the aircraft a handy multi-role fighter-bomber, the "Bombcat's" effectiveness in the air-to-ground mission could not then compare with that of the Intruder it was earmarked to replace. The F-14 was not compatible with any of the stand-off or guided weaponry routinely carried by the A-6E, namely the AGM-84 Harpoon and closely related Standoff Land Attack Missile, AGM-62 Walleye II, AGM-65 Maverick, AGM-88 HARM or laser-guided bombs (LGBs) such as the GBU-10, GBU-12 and GBU-16.

Furthermore, the Tomcat had no self-designating capability equivalent to the Intruder's undernose Target Recognition and Attack Multi-Sensor (TRAM) turret. This fully articulated sensor featured a laser range-finder and marked target seeker, a laser designator and

Lts Chris Blaschum (pilot) and Jack Liles (RIO) drop a Mk 83 bomb from VF-143's "Dog 106" (F-14B BuNo 161426) on the Lone Rock range at Fallon in October 1993. The weapon's BSU-85 Air Inflatable Retard ballute has already started to deploy, slowing the bomb sufficiently to allow the Tomcat to escape before it detonates – not an issue at this altitude, however. Note the orange Tactical Aircrew Combat Training System pod on the wing glove pylon, this store transmitting flight data to ground stations at Fallon to aid in the debriefing of the crew post-sortie. (US Navy)

F-14 "BOMBCAT": THE US NAVY'S ULTIMATE PRECISION BOMBER

Forward-Looking Infrared (FLIR). This sensor suite turned the A-6E into a deadly night-attack platform, and it also provided the jet with the ability to designate targets for its own LGBs. Although the F-14 was subsequently cleared to carry smart bombs, the jet initially had to rely on other aircraft with a laser designator pod to guide the weaponry to the target.

The addition of LGBs to the Tomcat brought with it an entirely new precision attack capability. The first such weapon approved for use was the 1,000lb (454kg) GBU-16 in early 1994. Its clearance coincided with the US Navy's decision that the F-14 should be able to designate its own targets, rather than relying on the A-6E or F/A-18C (the latter used the AN/AAS-38A Nite Hawk bolt-on FLIR pod). The Tomcat's night-attack capability also needed to be drastically improved. LGB integration progressed alongside an urgent program to find a basic FLIR and laser designator for the F-14 that would allow the jet to take over the precision heavy attack mission from the A-6E, which was now being withdrawn from service.

Left: Among the VF-143 crews to expend ordnance during the Fallon detachment in October 1993 were Lt Mike Davison (pilot) and Lt(jg) Stephen Davis (RIO), who took the aerial photographs of the bomb drop. They are posing with a Mk 83 "thousand pounder" prior to flying a strike mission over the live range. (US Navy)

Below: In November 1993, three months after completing its *WESTPAC 93* deployment with CVW-9 onboard *Nimitz*, VF-24 commenced its turnaround training with participation in Exercise *Coronet Sentry 94-1*. Working with USAF and US Army units, the squadron "promoted air-to-ground capabilities in joint operations." It then continued joint training with US Marine Corps assets shortly thereafter by performing CAS sorties at Camp Pendleton, north of San Diego. Photographed from a VMGR-352 KC-130T, "Renegade 205" (F-14A BuNo 161619) was involved in both events. (Rick Llinares)

LGB Pioneers

As detailed later in this book, VF-103 would play a major role in proving the worth of the Tomcat as a precision bomber with the advent of the modified AN/AAQ-14 Low Altitude Navigation and Targeting Infrared for Night (LANTIRN) pod, which the unit debuted operationally during CVW-17's Mediterranean/NAG cruise embarked in USS *Enterprise* (CVN-65) from June 1996. VF-103 had been chosen to introduce the targeting pod into fleet service because of the unit's experience delivering 1,000lb (454kg) GBU-16 Paveway II LGBs during its previous cruise onboard *Saratoga* in the first half of 1994. From January to late April of that year, VF-103 had racked up more than 320 CAP and TARPS missions over Bosnia-Herzegovina while helping to enforce a no-fly zone over the country as part of Operation *Deny Flight*. The LGB aspect of the deployment was detailed in the following account, which appeared in the fall 1994 edition of *The Hook*:

> The "Sluggers" left the confines of the Adriatic in late April for Palma, Spain. While en route, the nearness of the Capo Frasca target range on the western coast of Sardinia provided the opportunity to make Tomcat history. On May 2, 1994, two F-14Bs from VF-103 delivered three GBU-16 Paveway II LGBs to direct hits utilizing air wing F/A-18s and A-6Es for target illumination.
>
> The "Sluggers'" unqualified success in LGB deliveries came only after hard work by both aircrew and maintenance personnel. Technical data, including ballistic tables and profiles, were unavailable for F-14 LGB deliveries. Under the guidance of LT David Herringbone, VF-103 weapons training officer, delivery profile parameters were developed utilizing data developed by F/A-18s and A-6s. With the help of VA-35, VFA-81, SWATSLANT and VX-4, delivery tactics and profiles were developed and a series of test flights flown to prove proposed delivery parameters. Aircrew were briefed and maintenance personnel qualified on the handling and loading of Paveway II guided bombs. On May 2, 1994, the hard work paid off when "three for three" direct hits were scored on Capo Frasca.
>
> The first two GBU-16 releases, performed by Cdr Sandy Saunders/Lcdr Mike Ullrich, utilized an F/A-18 flying in a trail position for target illumination. After weapons release, the F/A-18 lased the target until impact, resulting in two direct hits. The third delivery, flown by Lt Dave Harrington/Cdr Steve Schlientz, employed an A-6E in trail formation to illuminate the target.
>
> The last days in the Mediterranean were highlighted by a second Tomcat LGB strike mission on May 31. Two more GBU-16 deliveries took place on the Capo Frasca target complex, with direct

Closing up behind the open ramp of the KC-130T, "Renegade 205" was armed with an unusual mix of "weapons" – single examples of the AIM-54C, AIM-9L and AIM-7M and two Mk 83 1,000lb (454kg) bombs. Only the latter were live, however, and they were soon to be expended on a bombing range in southern California. (Rick Llinares)

F-14A BuNo 162602 of VF-51 prepares to roll in at Fallon's Bravo 20 range during CVW-15's three-week stay at the naval air station in January 1994. "Screaming Eagles 102" is armed with four Mk 83s on its under-fuselage bomb racks. BuNo 162602 was written off on July 11, 1994 when it broke in half following a heavy landing onboard *Kitty Hawk*, the cockpit section of the jet sliding down the flightdeck and over the side of the vessel. The crew ejected, and the pilot was severely injured after landing in burning fuel on the flightdeck – the RIO was unhurt. Ten flightdeck personnel received minor injuries. (Cdr Tom Twomey)

hits to expand the "Sluggers'" record to "five for five." The second series of deliveries was designed to test the "buddy-bomb" concept between an F-14 and a laser-capable platform. The tactic was developed for night operations and situations where visual target acquisition by the F-14 would be difficult. In these deliveries, the Tomcat flew in formation on either an F/A-18 or an A-6E, releasing the weapon after verification that the laser platform had achieved an acceptable release solution. Upon weapon release, the F/A-18 or A-6E illuminated the target until impact.

The simple but tactically significant buddy-bomb concept provides greater capability in CAS operations, while giving strike planners added flexibility. If the Tomcat is upgraded to include a FLIR/laser illumination system, the F-14 will be completely autonomous in GBU operations. The Tomcat has long ago proved its proficiency in the air-to-ground arena, but as the "Sluggers" have proved the capability to deliver laser-guided munitions, the F-14 now has more punch in its role as a precision strike fighter platform.

Three months later, on August 23, 1994, VF-211 (which by then had reverted to the F-14A) became the first Miramar-based Tomcat unit to expend LGBs. The squadron proudly reported this event in an article boldly titled "Checkmates Unleash Weapons" published in the winter 1994 edition of *The Hook*:

> During a CAS sortie, the crew of LCDR J P "Bugs" Easterbrooke/Lt D A "Ogre" McGowen delivered a 1,000lb [454kg] LGB, with stand-off illumination provided by an F/A-18D Hornet from the VMFA(AW)-225 "Vikings." Two other VF-211 Tomcats crewed by Lts J K "Lefty" Hensley/S D "Cole" Porter and R J "Basher" Bello/R E "Mo" Vaught dropped a total of three laser-guided training rounds [LGTRs]. The event was an unqualified success as the LGB scored a direct hit on a truck parked at the base of a mountain. Two of the LGTRs made direct hits, with the third missing by only a few feet. Together, the "Checkmates" and "Vikings" proved that the Tomcat/Hornet team is a highly effective asset for ground forces support.

Having shown its ability as a precision bomber, the Tomcat was now on the cusp of becoming arguably the best strike aircraft the US Navy ever owned.

Lt Cdr Randy Parrish (pilot) and VF-51 CO, Cdr John Sill, release all four Mk 83 bombs in a diving attack on Bravo 20 range. VF-51 dropped 50,000lb (22,679kg) of live and inert ordnance during its air wing integration detachment in January 1994. (Cdr Tom Twomey)

CHAPTER 2
"BOMBCAT" BLOODED OVER BOSNIA

The long-running conflict in the Balkans provided the F-14 community with the opportunity to test the Tomcat in its new role as a bomber.

In the wake of *Desert Storm* and the fragmenting of Yugoslavia following the end of the Cold War, the US Navy, and more specifically carrier aviation, found itself heavily involved in multi-national operations. These established and then enforced no-fly zones backed by the United Nations (UN) that were aimed at protecting unarmed civilians in Iraq (Operations *Northern* and *Southern Watch*) and the Balkans. The conflict in the latter region soon escalated as Yugoslavia fragmented into separate states along ethnic, religious and political lines. Some of the worst fighting took place in Bosnia-Herzegovina, with powerfully armed Bosnian-Serb factions engaging the Muslim population in the newly created country.

Although it remained neutral as fighting wracked the region, the UN did recognise the new Muslim government in Sarajevo, and instigated a humanitarian airlift into the country while attempting to uphold an arms embargo against Serbia in an effort to stem the flow of weapons into neighboring Bosnia-Herzegovina. Aircraft flying these missions were principally based in Italy, although US Navy carriers assigned to Sixth Fleet

VF-41's color jets lead a pair of more mutely marked F-14Bs from VF-74 during the Oceana Airshow in September 1993. "Fast Eagle 101" (F-14A BuNo 162689) boasts nose art featuring the iconic "Anytime Baby" Tomcat leaning on titling that reads *BOMBCAT NO ESCORT REQUIRED*. This aircraft had left VF-41 by the time the unit embarked on its historic 1995 Mediterranean/OSW deployment. (David F Brown)

also played their part while sailing in the Adriatic. Operation *Sky Monitor* commenced in November 1992, and this gave way to Operation *Deny Flight* in March 1993, when the UN authorized North Atlantic Treaty Organisation (NATO) air commanders to forcibly prevent Serbian and Bosnian-Serb attacks.

Carrier-borne F-14s from COMNAVAIRLANT squadrons were heavily involved in the policing of the UN-imposed no-fly zones and in providing air cover for transport aircraft conducting humanitarian relief operations. Tomcats also flew regular CAPs looking for violations of the UN resolution, as well as unauthorized aerial activity. This had the desired effect, although USAF F-16s were forced to shoot down four Bosnian-Serb J-21 Jastreb attack aircraft that violated the no-fly zone when they bombed a factory in Bosnia-Herzegovina on February 28, 1994.

The Tomcat's role increased dramatically in scope after UN aircraft started attacking Serbian ground targets from August 1994, with aircraft flying pre- and post-strike reconnaissance sorties using TARPS. Nevertheless, a further 13 months would pass before the "Bombcat" was at last blooded in combat during Operation *Deliberate Force*, which was instigated on August 30, 1995 in response to a mortar attack by Bosnian-Serb forces on a Sarajevo market that had killed 43 civilians.

Lt Todd Parker was a RIO with VF-41 during the squadron's 1995 deployment embarked in USS *Theodore Roosevelt* (CVN-71), and he was one of a handful of Tomcat aircrew to drop ordnance in combat during the cruise. He recalled this event, and other aspects of the historic deployment, in *Grumman F-14 Tomcat – Shipborne Superfighter*, edited by Jon Lake and published in 1998:

A close up of the nose art applied by VF-41 to "Fast Eagle 101" in the late summer of 1993. Note the various weapons in the Tomcat's bandoleer – a Rockeye cluster bomb unit and a Mk 83 bomb flanking an AIM-9, AIM-7 and AIM-54. The "six-shooter" in the holster denotes that the aircraft was also armed with a 20mm cannon. (David F Brown)

Anyone familiar with Naval Aviation, and the continually evolving persona it undertakes, knows that change is not only constant but necessary. As theaters follow a sine-wave pattern of action and intensity, and corresponding threats upgrade and shift, the carrier battle group is the pre-eminent force to meet the challenges head on. Responding unlike any other force in the world in its combination of speed and intensity, the carrier air wing is the sword of the warrior in the battle group.

Accelerating the upcoming retirement of one of the distinguished members of this lethal team – the A-6 Intruder – identified a shortfall in the carrier air wing's capability. The carrier air wing projects power, and power projection equals bombs on target. All other missions support attack. The demise of the A-6, combined with the A-12 cancellation, meant that the number of aircraft capable of using air-to-ground ordnance had to be increased with existing jets. Thus began the five-year metamorphosis of the Tomcat from fleet air defender and strike escort to a multi-mission, self-escorted strike fighter unparalleled in Naval Aviation in its combination of performance, precision and payload, day or night.

As recently as six or seven years ago [this account was written in 1995], anyone who said Tomcats would be patrolling the skies of the former Republic of Yugoslavia loaded with laser-guided bombs awaiting strike, CAS or forward air control (airborne) [FAC(A)] missions would have been called insane. Yet here I was, just 20 miles [32km] south-east of downtown Sarajevo, en route to a target with two 1,000lb [454kg] GBU-16 LGBs.

Normally, at five months and one week into a cruise, one has seen and done just about everything there is. Not this time. The night before what was to be out last port visit, we felt that familiar rumble of the propeller shafts, and noticed that the ship's heading was due west, not the north-north-west heading that would have taken us to Rhodes. Several days earlier [on August 28, 1995], Bosnian-Serbs had fired mortar rounds into a market in Sarajevo in clear violation of UN restrictions. Once again, our presence in-theater was required. First though, a little background on how we got to this point.

I had been a RIO in VF-41 for three years when we deployed in March 1995 as a member of CVW-8 on board *Theodore Roosevelt*. This team had proved its lethality in the Gulf War, and those of us who had missed that opportunity were eagerly anticipating our shot. Things had been quiet in Bosnia for many months, primarily because of the winter weather, but we knew that the spring thaw

Above: By the time VF-41 visited Fallon during the final stages of CVW-8's pre-cruise work-ups in October–November 1994, the unit had replaced F-14A BuNo 162689 with BuNo 160394. As with the previous "Fast Eagle 101", this aircraft also had nose art in the form of the "Anytime Baby" Tomcat leaning on *STRIKECAT* titling. The jet was lost during a post-maintenance check flight over the Mediterranean on May 22, 1995, when its horizontal stabilators refused to work in unison. Both crewmen ejected successfully at 3,000ft and were quickly rescued. The RIO in the aircraft at the time was Lt Todd Parker, who would go on to drop ordnance on a target in Bosnia-Herzegovina less than four months later. (Michael Grove via Mike Crutch)

Left: A close-up of the *STRIKECAT* nose art and titling on BuNo 160394. This aircraft was the 250th Tomcat constructed for the US Navy. (Michael Grove via David F Brown)

Above: The pilot of "Fast Eagle 102"' (BuNo 160898) retracts the jet's undercarriage as it departs Oceana at the start of a bombing mission to the Dare County range in June 1994. The aircraft is carrying a pair of BRU-42 ITERs on its centerline, and they are carrying several Mk 76 Mod 5 blue bombs. BuNo 160898 had become "Fast Eagle 104" by the time VF-41 embarked 14 Tomcats in *Theodore Roosevelt* on March 22, 1995 and headed east across the Atlantic, bound for the Mediterranean. (Michael Grove via Mike Crutch)

Right: "Fast Eagle 102" goes in search of a target during pre-cruise work-ups in Nevada in late 1994. Such arid landscapes contrasted markedly with the lush forests and tree-covered mountains of the Balkans that VF-41 would patrol over when committed to Operation *Deliberate Force* in 1995. (US Navy)

usually meant increased hostilities. From the Tomcat standpoint, we were especially excited about the chance to employ a truly multi-mission capable aircraft. In addition to our standard CAP and fleet air-defense roles, we would be flying [On-Call] exercise CAS [XCAS], tactical reconnaissance and the newly emerging Naval Aviation mission of FAC(A).

After an uneventful six-week stint in the Red Sea and the Persian Gulf, we transited back through the Suez Canal to the Adriatic, with tensions in the area rising as expected. Once on station, we immediately began planning contingency operations, for clashes between the Bosnian-Serbs and Bosnian Muslims were already occurring daily. Unfortunately, the first four missions that launched off the carrier – all CAS and FAC(A) missions – were as anticlimactic as the three months that followed. In what became a routine drill for aircrew and planners alike, we would plan all night for a bombing mission we were certain would launch at first light, only to be shut down, sometimes literally, just before launch.

This is not to say there were not exciting moments, such as the shootdown of [F-16 Fighting Falcon pilot] Capt Scott O'Grady by a SAM, the one-night bombing of ammunition storage depots in Pale by the Marines and Air Force, the Bosnian-Serb takeover of Žepa and Srebrenica and the two-day offensive by the Croatians to reclaim the area in eastern Croatia that the Krajina-Serbs had held for two years. Each large-scale event brought an intense 24- to 36-hour planning period, followed by the wait for the order to launch, which never came. As might be expected, after doing this many times over three months, morale can suffer, and it became increasingly difficult to plan a mission that we doubted would ever launch. But plan we did.

The time between these large-scale events was by no means boring. The air wing flew daily missions as tasked by the CAOC [Combined Air Operations Center] in Vicenza, Italy. EA-6B Prowlers patrolled the sky providing electronic protection. E-2 Hawkeyes were the "eyes in the sky," acting as the command, control and communications link between Navy aircraft and the joint command and control system. S-3Bs and ES-3As provided electronic support and F/A-18 Hornets fulfilled the CAP, XCAS and HARM roles. Tomcats flew XCAS, TARPS and FAC(A) missions. Whenever possible, the carrier would pull in for a quick port visit in Corfu or Trieste, but there was always the tether of a short recall to worry about.

En route to what was supposed to be our second-to-last port visit in Marseilles, Saddam Hussein's two sons-in-law decided to defect [to Jordan], which proved to be the beginning of a dramatically altered last two months of cruise. Instead of pulling into port, we transited from the Adriatic to the eastern Mediterranean and stationed ourselves off the coast of Israel [for Operation *Infinite Moonlight* – a US/Jordan bilateral contingency plan to deter Iraqi military reprisals following the defections]. By August 28, the situation had defused, and we finally got the okay to start our trek west. The next stop would be our last port visit to Rhodes, followed by a turnover with USS *America* (CV-66) [with F-14B-equipped VF-102 embarked as part of CVW-1] in the western Mediterranean, and finally the trip home. But, as mentioned earlier, late that evening the propellers started churning the water faster than usual, and *Theodore Roosevelt* sped back to the Adriatic.

We assumed that – like in every other instance of increased hostilities – we would plan all night, only to see another opportunity lost. What made this instance different, and therefore even more difficult, was that we were supposed to be pulling into port and then heading home. Instead, we were returning to an uncertain area with even more uncertain intentions. However, it was not our job to question those intentions, so once again we began the now-familiar planning exercise.

VF-41's first taste of operational flying during its 1995 cruise on board *Theodore Roosevelt* took the form of OSW patrols over Iraq between April 29 and May 19. Typically flying in mixed packages that included both Tomcats and Hornets, the unit armed its jets with two GBU-12s. USAF tanker support was critical throughout OSW, particularly for short-legged F/A-18Cs – this example belongs to VMFA-312. Here, TARPS-capable "Fast Eagle 111" (BuNo 161864) holds off the right wingtip of a KC-135R awaiting its turn to refuel. (Capt Paul Filardi)

Devoid of any ordnance bar an AIM-9M acquisition round, "Fast Eagle 106" (BuNo 160407) undertakes a training sortie off the coast of southern Italy during a break from *Deny Flight* missions over Bosnia-Herzegovina in June 1995. During its 17 years of service with the US Navy, BuNo 160407 spent all bar four months of its time in the fleet assigned to VF-41. (Capt Paul Filardi)

Even though the first events manned up, and even launched, we still did not believe that anything would actually happen. So, when the first aircraft returned shortly after 0200hrs on August 30 with its racks and rails empty, we could hardly contain our surprise and excitement. Finally, after so much hard work and so little return, we were getting the chance to prove ourselves.

The main question on our minds was "How can we get the Tomcats to play?" At that point, our role of suppression of enemy air defenses [SEAD] was critical, and our TARPS flights were crucial for the intelligence picture, but the opportunity to be the first F-14 unit to drop live ordnance in combat was the pot of gold at the end of the rainbow. Once the CAOC was aware that Tomcats could indeed drop air-to-ground ordnance, the question became how and when it would be best to do so. At the time, we were not carrying the LANTIRN pod, which now provides Tomcats with a day or night, self-escort, self-lased precision-guided munitions [PGMs] capability. Nevertheless, we were still a lethal day iron-bomb dropper, and we could employ any of the LGBs as long as someone lased the target for us.

An initial target was finally identified – an ammunition dump in eastern Bosnia. Because of potential collateral damage issues, this was a precision-guided munition [PGM] target only, which meant that the Tomcats would drop their LGBs off a buddy-lase from a Hornet. We were not surprised that the "food chain" theory applied here, with this historic flight being top heavy rank-wise [in respect to the aircrew involved].

Two F-14s launched with the Hornet flight [on September 5], yet there was still the question of whether drop authority would really come for this unusual delivery tactic. Imagine the feeling of excitement when, two hours later, those same two Tomcats came back with their belly stations empty, except for the dangling arming wire that remains after the release of an electrically-fused weapon. The review of the FLIR tapes (the same tapes seen on television on CNN) proved that planning and diligence had paid off – two direct hits, and a history-making day for VF-41 and the Tomcat community at large.

Lts Todd Parker and Jesse Fox pose with a personalized GBU-16 1,000lb (454kg) LGB prior to flying a *Deliberate Force* mission over Bosnia-Herzegovina in September 1995. VF-41 dropped both LGBs and iron bombs (totaling (33,070lb/15,000kg)) during a week of combat in the Balkans. (US Navy)

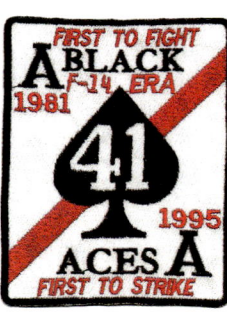

VF-41 (1995)

One particular ridgeline south of Sarajevo over which we were routinely tasked to fly was especially notable for the high surface-to-air threat activity coming from it. On one TARPS flight over this ridgeline, the crews noticed AAA being fired at them – a relatively common experience. What they did not see this time was the shoulder-launched SAM that was fired from behind them but did not reach them. It was not until they returned to the carrier that they saw their passage south-east of Sarajevo recorded live on CNN, complete with a missile smoke trail rising up toward their aircraft before falling away behind them. Despite these dangers, TARPS missions continued daily, and proved to be crucial for intelligence gathering.

In the meantime, the *America* battle group had arrived on station eager to turn over and participate in the action. The majority of the *Theodore Roosevelt* battle group had already begun its journey home, with only the carrier and its AEGIS cruiser escort remaining. The *America* battle group immediately assumed the duties of protecting both carriers, while *Theodore Roosevelt* and *America* began their turnover.

Normally, a turnover lasts only one to two days, but these were not normal circumstances. The CAOC wanted uninterrupted carrier air missions, yet was concerned about *America* picking up the full load of duties on its first day on station in an unfamiliar region. The turnover therefore lasted several days, with *Theodore Roosevelt* initially flying all of the missions and *America* picking up an increasingly larger share.

On the evening of September 11, the rumors began circulating that the next day *Theodore Roosevelt* would be released to transit back home. Even though we were already two weeks late for the scheduled start of our return, everyone had mixed feelings about leaving. We were anxious to return home, but no one really wanted to leave the theater. Several of us in the Tomcat squadron were especially disappointed, since we had not yet dropped live ordnance in combat, and it appeared we would not get the chance. So, I went to sleep that night with conflicting emotions.

I had only slept for two hours when the phone rang, and I was told to come up to the ready room immediately. We had just received word that the CAOC had ordered one more strike from the *Theodore Roosevelt* to be launched first thing in the morning. We had spent the last few nights attempting to beat down the aerial threats in the western and northern parts of Bosnia with HARM and Tomahawk cruise missiles, and now there was a priority target that was deemed reasonably safe to bomb. The request for a mixed F/A-18–F-14 strike package was approved, and we began flight planning well after midnight. By 0400hrs we were tired, but satisfied with the plan, and we retired to our staterooms for a few hours of rest before the mission launched.

I awoke feeling excited and alert, despite little sleep. We briefed for the flight, and as we manned up the aircraft we were relieved that there were no changes to the mission. The plan was relatively simple – eight targets at an ammunition storage depot, one for each aircraft. The Hornets would each lase their

The success of this mission set the stage for future ones. The next live-ordnance flight for the Tomcat was a dumb bomb target, selected carefully for its relative isolation and lesser potential for collateral damage. This particular target happened to be a radio-relay station on top of a mountain. Three Tomcats ingressed on a self-escort mission, and after making two passes each over the site, all tasked targets were destroyed with four Mk 83 1,000lb [454kg] unguided bombs each. Tomcat bombing missions continued throughout the week, until they became almost routine. Weather was a significant factor, and it became luck of the draw whether an individual Tomcat pilot or RIO got his chance to be part of history.

In the meantime, Tomcats continued to fly other missions in support of the effort. Most notable were the TARPS missions, generally flown in co-ordination with strike missions to provide immediate feedback on the success of the operations. The nature of these missions often caused them to be the most dangerous flown in-country. Tasking typically included targets just attacked, or significant threat locations, which meant that these aircraft were often tasked to fly into an aroused hornet's nest [of SAMs and anti-aircraft artillery (AAA)]. Combine this with the lower altitudes for TARPS flights, and you have a recipe for excitement.

Above: VF-41's officer cadre come together for a group photograph in front of the ill-fated "Fast Eagle 101" on the flightdeck of *Theodore Roosevelt* during the early stages of the 1995 deployment. The unit CO, Cdr Dick Bedford, is in the front row, ninth from the right. (US Navy via Capt Paul Filardi)

Left: USS *Theodore Roosevelt*/Operation *Deny Flight* (1995)

own target and then re-attack to buddy lase for their Tomcat wingman. I was the fourth aircraft in the lead division [of four jets], which would be followed shortly after by the second division [of four jets].

Approaching the target, the excitement grew as we realized our pre-flight chart study had paid off and we had found the target. The first three attacks went off without a hitch, scoring direct hits, with secondaries observed. We set up for our attack, ensuring that we were meeting our pre-planned parameters exactly. When we reached our release point, with confirmation of a good lase from the Hornet, my pilot pressed the pickle button. After what seemed like a millennium, but was actually only milliseconds, the "clunk-clunk" of two 1,000lb [454kg] LGBs coming off resonated beautifully throughout the aircraft. We jinked off target and looked down to check for threats and spot our hits. Fortunately, we were nearly directly abeam our target when the bombs hit, and two enormous explosions verified what we already sensed – this mission was a success.

We headed back to the carrier, hearing over our tactical frequency that the second division had achieved similar success. When we got back on board and reviewed the tapes, the success of the mission was overwhelming. Of the eight targets we had been instructed to hit, seven were completely destroyed, with no re-strike required.

CVW-8 and *Theodore Roosevelt* were replaced in the Adriatic by CVW-1 and *America* in early September, the latter vessel undertaking its final cruise prior to being decommissioned. The F/A-18Cs assigned to CVW-1 were in the thick of the action as soon as *America* was committed to *Deliberate Force*, expending LGBs and SLAM-ER. Although VF-102 did not get the chance to emulate VF-41's bombing success, it did fly plenty of CAPs, FAC(A) missions and TARPS runs, particularly after the Dayton Peace Agreement was signed by the warring factions in Bosnia-Herzegovina in December 1995. "Diamondback 104" (F-14B BuNo 162925) is undergoing its final weapons checks prior to launch shortly after VF-102 commenced *Deliberate Force* patrols, the aircraft carrying a single Mk 83 1,000lb (454kg) bomb on each centerline rack and an AIM-9L on its starboard wing shoulder pylon. (Alfredo Maglione via David F Brown)

After the excitement of the flight began to diminish, and we settled back into our routine, the ship's intercom whistled with the familiar sounds preceding a message from the captain. Instead of the usual, "Good evening, this is the captain," we heard music accompanying the words "Westbound and down, loaded up and trucking. We got a long way to go, and a short time to get there." Minutes earlier the captain had received the okay to pack up and ship out. The carrier, due home in eight days, would require almost record-breaking speed to get there on time. Miraculously, the approval came for "all ahead, warp speed." And without the rest of the battle group to slow the ship down, *Theodore Roosevelt* wasted no time.

It wasn't until it was all over that everyone really had a chance to reflect on what we had been a part of, and to realise that we in VF-41 had again made Tomcat history. The "Black Aces," with the first F-14 air-to-air kills in 1981, had been the "First to Fight," and now, 14 years later, we had been the "First to Strike."

VF-102 also routinely loaded its jets with two GBU-16s just in case the unit was required to drop PGMs. However, these could only be expended if laser guidance was available from a Nite Hawk pod fitted to one of CVW-1's F/A-18Cs. And with VMFA-251, VFA-82 and VFA-86 busy lasing targets for their own ordnance, VF-102 was not called upon to expend any LGBs. (Alfredo Maglione via David F Brown)

"Bombcat" XO over Bosnia

VF-41's XO on this historic cruise was Cdr Bob Brauer, and he was interviewed shortly after his return to NAS Oceana from the Adriatic by noted aviation historian Warren Thompson:

I've been flying the F-14 since 1982, and have accumulated more than 2,200 flying hours in the aircraft. My total time is about 3,800 flying hours. During the summer and fall of 1995 we flew a lot of missions operating off the coast of Bosnia as part of Operation *Deliberate Force*. We – VF-41 – were the only Tomcat squadron aboard *Theodore Roosevelt*, along with three squadrons [VMFA-312, VFA-15 and VFA-87] of F/A-18C Hornets.

In September 1995, we had the opportunity to make the F-14's first ever delivery of air-to-ground ordnance in combat. We dropped laser-guided munitions and free-fall Mk 82 bombs. I got to participate in the first strike, which involved one section of F/A-18s and one section of F-14s, flying as mixed sections. The F-14s each carried two 1,000lb [454kg] LGBs, and the Hornets carried the same load. That was the first time we had the chance to do the laser delivery in a mixed section, although we had previously practiced this a lot at NAS Fallon and while performing the *Southern Watch* mission in the NAG. The first time we actually got to do this with live bombs was in Bosnia. There was no air-to-air threat at the time. The surface-to-air threat was significant in some areas, but it had been pretty well beaten down by then.

On the day of the first strike the weather was a big factor, and it remained so into the winter months. As we launched off the ship and headed in over the coast, it didn't look like the strike was going to go because of an undercast. But about ten miles out from the target the weather cleared, and we could see it from some distance away. We rolled in at a very steep angle, from high altitude, against a munitions facility. The F-14s dropped while the F/A-18s lased. It was an incredible feeling as the 1,000lb [454kg] bombs kicked off the aircraft. Even in a 50-degree dive and at nearly 500 knots [926km/h], you could feel the release. I pulled the nose up, rolled and looked back over my shoulder to see the bombs' impact on the target. We then egressed the target area, having achieved absolutely superb results from direct hits that in turn caused impressive secondary explosions.

I can recall another mission over Bosnia in which some of our junior pilots were able to go in on the target and drop Mk 80 series dumb bombs, with excellent results. They picked up some very valuable experience.

One major benefit that the Tomcat brings to the air wing in its bomber configuration is the fact that the jet can launch with a heavy load of laser-guided munitions and bring it all back aboard if something like bad weather prevents us from

USS *Theodore Roosevelt* (CVN-71)/Med and Persian Gulf Cruise 1995 (1995)

reaching the target area. This is a distinct edge we have over the F/A-18 Hornet, which has similar carriage capabilities. Due to fuel and weight considerations, the Hornet has to get its bombs off, and if for some reason it can't drop its ordnance, it must jettison it before landing back on the carrier. Another plus for the F-14 is that it can carry a very significant air-to-air load – a Phoenix, a Sparrow and some Sidewinders – plus a couple of 1,000lb [454kg] LGBs. It can fight its way to the target, drop the bombs and fight its way back out. It's a self-escort mission for us!

During VF-41's week of bombing in the Balkans (from September 5 to 12, 1995) it delivered 33,070lb (15,000kg) of ordnance on Bosnian-Serb targets. The bulk of this weaponry had been guided using AN/AAS-38A Nite Hawk laser designator pods fitted to Hornets. More than three years would pass before Tomcats would again drop bombs in anger, by which time F-14 crews could self-designate.

Decoy Delivery

Although VF-41's "First to Strike" exploits in September 1995 rightly earned the unit plaudits for giving the Tomcat its operational debut as a bomber, the squadron also carried out a less glamorous, but no less important, mission during *Deliberate Force*. As VF-41's command history for 1995 notes, between "August 29 and September 11 the unit completed 100 combat sorties, totalling 170 hours, dropped 15 tons of air-to-ground ordnance and 39 TALD." As far as the author can ascertain, VF-41 was the only F-14 unit to ever employ the ADM-141 Tactical Air Launched Decoy (TALD) operationally.

Developed by the Brunswick Corporation in the 1970s for the USAF and the Israeli Defense Force Air Force, TALD was acquired by the former in significant numbers (around 2,000 examples) from 1987. Featuring a square fuselage and wings that popped out after deployment, and with a range of up to 70 nautical miles when dropped from high altitude, the ADM-141 was essentially a radar-reflective glider that mimicked the appearance of an aircraft. TALD had been given its operational debut with the US Navy on the first night of Operation *Desert Storm*, with more than 100 being launched by A-7s and F/A-18s as targets for Iraqi air defense radar. Thinking that the ADM-141s were enemy aircraft, the radar sites "flashed up" in an attempt to lock on to them, allowing HARM-equipped Corsair IIs, Hornets and Prowlers to knock them out.

TALD had only been cleared for use by the F-14 in the spring of 1994, with testing of the decoy being concluded at NAS Patuxent River on 28 April 1994. Fighter Wing, US Atlantic Fleet (FITWINGLANT) subsequently included TALD in its list of "approved for carriage that year" items, along with LGTRs. A VF-211 F-14A became the first fleet Tomcat to launch a TALD in September of that year, the decoy then being locked up by a second squadron aircraft and shot down with an AIM-7 Sparrow missile. VF-24 also downed an ADM-141 with a Sparrow shortly thereafter.

Two Brunswick Corporation ADM-141 TALD are seen here attached to the centerline stations of two BRU-42 ITERs carried by an F-14D of VF-31 at Miramar in 1995. Although only ever dropped operationally by VF-41, TALDs were occasionally employed by fighter units (and S-3 and F/A-18 squadrons) as targets for AIM-7s and AIM-54s during missile exercises. (Dave Roof)

Prior to the F-14 being given clearance to carry TALD, other US Navy aircraft, including the S-3 Viking, had occasionally launched them as targets for Tomcat units undertaking live-fire missile training. On July 24, 1994, for example, a Viking from VS-37 had deployed an ADM-141 that was engaged by an AIM-54 Phoenix missile fired by an F-14A from VF-111. Both aircraft were part of CVW-15 embarked in *Kitty Hawk*, which was sailing in the Sea of Japan at the time undertaking a six-month Western Pacific Ocean (WESTPAC) deployment.

Just over a year later, in early September 1995, NFO Lt(jg) (later Capt) Paul Filardi became one of just a handful of Tomcat aircrew to launch TALD operationally during *Deliberate Force*:

We could carry up to four TALD under the belly [a single decoy was attached to the centerline station of a BRU-42 Improved Triple Ejector Rack [ITER]. I don't recall any specific training we had on TALD, but I do remember that when I later became the unit's Air-to-Air Training Officer and was put in charge of missile shoots and co-ordination, we used them as air-to-air targets when there were no drones available. When we used TALD as a training target, it gave us the opportunity

The ADM-141 was trialed by the Naval Air Warfare Center (NAWC) and cleared for fleet use with Tomcat units following the conclusion of testing at Patuxent River on April 28, 1994. Intended to confuse and saturate enemy radar, AAA and SAM sites as part of an overall suppression of enemy air defenses strategy, TALD proved very successful during the opening stages of *Desert Storm* when employed primarily by A-7Es. Aside from its limited use in *Deliberate Force*, TALD were also used operationally by carrier-based F/A-18C units in the "Shock and Awe" phase of OIF I. (US Navy)

to release the decoy and then turn our aircraft away while other F-14s in the squadron ran intercepts and shot them down. We did this after the 1995 deployment in the W-72 operating area [off the coast of Virginia and North Carolina] when flying from NAS Oceana. This helped simplify the missile shoots, as we did not require BQM drones to be launched from their base at Dam Neck.

There were a few particulars about TALD employment that were important to note. It was a glide weapon with very small swing-out/pop-out wings that were not very strong or stable. I don't recall the exact parameters for employment, but TALD had to be launched in, or near, level flight, with turning forces not exceeding 1G. Lower airspeeds/mach numbers had to be observed or else the wings would deploy prematurely, creating unstable flight conditions that caused the decoy to tumble out of control and be lost.

Of course, TALD was made to be seen from the ground up, so it had a radar reflector in the nose. When being used in missile shoots, you wanted the decoy to have as long a glide-range profile as possible. This meant launching it from a higher altitude, well above the intercepting fighter, so that the guidance radar in the missile could "look up" at the TALD. Such an altitude advantage also allowed the decoy's radar reflector/cross section augmenter to "boost" the size of the TALD's radar return, thus making it look like a real aircraft. This concept of launching from higher altitude so as

to maximize the range profile also applied to TALD's use in the SEAD mission. TALD could also be pre-programmed to descend at different speeds (higher speed meant shorter range), as well as to turn and maneuver a little so the decoy looked like a real aircraft.

Due to its simplicity of operation, we never trained with TALD pre-cruise other than to study the tactics in which it could be employed and reading through the manuals associated with the decoy. TALD was a straightforward thing to use and drop since it didn't have any explosives in it. You just programmed the TALD on deck and released it at whatever altitude you had briefed in order to give it the maximum range desired.

I don't recall there being any special profile for our *Deliberate Force* mission. We were part of a [CVW-8] SEAD night strike, playing our part with TALD. The intent of this mission was to get a division of F-14s to drop TALD as decoys simulating a strike package heading in-country, and then F/A-18s and EA-6Bs [from CVW-8] would follow up with pre-emptive HARM shots [more than 25 AGM-88s were expended by CVW-8 squadrons during the 14 days of *Deliberate Force*]. The goal was to knock out enemy radar and SAM sites, with the latter posing a very real threat to us throughout our time in-theater. Mobile SAMs were our greatest concern, and we had to keep them firmly in mind when we executed missions further into Bosnia than normal [the TALD missions, for example], deploying weapons against known fixed, or expected, SAM and radar sites.

We briefed the operation as a later evening launch to have the greatest chance of surprise. It was a large strike package, with TALD droppers, HARM shooters, E-2 coverage, S-3 tankers, ES-3 ELINT

Above: Lt(jg) Paul Filardi poses with two TALD after checking that they have been correctly attached to his jet's BRU-42 ITERs on the flightdeck of *Theodore Roosevelt*. He successfully employed both TALD that night during a CVW-8 *Deliberate Force* SEAD strike. A total of 39 ADM-141s were expended by VF-41 between August 29 and September 11, 1995 in support of operations in the Balkans. (Capt Paul Filardi)

Left: In May–July 1996 CVW-8 joined *John F. Kennedy* for a brief 70-day NORLANT cruise that included port calls in Ireland and England. Most of the aircraft embarked in the carrier by VF-41 had been involved in the unit's *Deliberate Force* deployment of the previous year, including TARPS-capable "Fast Eagle 113" (BuNo 161161) complete with revised tail markings and two TALD. This jet served with VF-41 from October 1994 through May 1997, when it was passed on to VF-101 and dispatched to the Aerospace Maintenance and Regeneration Center (AMARC) shortly thereafter. (Capt Ken Neubauer via Warren Thompson)

[electronic signals intelligence] and helicopters doing their thing. Having briefed as a large force event, we were then put "on hold" until further notice. Something was causing a delay to the launch, so we just waited, either hanging around the ready room or heading up to our aircraft to chat with the flightdeck crews. This went on for hours and hours until finally, just after midnight, we were cleared to launch.

The launch and formation set up went smoothly, which was no small feat considering that the mission was being flown at night and there were a lot of aircraft in one relatively small area of sky. Our F-14s were not configured for night-vision goggles [NVGs] at that point, which made missions such as this one all the more difficult. Conversely, the NVG-equipped F/A-18 guys would often say they could see too much!

Although our strike was now running late due to the delays, at least we were all airborne, set up on the attack axis and heading to our launch points. We were "min[imum]-comms that night so as not to give away our presence to the enemy. As we hit our release point, my pilot, squadron XO Cdr Bob Brauer, launched the TALD, and we felt them drop away with a clunk under the jet. Shortly thereafter, the HARM shooters started launching their weapons, which was interesting because they were just in trail of us. Their HARM lit up the night sky, the missiles appearing to be perilously close to us as they streaked over the top of our formation on their preset routes. They would lock on to transmitting fire control radars searching for the TALDs.

Our part of the mission was now over, and we turned around and went back "feet wet" to our safe area over the Adriatic, before returning to the boat. We made a routine night recovery, which was followed by a debrief and then a late-night snack in the dirty shirt wardroom as we wound down from the multi-hour mission. The squadron completed another TALD mission the following night.

TALD continued to be used as a target for Tomcat units during their yearly missile shoots through to the aircraft's retirement, with decoys routinely meeting a fiery end during air wing integration deployments to Fallon. However, none were employed operationally by F-14 squadrons after *Deliberate Force*.

VF-41's "Fast Eagle 100" (BuNo 161607) had been one of two aircraft from this unit to give the Tomcat its operational debut in the air-to-ground role when they each dropped a single GBU-16 Paveway II on a Bosnian-Serb ammunition depot on September 5, 1995, the weapons being "lased" on to the target by accompanying F/A-18Cs from CVW-8. Featuring one of the smallest bomb symbols ever seen on a US military aircraft, BuNo 161607 approaches the runway at Oceana in May 1996 with an LGB silhouette on its port nose gear door just above the "100" modex. (David F Brown)

CHAPTER 3

LANTIRN REVOLUTION

By the time VF-41 blooded the "Bombcat" over Bosnia in September 1995, the aircraft was well on the way to becoming a precision strike fighter thanks to a program instigated literally "under the radar" within COMNAVAIRLANT. A number of key individuals responsible for providing the F-14 with the capability it needed to drop LGBs and, ultimately, GPS-guided weapons, recall how the aircraft's frontline fortunes were revitalized.

In a unique partnership between the US Navy and defence contractors, the Tomcat was made compatible with the AN/AAQ-14 LANTIRN pod. Martin Marietta's Dan Fischoff was in the vanguard of the upgrade from the beginning:

In the June 5, 1992 edition of *Aerospace Daily*, the lead article led with the headline "Navy pilots anxious to apply Gulf War lessons and expand plane's roles." In the article, the wish list included "LANTIRN pods for the F-14." Norm Augustine, the Chief Executive Officer of Martin Marietta, sent a note to his Vice President of Business Development stating "we need to make this happen." The note was in turn sent to me and I went to work. At the time, I was Manager of Advanced Programs for LANTIRN. This note started a three-year labor of love that resulted in a sole source procurement of LANTIRN pods for the F-14.

LANTIRN had undergone its operational test and evaluation in *Desert Storm* on the F-15E. During the campaign the USAF dropped more than 90 percent of the LGBs expended in the conflict. The A-6 was facing retirement and the [AN/AAS-38A] Nite Hawk pod on the F/A-18

Point Mugu-based VX-4 was heavily involved in missile testing with the F-14 throughout the jet's long frontline career until this role was taken on by VX-9 from September 1994. One of the last trials performed by the unit prior to its disestablishment was flight testing the AGM-84A Harpoon/SLAM air-to-surface missile. Only dummy rounds were fitted to the aircraft's shoulder pylon, the weapon being under consideration for the upgraded and ultimately stillborn F-14D Quickstrike program. Seen on the Point Mugu ramp, this F-14A is also carrying an inert Mk 84 2,000lb (907kg) bomb. (US Navy)

lacked long-range capability and most did not have laser designators. We saw the inherent beauty of the F-14 as a precision strike aircraft. It had outstanding range, loiter and weapons carriage capability, and it could easily carry the LANTIRN targeting pod on several stations. Many observers felt that the Navy would never really go forward with the program since the F-14 was planned for retirement starting in 2003, and that significant funding was going toward F/A-18 improvements. Others felt that even if the Navy program went forward, they would never put an "Air Force pod" on their jets.

Because of the importance of the program, we enlisted assistance from many people within Martin Marietta and outside consultants. We hired Whitney, Bradley & Brown [WBB] of Vienna, Virginia, to assist with a Concept of Operations for the F-14 with LANTIRN, and to help with obtaining support from the various stakeholders within the Navy and OSD [Office of the Secretary of Defense]. The core team for getting LANTIRN on the F-14 was myself, Monty Watson (a brilliant systems engineer), WBB's Capt Jim "Ruff" Ruliffson and Lt Cdr Dave "Hey Joe" Parsons – the former a Naval Aviator and the latter an NFO.

The F-14 Block 1 Strike upgrade program went through many different iterations during the three years between concept and contract. First, Grumman attempted to turn the desire for precision strike into new F-14D production, augmenting the jet with new computers and software. This program had a price tag in excess of US$3 billion, and was unaffordable within the limited budgets for Naval Aviation. Unsurprisingly, the program was never initiated.

Right: The most important weapons trials undertaken by the US Navy's flight test community in respect to the F-14's emerging role as a bomber took place in January–February 1994, when the Strike Aircraft Test Directorate of NAWC's Aircraft Division at Patuxent River performed a series of drops with inert GBU-16 1,000lb (454kg) Paveway I LGBs. NAWC NF-14A workhorse BuNo 159455 carried out the bulk of these flights, having also been involved in the original Mk 80 series GP bomb trials in the late 1980s, 23 air-to-ground sorties in November–December 1990 and the TALD clearance program in 1993–94. (US Navy)

Left: VF-21 Strike Fighter (1992)

By the time BuNo 159455 was tasked with conducting ordnance separation tests with the 2,000lb (907kg) GBU-24B/B hard target penetrator bunker-busting LGB in May 1996, the aircraft had swapped its long-lived DayGlo orange twin fins and NAWC titling for the striped fins, lightning bolt and SD codes adopted by the Strike Aircraft Test Directorate in the mid-1990s. This jet is fitted with FPU-1A/A external tanks that have been modified to carry cameras to record the weapon separation. Both the Tomcat and its ordnance are liberally covered with calibration markings to assist engineers with their analysis of the footage shot during the clearance trials. (US Navy)

In early 1994, the Navy decided that targeting pods would not be procured for the Tomcat. It came up with two "low cost" alternatives that required funding of about US$300 million. One was the integration of JDAM only, and the other was a radar upgrade called the "two box mod" that was intended to provide a modest targeting capability in the air-to-ground role.

Martin Marietta went on the offensive throughout 1994 to show that neither of these upgrades could compare to the true precision strike capability of the LANTIRN targeting pod. These points were effectively made by operations analyses and Concept of Operations [CONOPs] briefings, but we still had to make the system affordable. At this point, we made the connection with Fairchild, which had upgraded a small number of F-14Bs as part of the Multi-Mission Capability program. The key part of the upgrade was the installation of equipment that translated the analog signals of the existing AN/AWG-9 radar and AN/AWG-15 weapons control system into a format readable by a MIL STD 1553 digital databus.

Martin Marietta subsequently enlisted Fairchild to produce what we originally called the "Precision Strike Processor." This box was installed into the port console of the F-14 RIO station, and it featured a hand controller left over from the A-12 parts bin. We got permission from the Air Force to borrow two LANTIRN targeting pods, and came up with a plan to incorporate weapons

NAWC's NF-14A BuNo 159455 performed both level drop and dive-bomb separation delivery tests with the 2,000lb (907kg) GBU-24B/B hard target penetrator bunker-busting LGB in May 1996. The GBU-24 would earn a poor reputation for itself within the Tomcat community, being dubbed "pretty unreliable and a non-user-friendly weapon that doesn't have a high hit percentage" by VF-14 Operation *Allied Force* and OEF veteran Lt Cdr Van Kizer – he would get to drop a GBU-24 on a fuel/ammunition storage dump in Kabul on October 17, 2001. VF-41's Lt Cdr Scott Butler remembered that crews assigned GBU-24 missions had to spend "hours weaponeering and target planning in order to ensure the accurate delivery of the bunker buster. It was a labor-intensive weapon tailored exclusively for use against hardened targets, and fortunately for us, there weren't too many of those in Afghanistan." (US Navy)

delivery software into the LANTIRN pods. We obtained internal funding to modify the LANTIRN pod software and add Inertial Navigation System [INS] and GPS capabilities to wrap up the precision-strike solution.

On November 8, 1994, we briefed Vice Admiral Dick "Sweet Pea" Allen, COMNAVAIRLANT [and a former A-6 Bombardier/Navigator], on our concept. As a result of the meeting, he gave us the green light to proceed, and within six months we had a fully integrated precision strike solution on the aircraft. The Navy subsequently performed several LGTR and inert and live LGB deliveries with the modified F-14. We worked with the Navy to produce a videotape of the results, and we and the Navy started obtaining support from the Commanders in Chief [CINCs] and fleet commanders around the world.

In June 1995, a *Commerce Business Daily* announcement came out stating that a contract would be awarded to Martin Marietta to install LANTIRN on the F-14. On the cover of the June 30, 1995, edition of the magazine I wrote a note saying, "We Made This Happen", and sent it to Norm Augustine. This will always be the most important accomplishment of my career because of the positive impact that the LANTIRN capability had on the F-14 community and Naval Aviation.

Pilot's Perspective

As Dan Fischoff noted, one of his allies in this program was Capt Jim Ruliffson. A highly experienced fighter pilot, he too was committed to giving the Tomcat a new lease of life:

As a career F-4 Phantom II "driver," I never had the privilege of flying the Tomcat operationally. However, I did get to pick up aircraft No 21 [BuNo 158619, which was lost during a flight on February 22, 1977, when it entered a flat spin while testing its Pratt & Whitney TF30-P-414 engines] from the Grumman factory in the summer of 1974 while I was stationed at NAS Patuxent River after attending the US Naval Test Pilot School. Shortly thereafter, I got surprise orders to TOPGUN as CO in April 1975, which gave me both the opportunity and the obligation to truly understand the F-14 and its formidable weapons systems through the demanding ground and air syllabus at the Naval Fighter Weapons School.

Subsequent to a fleet CO tour of an F-4 squadron, I attended the Naval War College and used my TOPGUN background to co-author a classified study of the Tomcat weapons system versus the Tu-22M "Backfire" bomber threat, focusing on offensive and defensive Electronic Support Measures/Electronic Counter Measures [ESM/ECM] systems. This was during the height of the Cold War War-at-Sea tactical and training renaissance. The study was incorporated into TOPGUN's TOPSCOPE training as the primary text, which was very gratifying to me.

After that assignment, I had a tour in the Pentagon as the Fighter Analyst in Office of the Chief of Naval Operations [OPNAV], followed by the plum job as Deputy Program Manager [DPM] for Development

Although most LGBs would be dropped in horizontal flight once the F-14 was cleared to employ the weapon in fleet service, a number were also expended in dive-bombing attacks. This meant that the bomb had to first be cleared for delivery in angled flight by the Strike Aircraft Test Directorate. This photograph was also taken during the GBU-24 trials in May 1996. (US Navy)

in the F-14 Program Office [PMA-241]. This effectively put me in charge of both the F-14A+ and the F-14D, and their upgraded weapons system. After daily detailed engineering discussions of system capability and architecture with Navy and Grumman engineering staff, I felt that I knew the jet as well as anyone could without flying it for a living. When I retired in 1986 with the rank of captain, I had completed 24 years of service. The final three had seen me intimately involved with the Tomcat. Little did I know that this was to continue in my retired profession as well.

After I left the Navy, I became a defense consultant as a founding partner of WBB, and continued to work systems integration on the Tomcat in support of the ASW-27C digital data link and AN/AWG-15 and IRST [infrared search and track] upgrade. However, the most gratifying and transformational work came in 1994 when the LANTIRN "caper" began to unfold.

When word came that the F-14 Block 1 Strike upgrade would not be funded, I had already been working with Martin Marietta on integrating its USAF-developed LANTIRN pod on to the Tomcat on the presumption that a MIL STD 1553 digital databus would be present as part of the Block 1 Strike upgrade. Without the databus the pod was incompatible with the jet, making it a non-starter.

I specifically remember to this day the momentous meeting with Dan Fischoff of Martin Marietta in which he said, "We need to figure out a way to get LANTIRN on the jet without having to incorporate it into the software build." That was a "Eureka!" moment for me, as I realized that I knew exactly how to do it, and didn't know why I hadn't suggested it before. Fairchild had built a MIL STD 1553 translator for a classified program that I had been involved in as DPM, and it had all the relevant data needed by the LANTIRN pod for stabilization and pointing accuracy.

The tail fins of an inert 2,000lb (907kg) GBU-10 have just started to "pop out" as the weapon falls away from a NAWC F-14A that has cameras fitted to its modified port shoulder pylon. The jet also has camera-equipped FPU-1A/A external tanks. (US Navy)

Dave "Hey Joe" Parsons had just joined the WBB team with his extensive AIM-9X systems integration experience, and after hashing over the integration issues, we jumped in my car and raced up to the Fairchild plant near Germantown, Maryland, to chase down the actual translator and see if it was in working condition. It was, and we then proceeded to brief the integration concept to Capt Bob "Velcro" Riera, who had recently taken over the job as head of PMA-241. After we had explained about our unorthodox concept, he turned to his lead software engineer, Gary Stuart, and asked if it would work. His reply was "I'm afraid it will." That's when we knew we had a winning concept, but there was no money to demonstrate its feasibility. We convinced Martin Marietta to fund the effort, but we needed a Tomcat. And without funding, the program office was helpless to assist until OPNAV's cumbersome budget process could respond, which by then would be too late.

Vice Admiral Dick "Sweet Pea" Allen was an attack B/N who had seen combat in Vietnam flying A-6 Intruders. In his final US Navy tour, he served as COMNAVAIRLANT, and in that role he proved to be a staunch ally for those pushing to provide the F-14 with a precision-bombing capability that would allow the jet to take over the strike role following the retirement of the Intruder. (US Navy)

Former Crusader and Tomcat pilot Admiral Jay Johnson cited the delivery of the LANTIRN system to fleet F-14 squadrons in just 223 days from contract awarding as a primary example of innovation at work in the US Navy. He highlighted the "renaissance" of the F-14 in one of his first speeches after becoming the 26th CNO in August 1996. (US Navy)

In November 1994, we turned to COMNAVAIRLANT, Vice Admiral Dick "Sweet Pea" Allen, for support. He had been a staunch advocate for F-14 Precision Strike in his prior tour at OPNAV, and was now holding his former OPNAV staff cohorts to their promise to make sure the Tomcat was provided with this capability after the decision to retire the A-6 had been made. Allen jumped on the chance to demonstrate the capability to OPNAV. As COMNAVAIRLANT, he could authorize use of one jet for this purpose, and he directed his staff to co-operate with us to demonstrate LANTIRN on the Tomcat.

By the spring of 1995, a VF-103 jet was getting "shacks" with LGTRs and inert LGBs. By any measure, the demo was wildly successful, and OPNAV and NAVAIR were able to respond accordingly by making it an official program in record-breaking fashion and fielding the first LANTIRN pods with VF-103 in just 223 days from the contract being issued. We didn't stop there, as additional capability was subsequently inserted through Tomcat Tactical Targeting [T3] and Fast Tactical Imagery [FRI], followed by JDAM integration – all at relatively low cost and also in record time. The ultimate proof of success was on deployment, as the Tomcat quickly became the heavyweight strike capability for the air wing first over Iraq in OSW, then in Bosnia, Kosovo, Afghanistan and, finally, over Iraq once again.

I hold to this day in high esteem a letter from the then-Commodore of FITWINGLANT, Capt Mark "Clem" Clemente, who stated I had done more for the F-14 and Tomcat community after I retired then I ever did on active duty!

PMA-241 Boss

Heading up the F-14 Program Office during this critical period in the "Bombcat's" development was Capt Bob "Velcro" Riera. His support for LANTIRN was crucially important in getting the targeting pod into fleet service so expeditiously. Years later, he recalled:

In one of his first major speeches to past and present leaders of Naval Aviation, newly appointed CNO Admiral Jay Johnson [a former F-8 and F-14 pilot], presented his vision for the Navy while under his command. He stated that he would use four principal stars to set the Navy course, and to prioritize his day-to-day activities. Similar to his predecessor Admiral J Michael Boorda, Johnson's first guiding star was "our people first." His second star, continuing in order of prominence, was "innovation in the way we do business." "Innovation," he said, "is crucial for the Navy if we are going to be able to survive in this era of drastically declining budgets."

Johnson's primary example of innovation was the F-14 precision-guided weapon system, or LANTIRN, delivered fully operational to the fleet in just 223 days from contract award. Johnson went on to talk about the F-14 "Renaissance" and the remarkable resurgence of morale within FITWINGLANT. He credited this to the delivery of new capabilities such as LANTIRN and the highly effective digital reconnaissance system, TARPS-DI [Digital Imaging]. "No other aircraft," Johnson said, "at this stage of its life, had ever been able to regain such prominence and critically needed capability."

What made his comments so notable was not the fact that he was highlighting the F-14's incredible performance. Rather, it was the fact that just 12 months earlier [1995], the Tomcat community had suffered one of the worst years ever. It had lost three F-14s in less than one month, and two more were to follow just a short time later. Moreover, having been designed in the late 1960s and built in the early 1970s, the aircraft was being plagued by an ever-increasing number of prohibitively expensive reliability, maintainability and safety problems. These problems were threatening the existence of the jet because, simply, the Navy could no longer justify the F-14's high cost for its marginal return on investment. Correspondingly, Tomcat community morale was at an all-time low because *Desert Shield/Desert Storm* had confirmed that the F-14, without precision strike capability, lacked a meaningful "mission" for the future. So, how did we go from this crisis state to a CNO-described "renaissance" in just one year?

I had just taken over as PMA-241 in 1995, with a prior tour there and in the fleet as a fighter pilot. With probably more bravado than the current situation justified, we fashioned a comeback strategy around the motivational rallying cry "The Cat is Back." Seizing the opportunity to introduce LANTIRN to the Tomcat after the successful demonstration of an industry-funded concept demonstration in early 1995, the real work to formally build a Navy-sponsored "program" began in the fall of 1995 with the Operational Advisory Group.

In just two days, this team of select senior and junior officers representing the entire Tomcat community, including TOPGUN, put together a realistic F-14 "Roadmap for the Future," principally utilizing the premise "what's good enough?" The roadmap fully balanced operational requirements with safety and logistic improvements, thereby sending the clear message that the F-14 community understood full well that operational capabilities such as LANTIRN and TARPS-DI were worthless if the jet was down for maintenance.

After the Tomcat Roadmap was completed, the F-14 Program Office, acting as the fleet's trusted (but as of then still unproven) agent, presented it to the leaders of Naval Aviation for their approval. The opening and closing pages of the Roadmap brief were titled, "The cheapest car you'll ever own is the one you're driving," and "A wise man makes sure he has reliable transportation, until he has his new car in the garage." Although the many Flags [admirals] present may not have fully appreciated our thoughtful humor, they did appreciate our plan. What really got their attention, however, was our proposal that if they would protect our current budget, we would fund the entire Roadmap requirements – a total of US$200m – from entirely within our budget. Initial skepticism was soon replaced by true belief and acceptance after the Flag Board reviewed our US$200m of proposed offsets such as F-14D AMRAAM and 14 F-14B upgrades.

Additionally, we provided written (a small miracle, and testimony of the quality of the Roadmap) approval of both the F-14 Roadmap and offsets by FITWINGLANT, TYCOMS [type commanders] and CINCs. With the full blessing of the Flag Board – meaning "this idea had better work, Captain" – we were on our way.

Relying on the incredible power of a fully, and truly, integrated Fleet, Industry and Program Office Team, we started seven new programs and restructured nine existing programs for effectiveness and efficiency. At the start, the team determined that both the new and existing programs needed to be run with a new "game plan." To constantly remind us of the new rules of the game, we created and followed imperatives such as "Failure is not an option," "Do what you say you are going to do," "The program has failed if it's even one day late" and "If you say it's going to cost X, it better cost X, or less." In other words, in the new game, we needed to restore and maintain one principal thing – program credibility!

In this new game, we adopted modern business processes common to industry but new to government such as earned value management, concurrent modifications and inventory control. We instituted our own new program like zero-sum program-operating memorandums and targeted force structure reductions. The pay-off was tremendous, and well worth the considerable effort. In addition to providing the Fleet with proven capabilities such as LANTIRN and TARPS-DI, the team delivered every program it started on-cost, schedule and performance. These included programs critical to the effectiveness, safety and protection of our aircrews such as Joint Tactical Information Distribution System [JTIDS], Airborne Self-Protection Jammer [ASPJ], AN/ALR-67 [Radar Warning Receiver], BOL Chaff, TF30 Breather Pressure Indicator, Night Vision Devices and the much needed safety improvements provided by the Digital Flight Control System [DFCS].

We, the team, were truly proud of what we had done in the defense of our country. We believe that the F-14 retired with the sentiment that it was and still would be the battle group commander's weapon of choice for strike and air superiority missions because we had honored the tenets of "set your sights high," "be innovative" and, most importantly, "do what you say you are going to do" and more. "THE CAT WAS BACK!"

"Bombcat" Proponent

A veteran of *Desert Storm* and a staunch Tomcat advocate, Lt Cdr Dave "Hey Joe" Parsons of WBB also played a key role in the evolution of the "Bombcat." The following account of the program from his perspective was originally published in Erik Hildebrandt's outstanding *Anytime, Baby!* volume on the F-14, released in 2006:

Since its introduction to the Fleet Air Defense mission, the Tomcat set new standards for range, endurance, radar detection and multi-mission launch capability. The Tomcat performed this mission well throughout the Cold War, but was never called upon to show its claws in air-to-air combat in naval service except for a few isolated instances. It is perhaps the ultimate irony that the Tomcat, designed to be the supreme air superiority machine, left a much larger combat legacy as a precision strike fighter – a role it was never envisioned to perform. Indeed, it very nearly missed the opportunity altogether. As the Cold War abruptly ended and *Desert Storm* concluded, the Tomcat was looking like a prizefighter past its prime, and in the Pentagon budgetary debates of the early 1990s it came under scrutiny for early retirement.

Photographed during VF-32's *Desert Shield/Desert Storm* cruise in 1990–91, Lt Cdr Dave "Hey Joe" Parsons was a NFO for 15 years – he had previously served for five years in the US Marine Corps. Having joined WBB immediately after leaving the US Navy in March 1994, Parsons was a key facilitator in the "great LANTIRN integration caper" that turned the F-14 into a precision bomber. (via Lt Cdr Dave Parsons)

At the time, the Tomcat was slated for a multi-billion-dollar program initiative called Block 1 Strike to upgrade the F-14 to the strike fighter role with precision strike capability. However, the defense budget was under attack and demands for a peace dividend resulted in many program cancellations or severe reductions. It came down to eliminating a type/model/series from the carrier deck. The Navy pitted the Tomcat against the A-6 Intruder, another Grumman product, and itself proposed for the wholesale upgrade/modernization of all remaining examples to A-6F configuration. In the Navy program inner sanctum, the contest was almost too close to call.

The A-6 was older, but had proved itself in combat time and time again as a world class all-weather precision strike platform with long legs and unrivalled payload. The Tomcat was a few years younger, but had not proven itself at all in the strike role. In fact, some career attack aircrew doubted that the Tomcat community had the maturity and resolve to transition into the strike role, while a member of the Hornet "mafia" snorted, "Any dollar spent on a Tomcat is much better spent on a Hornet!" The tension was palpable, and when it finally came to a decision, Rear Admiral "Sweet Pea" Allen was in a very influential position as head of the Navy's Program Analysis Division. The former A-6 Bombardier/Navigator, who was slated to pin on a third star and become COMNAVAIRLANT, had participated in the many debates about which aircraft would remain in fleet service.

In the last rounds of deliberations Allen relented to let the Tomcat prevail, but only if it was upgraded with precision strike capability. He left the Pentagon with a final admonition to his fellow admirals in the Office of the CNO that he would come back and haunt them if they waylaid the funds designated to provide the Tomcat with a reduced scope precision strike capability.

In the early months of 1994, the Tomcat program was in a state of gloom and disconsolation. *Desert Storm* had relegated the F-14 to second string status, and without a robust upgrade there wasn't much chance of making the first string. There wasn't enough funding to fully integrate a FLIR targeting system, so it appeared that JDAM would be the only weapon available to the Tomcat to usher it into the strike role, and that would be years away. Potentially more troubling was the decision to equip each carrier with only a single Tomcat squadron, and transition a third of the Intruder Naval Aviators and NFOs into F-14s. Tomcat aircrew in all ranks instantly found themselves facing a reduction of available cockpits and in a virtual game of musical chairs.

The Pentagon budgetary drills in the fall of 1994 didn't look promising at all, and there were rumors of a raid on the remaining Tomcat modernization funding and retirement as early as 2003. With traditional research, development, test and engineering taking anywhere from five to 12 years to integrate and field new capability, the Tomcat needed a breakthrough, if not a miracle, to get precision strike capability and revive the flagging spirits of the once proud community.

F-14 Supersonic Attack – Bombcat (1994)

In spring 1994, a Martin Marietta (now Lockheed Martin) representative by the name of Dan Fischoff sat with Jim "Ruff" Ruliffson to discuss the state of affairs with the Tomcat strike upgrade, and whether a USAF LANTIRN targeting system could be integrated with the F-14. LANTIRN was the key to the USAF F-15E Strike Eagle's impressive precision strike capability, and had debuted in combat during *Desert Storm*. It had taken years to integrate the LANTIRN pod with the aircraft's weapons system software, followed by many months of testing for weapons compatibility and clearance before it was fielded.

Undeterred, Dan had enlisted the assistance of talented systems' engineer Monty Watson, who began tearing into the wiring diagrams of the Tomcat to see what made it tick.

Right: LANTIRN Tomcat (1996)

Below: Martin Marietta's Dan Fischoff (second from left) and systems engineer Monty Watson (center) pose with civilian contractors and a VF-103 maintainer during the critically important, and highly successful, LANTIRN test detachment to NAVSTA Roosevelt Roads. Fischoff was Manager of Advanced Programs for LANTIRN at Martin Marietta, and he spent three years overseeing the US Navy's sole source procurement of the targeting pod for the F-14. It was Monty Watson's task to virtually hotwire a LANTIRN pod on to the Tomcat, thereby eliminating a lot of the complications of a full-up integration and the requisite months of testing that went with it. (Capt Alex Hnarakis)

He was looking for a way to virtually hotwire a LANTIRN pod on to the Tomcat, thereby eliminating a lot of the complications of a full-up integration and the requisite months of testing that went with it.

"Ruff" brought me in as his new hire, and I was tasked with looking at the weapons integration issues and to work out the best way to approach the Pentagon for support. I had just left the Pentagon, having worked as a requirements officer who oversaw weapons integration on TACAIR [Tactical Air] platforms. I also had valuable insider knowledge of the latest Pentagon budgetary battles. Fischoff, Ruliffson, Watson and I became the nucleus of a team of experts in weapons integration, flight-testing, program management and Pentagon programmatics that began tossing around ideas, as well as discussing issues that would have to be overcome.

Jim Ruliffson had been involved in the testing of F-14D digital components on the analogue F-14A, and he knew that Fairchild still had translators and test equipment in a locked and sealed cage in its Germantown facility. He also knew the right people there who would donate some of their time to looking at the issue of getting the LANTIRN pod access to the aircraft weapon/navigation system data without a full integration. Meanwhile, I led the team to NAS Oceana to garner interest from Capt Dale Snodgrass, the FITWINGLANT commodore, who assigned Cdr Dana Dervay, a former VX-4 Operational Test Director and his staff Readiness Officer, to work with the burgeoning team of non-traditionalists who were beginning to believe that they could make a miracle happen. Monty began debating with the NAVAIR Tomcat Point Mugu crowd who weren't convinced of his approach and preferred the traditional method, but there wasn't enough time or funding for that path.

Once the issues were laid out, it appeared that not just one miracle would be needed, but a succession of miracles. However, there was the solid foundation of the LANTIRN pod to work with, and some fortunate breaks. One of the Tomcat initiatives cancelled in the downsizing was HARM integration testing. Luckily, it had produced a HARM adaptor for the Tomcat that had received flight clearance. It proved to be just the right configuration to mate to the LANTIRN pod, and the testers at Patuxent River were able to fly the latter to establish flight clearance on the F-14. Other major issues that needed miracles were the hand-control integration scheme, physical boresighting requirement and LANTIRN alignment.

Monty was able to design a hand-control unit that dropped into the rear cockpit console in place of the TARPS control panel. Using the A-12 hand-control stick, the unit hosted cards that allowed the LANTIRN pod to listen in on the Tomcat's existing INS alignment and the RIO's computer address panel actions, thereby providing the LANTIRN pod with the information it needed without actually interfacing with the AN/AWG-9 radar. This was a major breakthrough. Monty also designed the cards so they were a single configuration for all Tomcat models, despite insistence by NAVAIR experts at Point Mugu that it couldn't be done.

The physical boresight to the aircraft armament dataline required a large boresight rig for the F-15E, and there wasn't time or funding to duplicate that approach for the Tomcat. Borrowing

Right: Legendary high-time F-14 pilot (4,900 flying hours in the jet) Capt Dale "Snort" Snodgrass was the FITWINGLANT Commodore at NAS Oceana when the LANTIRN pod was being trialed for use in the front line. A staunch supporter of the upgrade, he flew nine sorties in a LANTIRN-equipped F-14B during VF-103's test program with the pod in the spring of 1995. Snodgrass was killed in a flying accident on July 24, 2021. (US Navy)

Below: The AN/AAQ-25 LANTIRN pod was exclusively responsible for revolutionizing the F-14's combat employment in the jet's final decade of frontline service with the US Navy. Featuring an integral GPS for position information and an IMU for improved stabilization and accuracy, the pod also boasted an internal computer with ballistics data for the various PGMs carried by the F-14. Each pod cost the US Navy around $3 million apiece, which meant that only 75 were bought for fleet use. Typically, a squadron would take six to eight pods with it on deployment, and these would be permanently fitted to non-TARPS jets. (Danny Coremans)

slew the pod to selected waypoints without having to use radar cueing. This feature, which was not resident in the F-15E, would give the Tomcat an important capability when it arrived in-theater.

By the late summer of 1994, the NAVAIR F-14 program office wasn't keen on the seemingly radical integration being developed by Monty and the crew, and even worse, it would not provide an aircraft for testing once Monty had a working solution to the LANTIRN integration. NAVAIR Tomcats do not fly without program support (funding coming from a program of record – the LANTIRN caper was merely a contractor "proposal"). By the fall, we had solved the major issues and now needed a Tomcat to prove that the unconventional integration scheme would work.

Cdr Kelly "Psycho" McBride, an experienced test RIO, was in charge of advanced development in PMA-241 at the time. Although he knew what had to be done, he had no money, and time was running out. The Tomcat was in real danger of being retired altogether unless a miracle happened before the next round of budget cuts that were sure to come. Then the start of the miracle began.

As previously noted, it was centered around the notion of integrating a USAF-developed LANTIRN pod on to the Tomcat without undergoing formal software development or integration with the aircraft weapons system in the traditional sense, which would take years and tens of millions of dollars. It started as an unsolicited proposal from a contractor team comprising three companies (Martin Marietta, WBB and Fairchild), and was dependent on Martin Marietta's management to

The multi-purpose pylons located under the Tomcat's wing glove were initially built for missiles, with a combination of AIM-7s, AIM-9s and AIM-54s being fitted on the two stations. However, from 1996 onwards, the lower pylon on the starboard side of the jet was rewired to accept the fitment of the most important store carried by the F-14 in its final decade of fleet service – the LANTIRN pod. The following year, the Swedish-built LAU-138 weapons rail/countermeasures dispenser, seen here, was also introduced. Used for the carriage of AIM-9Ms, the rail could also be filled with 160 rounds of either BOL chaff or BOL infrared flares. (Danny Coremans)

The US Navy's LANTIRN pods were delivered in Gunship Gray, with some examples (like this one) subsequently being repainted light gloss gray. To keep as many pods serviceable as possible, pod sections were swapped out during maintenance and whole pods transferred between squadrons – units going on deployment had priority for serviceable pods. This externally mounted scoop provided cooling air for the avionics within the pod. (Danny Coremans)

from my AMRAAM and AIM-9X weapons integration experience, I recommended using an internally mounted Inertial Measurement Unit [IMU] that both missiles employed. This not only eliminated the need for the bulky and single point Air Force solution, it provided continuous boresight alignment and, surprisingly, helped in image stabilization. More importantly, when Litton was contacted for a bolt-on IMU, the team discovered it was available in an integrated GPS configuration that provided the Tomcat LANTIRN integration with an unprecedented ability to

approve the requisite investments on their part to prove the somewhat unorthodox, if not radical, integration scheme to provide precision-targeting capability that would work.

The "little hen that planted the wheat" was Dan Fischoff from Martin Marietta, and he had an uphill challenge with his own company and the Navy. WBB had the systems integration expertise and insider knowledge of the Navy requirements and budgeting process to help lay out the strategy. Jim "Ruff" Ruliffson, a WBB partner, retired as the F-14D program manager after a sterling fighter community career and enlisted my assistance. Fresh from an OPNAV requirements tour, I was an expert on systems integration from my role in AIM-9X development. "Ruff" knew the Tomcat inside out from the development perspective, and had an idea how the LANTIRN could be rapidly integrated on to the jet. I also knew how Washington and the Pentagon worked, having taught this subject as an instructor while still in the Navy. All of our collective knowledge would come into play.

Without official OPNAV or NAVAIR sponsorship, there was some skepticism outside the team. Indeed, many thought that Dan was literally putting his career on the line. The F-14 Requirements Officer, Capt Stan "Steamer" O'Connor, was definitely interested in seeing the concept achieve success, but he could not openly endorse the efforts of the team as it was not an official Navy initiative. In fact, OPNAV was on record supporting the JDAM integration instead, which would take years and likely have the funding pulled anyway. The Tomcat needed an instant solution, or something within a year or it may not survive.

The team now saw Vice Admiral Allen as a potential ally, as we knew that he had previously taken a positive stand on the development of the Tomcat's precision targeting capability. As COMNAVAIRLANT, he had both a strong voice and half of the fleet's operational Tomcats. The team only needed access to one, and although fleet aircraft are not part of the test community, a jet could be used for rapid prototype testing under certain conditions. During an audience with Vice Admiral Allen, he was briefed on the proposal by one of Martin Marietta's vice presidents. To their great delight, and as they had hoped, Allen directed his staff to provide an aircraft and whatever support the team needed.

The team huddled with Capt Snodgrass, who determined that VF-103 would be the squadron to participate in the testing, which was planned for the spring of 1995. A pilot and RIO, VF-103 XO Cdr Alex "Yogi" Hnarakis and Lt Larry "Rat" Slade, were designated as the test crew. Both had prior tours in the Development and Operational Test communities, respectively, which was a huge plus both then and later, when the case was made to forego formal testing and allow LANTIRN to deploy immediately.

A USAF LANTIRN pod was shipped to the Fairchild Germantown facility, where "Rat" and I checked out the displays on a Tactical Information [TID]. Symbology from the pod and an AN/AWG-9 simulation were interleaved on the TID to avoid any overlap between the two systems. To the uninitiated eye, it appeared that the LANTIRN pod and the aircraft's AN/AWG-9 had been integrated seamlessly, when there were, in fact, two independent sources of information being presented on the TID. By March 1995, the pod was back at NAS Oceana and mounted on a VF-103 F-14B appropriately adorned with *FLIRCAT* nose art.

During the first round of testing the aircraft was scheduled to fly out of Oceana to nearby Dare County, where it would drop LGTRs. The second round in the program involved live LGBs at the Vieques Island range in Puerto Rico. Both rounds were unqualified successes, and the team felt a deep sense of satisfaction. It was obvious that the miracle had indeed come true and the Tomcat had been transformed into a potent strike fighter. The major hurdle now was to make the LANTIRN integration an officially sponsored program with the involvement of NAVAIR and the F-14 program office, which would take on the task of procuring Navy LANTIRN targeting pods and the associated hand-control units and then work with the fleet to conduct installations.

Right: This rear view of a LANTIRN pod in situ on an F-14 at NAS Oceana reveals the store's exhaust vents. The rear of the LAU-138 is also open, ready for a cartridge of BOL chaff or flares to be installed. (Danny Coremans)

Far right: The starboard wing glove multi-purpose pylon could be configured with four different adapters depending on the ordnance to be carried – AIM-7, AIM-9, AIM-54 or, from 1995, the LANTIRN targeting pod. As previously noted, only the starboard pylon was rewired for pod fitment. The shape of the vertical pylon differed depending on the store uploaded. (Danny Coremans)

Capt Bob Riera, the newly installed F-14 Program Manager, saw the value of LANTIRN and what it meant to the community. He assigned Jim Blackmon the task of making it all happen. Capt Snodgrass now brought in his maintenance experts, led by Cdr Pat Taylor, to align the fleet Tomcats with the NAVAIR effort. The Navy and Industry team were now united at all levels to make sure contracting, logistics, training and hardware all came together in rapid fashion. Years of traditional procurement and development actions were compressed into months and even weeks.

Although a consequence of timing, it came as a pleasant and just reward that Martin Marietta would be able to deliver enough LANTIRN pods to enable VF-103 to depart on cruise in June 1996 with the precision-strike capability. It had been a whirlwind year to get the LANTIRN integration concept proven as a viable capability, but it was now in fleet use. The excitement rippled

Left: The crowded cockpit of the F-14A, with its mix of multi-function displays and conventional strip and dial gauges. Note the lack of a HUD atop the cockpit coaming, vital inflight information being projected directly onto the windscreen immediately ahead of the pilot. Many of the key instruments and systems seen here were unique to the front cockpit, including the control column, wing sweep handle, throttle and landing gear actuator handle. (Danny Coremans)

Below: The RIO's cockpit in the Tomcat shared little in common with the space occupied by the pilot – it was also less cluttered. The central console in the F-14A/B was dominated by the 9in-diameter TID, with its dedicated hand-control unit mounted on a pedestal immediately in front of it between the NFO's legs. The rectangular panel immediately above the TID was the Detailed Data Display console, and at its center was a small cathode ray tube display. (Danny Coremans)

LANTIRN REVOLUTION

through the community. Many former A-6 aircrews were by now part of that community following the disbandment of virtually all the Intruder units. They brought with them their extensive strike planning and weapons execution expertise.

The Tomcat community did not forget the crucial role Vice Admiral Allen played in allowing use of a fleet F-14 to conduct the proof of concept. The contractor team got together and endowed an annual award in his name for the best Tomcat squadron contribution to precision strike. Individual awards were to be presented by FITWINGLANT at the annual Tomcat Fighter Fling event, where traditionally all other types of awards were presented. Vice Admiral Allen was invited to present the first such award [in 1996], which went to VF-103 for its significant role in demonstrating that LANTIRN worked.

A little over a decade after test community Tomcats had conducted intensive trials that cleared a range of dumb and laser-guided bombs for fleet use, the aircraft commenced similar testing with the 2,000lb (907kg) GBU-31 JDAM. Naval Weapons Test Squadron NF-14A BuNo 161609 participated in the weapon carriage phase of the evaluation program, with the jet being seen here starting its engines at a wintery Point Mugu. A-model Tomcats were unable to drop JDAM operationally, as the aircraft could not generate computed aim point co-ordinates for the weapon due to the jet lacking a MIL-STD 1760 data bus and associated onboard GPS system. NF-14As were, therefore, the only A-model jets to ever carry the weapon, albeit inert examples like the one seen here. (US Navy)

CHAPTER 4
LANTIRN DEMO AND OPERATIONAL DEBUT

The late Capt Alex "Yogi" Hnarakis (who sadly passed away on May 3, 2018) was tasked with proving that the LANTIRN pod would indeed turn the F-14 Tomcat into the "multi-mission strike fighter and FAC(A) platform it was always capable of being!"

In December 1994, I [Capt Alex "Yogi" Hnarakis] joined VF-103 as XO for CO, CDR Steve "Snotty" Schlientz. Among the first things he asked was, "'Yogi', I have something I need you to do."

"Sure Skipper, I'm your XO. What do you need?"

"COMNAVAIRLANT, Vice Admiral Allen, has arranged for and secured NAVAIR approval for a demonstration of the Martin Marietta LANTIRN Targeting FLIR pod on a Tomcat. Martin Marietta is doing this on its dime, outside the Navy's normal acquisition process, but the company does not have a jet or crew. FITWINGLANT has asked VF-103 to provide the F-14, crew and fuel for some flying to work with the Martin Marietta engineers on this 'science project.'

"Lt Cdr Larry 'Rat' Slade, one of our super RIOs with VX-4 operational test experience, has been working for a couple of months on this, and there are a few more months of planning and

VF-103 was one of the first F-14 units to complete the TASS course, which had been established in fall 1991 to replace the FRS bombing course that had initially introduced fighter aircrew to the air-to-ground mission. F-14B BuNo 161601 was photographed at Oceana in early September 1993, carrying a 1,000lb (454kg) Mk 83 bomb and an AIM-54C (hidden from view) on its under-fuselage stores racks, plus an AIM-9L on its shoulder pylon. BuNo 161601 was lost just days after this photograph was taken when its crew was forced to eject 40 miles (64km) west of Cape Hatteras, in North Carolina, on September 13 after the jet became uncontrollable. Both the pilot and RIO were plucked from the water by fishermen. (David F Brown)

aircraft modifications to go before first flight. I specifically want you as the pilot because of your developmental test experience from your Patuxent River tour."

"Sounds great," I replied. "I'll start ramping things up with 'Rat.'"

"Snotty" then told me, "I chose you and 'Rat' because if this works, I don't want the acquisition system to claim that 'fleet guys can't test,' and then do years of testing – including the duplication of what you guys will do in this demonstration. I figure you guys will apply the planning, methodology, logical build-up and risk mitigation you learned in the test world."

Unsurprisingly, shortly after we had had this brief discussion, we were told that somewhere higher up in the chain of command it had been determined that we would not be allowed to call our well thought out and concisely written plan of action a test plan. Furthermore, the sorties we made with the LANTIRN pod could not be called test flights. We could care less! Unperturbed, we pressed ahead and called our test program a "demo plan," and the hops we made "demo flights." "Snotty" was both a visionary and clairvoyant!

After a few months of hard work by VF-103, led primarily by "Rat" Slade, and a very small Martin Marietta team led on the scene by Monty Watson and co-ordinated with the company by Dan Fischoff, we started flying F-14B BuNo 161608 in March 1995. Our pod differed from a standard USAF LANTIRN pod through the addition of a GPS/INS for cueing of the FLIR "soda straw"' [the image created by the sensor and displayed on the screens in the cockpit], since the F-14A/B/D air-to-ground radar was not good enough to use for target cueing. The jet's INS wasn't up to the job either, since it drifted too much and was not GPS-aided in those days.

Sluggers. (VF-103, 1995)

Ballistics for the GBU-16 1,000lb [454kg] LGB and the LGTR were also added [to the pod], as were display signal paths between the pod and the RIO's new 8 x 8in Programmable Tactical Information Display [PTID], which had to be used instead of the then standard fishbowl TID – the PTID displayed both radar and LANTIRN data. Paths were also established to the pilot's Vertical Display Indicator, in place of the Television Camera Set. F-14 aircraft modifications included dedicated "hard wiring" from the LANTIRN pod to a GPS antenna on the jet's turtleback and from the LANTIRN station to the cockpit displays.

There was no integration between the pod and the aircraft's weapon system, sensors or stores management system, however. This meant that the F-14's software remained untouched, thus preventing the costly and time-consuming changes and tests needed when an aircraft's weapons system code is modified. The F-14 did not even "know" the LANTIRN pod was on the jet! GBU-16 and LGTR release was performed manually by the pilot when the LANTIRN ballistics software displayed to the crew the countdown to release.

Prior to our first flight with the pod [on March 20], "Rat" pressed F-15E pilots and Weapon Systems Officers that he knew who had LANTIRN experience for as much information as he could get out of them, and we got one F-15E simulator "ride" together to practice procedures, even though the "knobology" and displays were different in the Tomcat. The initial F-14 hops that "Rat" and I did were medium altitude, round-robin flights during which we learned how to use the LANTIRN pod and practice simulated attack procedures – no weapons were aboard and eye-safe laser mode was used. Four flights [on April 3, 1995] were then flown to Dare County Bombing Range, where we released a series of LGTRs – SHACK! The final Dare County training runs [on April 5] were for two inert GBU-16 LGBs – SHACK again!

We then flew off to Puerto Rico [on April 8] to drop on the Vieques range, delivering the ordnance from our demo Tomcat. On this occasion we were accompanied by a photo/safety chase Tomcat

"Super RIO" Lt Cdr Larry "Rat" Slade and VF-103 XO Cdr Alex "Yogi" Hnarakis were the first US Navy crew to self-designate an LGB with the LANTIRN pod. The pair initially dropped four BDU-59 LGTRs (April 3, 1995) and two inert GBU-16s (April 5, 1995) on the Dare County range before flying off to Puerto Rico to expend live ordnance on the Vieques Range. (via Capt Alex Hnarakis)

and two F/A-18Cs from our air wing [CVW-17]. Both Hornets were equipped with AN/AAS-38A Nite Hawk FLIR pods and AN/ASQ-173 Laser Spot Trackers (LSTs). Since LANTIRN had no LST capability, the Hornet's job was to fly a couple of miles in trail to confirm our LANTIRN laser was indeed over the intended aim point, and that there was enough laser energy being emitted by the pod for the GBU-16 to be guided to the target.

We were confident based on our experience at Dare County that the Tomcat/LANTIRN combination would continue to work well, but "Rat" Slade was concerned that if we flung a bomb off target, we might not know if there was a LANTIRN problem or if it was merely a malfunctioning GBU-16 guidance unit or a bad fin kit. Should a bomb miss the target, we did not want to chase LANTIRN "ghosts" if they did not really exist.

The first ordnance dropped at Vieques [on April 10] consisted of two more inert GBU-16s – SHACK, with video! We followed this up with four live GBU-16 drops [on April 11–12], one of which failed to hit the aim point but landed safely within the target range and planned hazard pattern. The post-flight review of our LANTIRN tapes and the Hornet LST video quickly showed that the laser spot was on the intended aim point the whole time, and with plenty of laser energy, so we confidently attributed the failure to a bad GBU-16. We were immediately able to get back on to the preparation for the final drops. Like "Snotty," "Rat" Slade was also a visionary and clairvoyant with his plan to incorporate our Hornet buddies into the Vieques evaluation.

The remaining three live drops went to plan – again, SHACK with video! It was great watching the on-target, high order explosions blowing up [M60] tanks. We were laughing like kids with new toys! We got Capt "Snort" Snodgrass, the FITWINGLANT commodore, out on one of the drops with "Rat," as well as a LANTIRN familiarization flight for the VF-103 chase pilot, Lt Otto "Lechter" Sieber, again with

Left: Cdr Hnarakis and Lt Cdr Slade prepare to taxi out at NAVSTA Roosevelt Roads during LGB trials over the Vieques Range on April 10, 1995. Their F-14B, BuNo 161608, "armed" with two inert GBU-16s, carries their names on the cockpit rail as well as specially applied *FLIRCAT* titling and nose art. (via Capt Alex Hnarakis)

Right: Sluggers Fighting 103. (VF-103, 1993)

Above: "Clubleaf 213" cruises along the coast of Puerto Rico, inbound to the Vieques Range. Two GBU-16s are nestled beneath the fuselage of the jet on the centerline BRU-32 bomb racks, and the all-important LANTIRN pod can just be seen on the specially modified starboard shoulder pylon. This photograph was taken by a second VF-103 jet that acted as a photo/safety chase aircraft for the trial flights. (via Capt Alex Hnarakis)

Right: Cdr Hnarakis and Lt Cdr Slade release the first live LGB to be self-designated by an F-14 as they approach the target area on the Vieques Range on April 11, 1995. (via Capt Alex Hnarakis)

"Rat" in the back seat. I took the chase RIO, Lt Lonn "Lumpy" Larson, out for a LANTIRN familiarization flight too.

The Hornet guys requested some side-by-side runs, with ranges called out by voice on the tapes and qualitative descriptions by Tomcat and Hornet crews on when they could identify target features on their cockpit displays via their respective pods. We were a little concerned that this information would get twisted, but our F/A-18 buddies wanted to take the side-by-side video to the Hornet Operational Advisory Group as ammo to get something better for the Hornet than Nite Hawk. On the way home, I was able to give Lt Cdr Dana "Devo" Dervay a LANTIRN familiarization [fam]. He was our main FITWINGLANT co-ordinator and heavy lifter for LANTIRN.

Returning to Oceana [on April 14] with lots of great video and a demonstration successful beyond our earlier estimates, we flew a few more fam flights locally with LGTRs, and some without ordnance, in order to give "Snotty" a chance to see LANTIRN for himself. We also undertook fam flights for CVW-17's DCAG, Capt James Zortman, CVW-7's CAG, Capt Ron McElraft and CVW-3's CAG, Capt Tom Zelibor. Then we had to de-configure the jet, since the mods were not cleared for use when the aircraft was embarked on a carrier, and the only LANTIRN pod modified for the Tomcat had to go back to Martin Marietta.

Left: The GBU-16 dropped by "Clubleaf 213" zeroes in on its target (an old M48 Patton tank), which has been painted in laser energy by the LANTIRN pod controlled by Lt Cdr Slade. (via Capt Alex Hnarakis)

Below left: "Shack!" The LGB scores a direct hit on the Patton tank. The VF-103 detachment completed a further three live drops on the Vieques range. Three LGBs hit their aim points and the fourth weapon landed safely within the target range, but missed its target. In respect to the latter bomb, Cdr Hnarakis explained, "The post-flight review of our LANTIRN tapes and the Hornet LST video quickly showed that the laser spot was on the intended aim point the whole time, and with plenty of laser energy, so we confidently attributed the failure to a bad GBU-16." According to the Joint Munitions Effectiveness Manuals issued to the US military, an impact inside 20ft is considered to be a bullseye. (via Capt Alex Hnarakis)

Below right: Both M48s suffered significant damage from the LGBs expended by "Clubleaf 213." Here, the tanks are being closely examined by range personnel. (via Capt Alex Hnarakis)

"Clubleaf 213" provides an impressive backdrop for six Naval Aviators and NFOs that were heavily involved in the LANTIRN/LGB trial in 1995. They are, from left to right, Cdr Alex 'Yogi' Hnarakis, Lt Cdr Dana "Devo" Dervay (who was the FITWINGLANT co-ordinator and a "heavy lifter" for LANTIRN), Lt Cdr Larry "Rat" Slade (the key VF-103 Project Officer for LANTIRN), Lt Lonn "Lumpy" Larson (RIO in the chase F-14), Capt Dale "Snort" Snodgrass (FITWINGLANT CO) and Lt Otto "Lecter" Sieber (pilot of the chase F-14). All six men guided ordnance on the Vieques Range using the LANTIRN pod, with Hnarakis, Slade and Snodgrass dropping live LGBs on April 11–12 and Dervay, Larson and Sieber expending BDU-59 LGTRs on April 12–13. (via Capt Alex Hnarakis)

I thought that LANTIRN gave the F-14 an exceptional capability, being a real force multiplier for the jet. I also figured it would take too long for the Navy to make a decision to buy, complete the development and then field Tomcat LANTIRN – despite the documented demonstration work we had completed. I never figured I would see it fielded in my remaining flying lifetime, and at that point I still had more than two years ahead of me in my CO/XO tour. However, "Snort" immediately went to work with the video we brought home, showing it to combatant commanders and Naval Aviation and NAVAIR leadership. In June 1995, we got word that the Navy was going to buy LANTIRN for Tomcats.

The Navy and Martin Marietta felt that they would not have enough modified LANTIRN pods or Tomcats to support the next deployment, planned for December 1995, but thought they could have six modified pods, nine modified Tomcats and minimum essential further testing complete in about a year for a June 1996 deployment. Not only was that "light speed" for the acquisition system in those days, but the

Above: Having returned to Oceana, "Clubleaf 213" performed a handful of familiarization flights for senior officers within the East Coast air wing community, specifically Capt Jim Zortman (DCAG of CVW-17, and an A-6 B/N), Capt Ron McElraft (CAG of CVW-7, and an F-14 RIO) and Capt Tom Zelibor (CAG of CVW-3, and also an F-14 RIO). The aircraft, seen here in June 1995, remained configured with the LANTIRN pod and its special Station 8B Adapter fitted to the multi-purpose shoulder pylon into the fall of that year. Delivered new to VF-124 as an F-14A in August 1983, BuNo 161608 later served with VF-21 and then VF-103 following its upgrading to F-14B specification. Subsequently passed on to VF-102, the aircraft saw combat over Afghanistan with the unit during OEF in 2001–02. (via Mike Crutch)

Left: A close-up of the *FLIRCAT* nose-art worn on both sides of "Clubleaf 213's" radome. The artwork featuring a muscular tom cat holding an illuminated lantern while riding an LGB disappeared when VF-103 "Sluggers" became VF-103 "Jolly Rogers" in October 1995, with the majority of the unit's jets being repainted with the skull and crossbones at this time. (Don Linn)

Tomcat squadron deploying in June 1996 [onboard *Enterprise*] was us in VF-103! Lots of behind-the-scenes work began, including required "shake, rattle and roll work" of LANTIRN by NAVAIR and expanding the Tomcat release envelopes for various precision air-to-ground and mixed air-to-air and air-to-ground load-outs.

In January–February 1996 during COMPTUEX, as we were leaving the Puerto Rican Operating Area, CAG Zortman and I were giving the embarked Flag, Commander, Carrier Group Four, a "backfill" on how the NVG training det at Roosevelt Roads, in Puerto Rico, had gone. VF-103 was also the first East Coast Tomcat squadron to deploy with NVGs, and six of the nine LANTIRN jets had their cockpits modified to make them compatible with the goggles. Some of our crews received their initial NVG training during a brief shore det because the moon cycle required flying later at night than the cyclic ops time *Enterprise* was operating to during COMPTUEX.

As we were wrapping up the update, we also mentioned to the admiral that during the final set of work-ups – CJTFEX [Combined Joint Task Force Exercise] in April–May [1996] – VF-103 would have the first three of six LANTIRN pods we would use on deployment, with the remaining pods due to reach us prior to our departure on cruise. The admiral, formerly a Tomcat RIO, mentioned as CAG and I were heading out the door that "we in the Tomcat community had not really embraced the air-to-ground mission, and still had a way to go before NVGs and LANTIRN pods would make that much of a difference." When we got into the passageway, CAG asked me what I thought about the admiral's parting comment. I told him that the admiral was quite misinformed, and I was not pleased with the use of the word "we." CAG said he figured I'd say as much.

I confirmed with CAG that we were still scheduled to fly the admiral in one of our jets in a couple of days' time as planned when we got nearer to the Cherry Point Operating Area [off the coast of North/South Carolina]. I then cleared with CAG my "on the spot" plan to schedule the admiral as wingman on a typical CAS training mission, with Lt Cdr "Morty" Moradian as his pilot and a couple of junior officers [JOs] leading, including Lt Mike "Tung" Peterson as the briefer. On the day of the flight, I watched from the back of the ready room as a standard CAS scenario, with a forward line of troops, 9-line briefs, Joint Munitions Effectiveness Manual [JMEM] weaponeering and air-to-ground deliveries was professionally covered as was typical for VF-103.

A couple of hours later, I watched the debrief too, and afterwards asked the admiral if he still thought "we in the Tomcat community had not yet embraced the air-to-ground mission?" He replied that his eyes had indeed been opened by this sortie, and commented that he didn't realize that "we" in the Tomcat community knew how to do what he had just witnessed from the back seat of an F-14.

I replied that 70 percent of the squadron were JOs who had come through VF-101 since 1993, when strike was added to the syllabus, and that they had never known a single-mission

Fighting 103 (VF-103, 1996)

Lt Michael "Tung" Peterson stands in his flight gear alongside "Victory 203" in the NAG in October 1996. He would become one of the leading exponents of LANTIRN use in the Tomcat fighter community, seeing combat in OSW, OEF and OIF I. (via Cdr Michael Peterson)

No need for laser guidance when employing the weapons photographed on the VF-103 flightline in September 1995. All destined for the Dare Country range, these 500lb (226kg) BDU-45/B practice bombs have been fitted with BSU-86/B retarded (high-drag) fins. Parked behind the bomb trolleys is F-14B BuNo 163215, which was delivered new to VF-103 in December 1988 and later served with VF-32 (participating in Operation *Desert Fox*), VF-143 and VF-101. (David F Brown)

On June 10, 1996, VF-103's "Victory 202" (BuNo 161873) successfully dropped the unit's first 2,400lb (1,088kg) GBU-24 "bunker-busting" bomb on the NAWS China Lake ranges. This gave the squadron approval to use the weapon on its next deployment, which saw VF-103 embark in *Enterprise* as part of CVW-17 just 18 days after the Paveway III bomb drop trial. (US Navy via David F Brown)

jet. They were the same air-to-air killers the Tomcat community had always produced, but they were now also pros at air-to-ground, including CAS for all aircrew in VF-103 and FAC(A) for eight Naval Aviators and NFOs. The only thing they needed were the tools to do it well at night, with precision, and that is where NVGs and LANTIRN were going to make the difference.

When the admiral flew with us in CJTFEX, he had experienced a LANTIRN flight. I told him that the pod was user friendly enough to learn with an extended flight brief and the aid of a pilot already familiar with the system. He had accepted the challenge, saw the capability for himself during JTFEX and wrote a great P4 [a designator for communications between the US Navy's highest ranking officers] message to the world immediately afterwards.

In the month prior to deployment, VF-103 sent a jet and crew to NAWS China Lake [in California] to drop a 2,400lb [1,088kg] GBU-24 Paveway III "bunker buster" LGB. Lt Cdr Pete "Skids" Mathews was the RIO for the flight, and he had VX-9 pilot Lt Pete "Pester" Hooper up front. That way, if needed on deployment, we had an Operational Test-proven LANTIRN Tomcat "bunker-busting" capability, vice just an "on paper" capability. VF-103 and CVW-17 also got in "Skids" a RIO who had actual experience dropping a GBU-24 – something few other aircrew in any community had done at that time.

The first ever Tomcat LANTIRN deployment [and east coast F-14s with NVGs] went well thanks to the efforts of many unnamed personnel – especially the VF-103 sailors who bore the brunt of the additional work required to modify aircraft to make them NVG and LANTIRN compatible. VF-103 crews formulated the baseline LANTIRN procedures and tactics that subsequent squadrons built on and further improved. NAVAIR continued to expand the envelope, and more flight clearances arrived shortly after deploying that included catapulting Tomcats with GBU-24s. Always thinking ahead, the JOs also found innovative ways to use LANTIRN that went beyond the original intent.

Early in the deployment, a division of Tomcats did a training event that involved a 600-nautical-mile [1,111km] round-trip precision strike at night against simulated targets in southern Israel, fighting through air-to-air opposition and bringing back hit assessment video on all four planned aim points – all the hits were within ten seconds of the planned time on target (GPS time on the recorded video display kept everyone honest). Even more interesting was the fact that all but the initial 50 nautical miles [93km]

By the time "Victory 204" (BuNo 163215) and VF-103 embarked in *Enterprise* with the rest of CVW-17 on June 28, 1996 for the carrier's scheduled Mediterranean cruise, the "Sluggers" were no more. In October of the previous year, following the disestablishment of fellow FITWINGLANT unit VF-84, VF-103 took over the insignia and traditions of the "Jolly Rogers." Armed with Mk 76 Mod 0 training bombs, "Victory 204" has crouched in preparation for attachment to the launch shuttle of waist catapult two. This aircraft was one of nine LANTIRN-capable Tomcats deployed by VF-103 – note the targeting pod (one of six supplied to the unit) on the shoulder pylon. The remaining five jets flown by the squadron were wired up for TARPS employment – the LANTIRN hand controller initially replaced the TARPS panel in the first aircraft to be modified, although a work-around for this mission restriction was quickly found. CVW-17's 1996 Mediterranean/NAG cruise was the only time LANTIRN-capable Tomcats flew from the same flightdeck as the precision strike platform they were destined to replace, the A-6E. (US Navy)

historic deployment with CVW-17 embarked in *Enterprise*. His account was published in the winter 1996 edition of *The Hook*:

I joined VF-103 in April 1996 after two back-to-back summer cruises – a 1994 Med cruise with VF-142 and a 1995 WESTPAC cruise with VF-21, based in Japan. Sadly, both squadrons were then disestablished. On these two cruises, the F-14s were relegated to sweep missions, high-value-unit CAP and an occasional strike that allowed daylight "dumb bomb" deliveries. If you didn't have NVGs on night strikes, you weren't a player. That's why I jumped at the chance to cruise with the "Jolly Rogers," the first squadron in the Navy to deploy with LANTIRN infrared and a laser-guided bombing system, and the first east coast Tomcat squadron with NVGs.

When I checked into the "Jolly Rogers," they were getting ready to deploy for CJTFEX, a large, combined exercise held off the coast of North Carolina with all US services participating, along with the British Royal Navy [VF-103's LANTIRN-equipped jets lased targets for Fleet Air Arm Sea Harrier FA 2s]. The pace throughout work-ups had been more demanding than usual due to the training necessary for NVG and LANTIRN proficiency. In that short time, the "Jolly Rogers" successfully demonstrated the significance of the F-14's night precision-bombing capability.

Additionally, the squadron spearheaded efforts to accelerate clearance for the Tomcat to drop the GBU-24 "bunker buster," including dropping and guiding the first from a VF-103 Tomcat to a direct hit at NAWS China Lake [on June 10, 1996]. To further demonstrate the Tomcat's awesome strike fighter capability, the "Jolly Rogers" flew a 540-knot [1,000km/h], 600-nautical-mile [1,111km] unrefuelled LANTIRN/NVG opposed night division strike into the BT-9 Target Complex in North Carolina. Results – hits on all targets (confirmed on video) and all bogeys killed.

From my first flight with the LANTIRN pod I knew that the Navy had received its money's worth, and that Capt "Snort" Snodgrass and the boys at FITWINGLANT had scored a direct hit. Throughout work-ups and through three months of cruise, it is apparent that LANTIRN has added an enormous punch to the air wing's strike capability. The Tomcat has become the platform of choice for many strike scenarios.

The "Jolly Rogers" demonstrated an impressive night FAC(A) capability during Operation *Decisive Endeavor* in the skies over Bosnia-Herzegovina. In recent surge ops during Exercise *Juniper Hawk*, the "Jolly Rogers" flew another 600-nautical-mile [1,111km] strike, this time from the carrier, against simulated targets in southern Israel. Once again, the results were perfect. RADM [Martin J] Mayer [Commander, Cruiser Destroyer Group 12] reported that each additional LANTIRN-configured F-14 added significantly to the air wing's day/night PGM strike capability, and he pointed out the necessity to augment additional pods so that all LANTIRN-capable F-14s on board could deliver PGMs.

Above: One of the six LANTIRN pods issued to VF-103 for its 1996 deployment. The unit received its first three in April 1996, with the remaining three arriving at Oceana just two weeks prior to VF-103 embarking in *Enterprise* and commencing a six-month deployment. (Marc Chiabaud)

Right: Fighting 103 Med-Persian Gulf (1996)

on ingress and final 50 nautical miles on egress were flown at 540 knots [1,000km/h]. That's 500 nautical miles [926km], or almost an hour of flight time, at 540 knots on a normal night cycle with zero tanking!

No weapons were dropped or fired in anger on our 1996 deployment, and at the end of it our LANTIRN pods were cross-decked to our VF-32 relief as we headed home. All subsequent Tomcat deployments were undertaken by units that had aircraft equipped with LANTIRN. It was fun to be among those involved on the ground floor with the LANTIRN integration, and to see where it has led since. It also set the chinning bar higher for subsequent FLIRs employed by both the Navy and Marine Corps.

"Jolly Rogers" Make History

As a postscript to Capt Hnarakis' detailed description of the crucial role VF-103 played in getting LANTIRN cleared for fleet use, Lt Cdr Tom "D-Day" Lucas provided the following insight into his experiences with the unit's cutting-edge F-14Bs while participating in VF-103's

Fighting 103 Med Cruise '96 (1996)

Although VF-103 made Tomcat history by being the first fleet unit to deploy with the LANTIRN pod, thus turning the jet into an all-weather precision strike fighter, the squadron also remained dedicated to its traditional air superiority role. Proving this point, "Victory 201" (BuNo 163224) conducts an OSW patrol over southern Iraq in October 1996 armed with an AIM-54C, an AIM-7M and an AIM-9M. (VF-103)

With its tailhook down and wings swept forward, "Victory 203" (BuNo 161433) circles overhead *Enterprise* in typically hazy weather in the NAG in the fall of 1996. The aircraft had just completed yet another uneventful OSW mission, escorting one of five TARPS jets assigned to VF-103. (VF-103)

The most impressive part of the system is its reliability – the pods are left on the aircraft and are used on all missions, day and night. They've taken the punishment that only Naval Aviation can dish out [heat, jet exhaust in close proximity, salt spray, catapult launches and arrested landings], along with Air Combat Maneuvering [ACM] flights. LANTIRN functioned perfectly for me on every mission I flew on cruise [VF-103 experienced only seven LANTIRN failures in more than 460 missions during the first four months of the deployment].

LANTIRN also gives us capabilities that aren't so apparent. How many times on my last two cruises have I been vectored on to a bogey at night and been asked to identify it? My first reaction was a strong desire to strangle the controller – FLIR identification at night is easy now, however. It can also be used for raid counts and for keeping a bogey "locked up" during pre-merge maneuvers. The GPS in the LANTIRN is an outstanding navigational tool, and although not fully integrated with the INS, it's a helpful addition. The real "bread and butter" of the system is the IMU stabilized laser-point tracker – something that no other precision strike fighter presently has. This gives us better tracking capability and more precision. The Tomcat also has better definition for target selection due to the RIO's larger display screen, which means earlier target detection.

The officer cadre of VF-103 pose with "Victory 201" and the skull and femurs of Ens Jack Ernie, which have been mounted in a glass case in memory of his alleged service with VF-17 "Jolly Rogers" in World War Two. He was posted missing in action during the invasion of Okinawa in April 1945, and his remains were located in the late 1950s. Ernie's family suggested that the "skull and crossbones" be presented to the unit as a way of preserving his memory. (VF-103)

"Victory 102" departs Yuma in afterburner at the start of a training mission in April 1997. By then, VF-103 had switched to the traditional "100" modex usually seen on fighter aircraft within an air wing. CVW-17 had also recently swapped carrier assignment to *Dwight D. Eisenhower*, hence the titling on the wing glove leading edge. "Armed" with an inert Mk 83 bomb and carrying a TACTS range instrumentation store, the aircraft is noticeably devoid of any LANTIRN pod. All six assigned to the unit when on deployment had been transferred to VF-32 embarked in *Theodore Roosevelt* when the latter carrier replaced *Enterprise* in the Mediterranean on December 8, 1996. (Dan Stijovich)

In short, the Tomcat adds a precision strike capability to the CAG's arsenal that is capable of flying long ranges with higher speeds on and off the target. Combine all this with NVGs and two crewmen, and you have an extremely capable weapons delivery and FAC(A) platform. With the recent clearance to carry two GBU-24s, the Tomcat is now an ideal platform for "hard targets," and the only platform with GBU-24 "bring back" capability [the ability to land on the carrier with unexpended GBU-24s – critically important, given the bomb's $600,000 price tag]. Armed with a deadly air-to-air loadout and LGBs, the F-14 with LANTIRN and NVGs is the Navy's best choice to fly night, long-range, self-escort precision strike missions well into the 21st century.

CHAPTER 5

BOMBCAT CO

A veteran TOPGUN instructor and highly experienced F-14 RIO, Cdr Dave "Bio" Baranek returned to the fleet for his command tour just as the "Bombcat" came of age with the advent of the LANTIRN pod.

Before I started at VF-124 (the F-14 FRS on the west coast) in 1980, I had seen the Grumman promotional photographs of a Tomcat on the ramp with an impressive variety of ordnance arrayed in front of it, including bombs. But as we quickly learned in the "RAG" [the nickname for the FRS, derived from the defunct term Replacement Air Group], our mission was air-to-air. Officially, the F-14's air-to-ground capability at that time was limited to its 20mm cannon. In my first fleet squadron [VF-24] we flew training flights that taught us how to effectively strafe targets, but any mud-moving ordnance larger than 20mm shells would be delivered by the air wing's A-6 and A-7 squadrons.

In my second F-14 squadron [VF-2], which was in CVW-2 in the late 1980s, CAG [Capt Christopher T] "Boomer" Wilson had attempted to get clearance for his Tomcat squadrons (VF-1 and VF-2) to carry bombs during the 1989 WESTPAC deployment onboard *Ranger* – permission was denied, as I recall, due to non-certified bomb racks.

Then I did several staff tours – payback for ten years of flying at Miramar – during which time the Tomcat community underwent a huge change, as did Naval Aviation as a whole. By the time I returned to the fleet it was 1996. The A-6 and A-7 were gone and most carrier decks hosted three F/A-18 Hornet squadrons, along with one F-14 unit. The Hornet was a strike fighter from the start, with excellent capability in air-to-air and air-to-ground roles, and by this time the Tomcat's nascent strike capability had been developed. Each carrier now effectively sailed with four squadrons of strike fighters embarked.

Things had also changed within F-14 squadrons, with the Strike Fighter Weapons and Tactics (SFWT) program having raised the bar on aircrew professional knowledge and performance. When I joined VF-211 as XO in 1996, I found eager JOs who had embraced the SFWT program and worked hard to complete their flight quals [qualifications] and written tests. Most of the department heads (lieutenant commanders) had already completed their own SFWT quals at the highest levels. The program addressed both air-to-ground (strike) and air-to-air (fighter) roles, producing versatile aviators in the process.

With the demanding performance goals and expert oversight by former TOPGUN instructors who were assigned to the squadron, aviators were not only versatile but also lethal. In addition, each

Former TOPGUN instructor Cdr Dave Baranek joined VF-211 as its XO in 1996, just as the unit was preparing to receive its first LANTIRN pods. New to the world of the F-14 as a precision strike fighter when he was posted to the "Fighting Checkmates," Baranek had come close to dropping ordnance from the jet when with VF-2 some seven years earlier. The "Brutus" character on the tail of "Nickel 101" (F-14A BuNo 161856) dates back to VF-211's establishment as VB-74 in 1945. Fittingly, he is firmly grasping a rocket as if to underline the unit's air-to-ground mission. (via Cdr Dave Baranek)

F-14 squadron had a handful of former A-6 pilots and Bombardier/Navigators [B/Ns] who were worth their weight in gold as we took on the strike mission. To their credit, they in turn embraced the air-to-air role, which was totally new to them. Many guys wore patches that indicated the Tomcat's multi-mission capability, and although they looked sharp, I stuck with the classic "leaning Tomcat" patches I already had from the 1980s.

One of the most obvious changes from the old days was found in the flight brief. Preparing for an air-to-air training sortie in the early 1980s, we could cover the essential information in 20 bullet points or less on the whiteboard. For a typical strike fighter training mission in the late 1990s, the large whiteboard in the squadron ready room would be crammed full of important information – weapon type, delivery profile, cockpit switch settings, release parameters (airspeed, dive angle, release altitude) and many more details. Furthermore, the air-to-air threat had become more challenging with the proliferation of fourth-generation fighter aircraft.

From my perspective, then, the air-to-ground mission was totally new, and air-to-air combat was appreciably more complex than it had been in 1982. This reality was all the JOs knew, and they had no doubt that their training and skill would bring success to us in our A-model Tomcats.

VF-211 moved from Miramar to Oceana in 1996, and we soon became well acquainted with the Dare County Bombing Range. During work-ups, VF-211 returned to southern California on a training detachment to [NAF] El Centro for SFARP [Strike Fighter Advanced Readiness Program],

Above: VF-211 (1997)

Left: VF-211 was the first Pacific Fleet Tomcat unit to complete the AARP in June 1992, which gave crews exposure to the same kind of ground attack training undertaken by Naval Aviators and NFOs flying the A-6 Intruder and F/A-18. F-14A BuNo 162600, armed with four 500lb (226kg) Mk 82 bombs, was photographed commencing its attack run on the range at MCAS Twentynine Palms, in California, in July 1996. This base is home to the Marine Air Ground Task Force Training Command and Combat Center. Issued new to VF-2 in November 1985, BuNo 162600 joined VF-211 in 1992 when the unit replaced its F-14Bs with A-model Tomcats. The aircraft remained with the "Fighting Checkmates" until it was transferred to VF-101 in November 1999. (US Navy)

which had replaced FFARP [Fleet Fighter ACM Readiness Program] of the 1980s. This was an intense squadron-level program where all aircrew were challenged. Those who had been here before knocked off the rust, while "newbies" stepped up their abilities.

Meanwhile, squadron maintenance personnel met every challenge thrown at VF-211, working long hours whenever necessary to keep our old F-14As in the sky to support training. As with the aircrew, I saw a high level of confidence and enthusiasm among the sailors working on our jets. I wondered about the "ordies," who in the past had mostly lifted 200lb [90kg] Sidewinders, 500lb [226kg] Sparrows and the occasional half-ton [454kg] Phoenix. Now they often lifted 1,000 [454kg] and 2,000lb [907kg] bombs, in addition to the missiles. But in typical Navy fashion, the harder they worked, the more pride they displayed.

Later in work-ups we deployed to Fallon for air wing training, where Tomcats and Hornets [VMFA-314, VFA-146 and VFA-147, all equipped with F/A-18Cs] flew many of the same missions. The intensity and quality of training at Fallon was an order of magnitude higher than it had been in the 1980s, based on my recollections. Of course, the entire air wing was involved, and I could see that the performance of E-2C Hawkeye crews – CVW-9 had VAW-112 assigned to it – had improved like everything else.

I haven't yet mentioned the LANTIRN Targeting System [LTS]. The vast majority of our bomb deliveries during pre-deployment training were visual, as we had only one LANTIRN pod for the latter part of work-ups and just a few select aircrews flew with it. In my case, during work-ups, the only exposure I had to LANTIRN was a brief qualification session with a "table-top trainer," which consisted of a LANTIRN hand-control unit and a tutorial program run on a computer. Fortunately, LANTIRN was fairly straightforward to operate, and several weeks later, when I finally flew with a real pod, I quickly felt comfortable. I subsequently received my LANTIRN patch.

VF-211 deployed with CVW-9 on board *Nimitz* in September 1997. We crossed the Pacific, made a few port visits and then entered the Persian Gulf to enforce the southern no-fly zone over Iraq as part of OSW. Aircraft carriers sailing within the Persian Gulf, which had become commonplace

Left: VF-211 "Brutus" (1997)

Right: *Nimitz* underway in the NAG in October 1997, shortly after the carrier had arrived in-theater. CVW-9 commenced OSW patrols within days of coming under Fifth Fleet control. Seven of VF-211's ten Tomcats can be seen chained down on the flightdeck as aircraft are worked on between missions. *Nimitz* was on deployment for six months between September 1, 1997 and March 1, 1998, the vessel undertaking a world cruise that saw it depart NAVSTA Bremerton, in Washington state, and return to NAVSTA Norfolk. The carrier subsequently underwent a three-year Refueling and Complex Overhaul in Northrop Grumman's Newport News shipbuilding yard. (US Navy)

from the summer of 1990, following the instigation of Operation *Desert Shield* in response to the Iraqi invasion of Kuwait, was something we could have hardly imagined back in the 1980s – yet another example of the changes affecting Naval Aviation.

The major missions that supported OSW included:

- **Surveillance** – critical to effective no-fly zone enforcement, and usually provided by a USAF E-3 AWACS aircraft. It could also be undertaken by Navy E-2 Hawkeyes, if necessary. In addition to radar surveillance of the no-fly zone, they provided command and control that was essential for keeping these complex operations on track.
- **Counter-air** – an armed response in the form of coalition fighters on patrol over southern Iraq that were ready at a moment's notice to attack IrAF aircraft that flew into the no-fly zone.
- **Interdiction**, known to us as Strike Familiarization [SFAM] – strike and electronic warfare aircraft on patrol always carried a variety of air-to-ground weapons to respond immediately to any Iraqi attempt to shoot down a coalition aircraft. These flights usually performed a simulated air strike, so we called them SFAM. Several Iraqi SAM and AAA sites remained intact on the ground under the no-fly zone.
- **Refueling** – the airborne armada was supported by USAF KC-10s, RAF VC10 tankers and air wing S-3s.
- **Reconnaissance** – USAF U-2s performed reconnaissance from 70,000ft while TARPS F-14s took photos from appreciably lower altitudes.
- **FAC(A)** – required incredible skill and crew co-ordination. FAC(A)-qualified crews would direct bombing attacks on enemy forces in close contact with friendlies.

An event launched from the ship included some or all of the above mission elements. F-14s flew counter-air, interdiction and TARPS. Due to realities such as flying over hostile territory, OSW flights were designated as combat operations – missions were recorded in green ink in aviators' logbooks.

We flew with a mixed ordnance load, carrying multiple air-to-air missiles to back up our no-fly zone patrolling duties and air-to-ground ordnance for quick reaction in case anything happened while we were aloft. Our Tomcats usually carried two cluster bomb units [CBUs], and these were sometimes replaced by a pair of LGBs. The "Checkmates" rarely loaded four bombs (on under-fuselage weapon rails in what we referred to as a super-bomber load-out) because such a configuration reduced the amount of fuel that could be brought back for a safe landing on the ship. However, if a strike seemed likely, we would set up several super-bombers.

"Nickel 110" (BuNo 162606) heads for *Nimitz* after completing an OSW mission in late 1997. The aircraft features a common loadout for the F-14 during VF-211's 1997–98 deployment – an AIM-54C and an AM-9M on the port shoulder pylon, an AIM-7M in the aft belly station, two Mk 20 CBUs on the forward BRU-32 racks, a LANTIRN pod on the starboard shoulder pylon and, of course, 600 rounds of 20mm ammunition. Following fleet service with VF-2 and VF-211, BuNo 162606 was assigned to NSAWC at Fallon. (Cdr Dave Baranek)

"Nickel 102" (BuNo 161274) keeps station off to the right of a KC-10A from the 6th Aerial Refueling Squadron (ARS)/60th Air Mobility Wing (AMW) while three F/A-18Cs from VFA-147 (the two jets closest to the camera) and VFA-146 top off their tanks over Kuwait. The 400th F-14 delivered to the US Navy, BuNo 161274 was issued new to VF-1 in May 1981. It later served with VF-154, VF-213 and VF-24, prior to joining VF-211. Passed on to VF-41, with whom it saw combat in OEF in 2001, the fighter then briefly served with VF-101 before rejoining VF-211. It was flown by the unit to AMARC in August 2004 and sold for scrap to HVF West in June 2008. (Cdr Dave Baranek)

"Nickel 103" (BuNo 161850) has the same stores' configuration as "Nickel 110" seen earlier in this chapter. Despite flying more than 370 OSW sorties during the four months *Nimitz* was in the NAG, VF-211 did not get the chance to expend any ordnance in anger. Indeed, CVW-9 was not called on to drop any weaponry on targets in southern Iraq. Delivered new to VF-101 in April 1984, this aircraft later served with VF-31 before being transferred to VF-211. It remained with the unit until stricken in September 1999. (Cdr Dave Baranek)

The CBUs were part of the default OSW load-out. The thought was that the initial air-to-ground action would be quick-response, and initially we would probably be tasked with hitting a missile site or similar soft target. Being an area weapon, the CBU would give us the best chance of causing damage [the Tomcat's history with the CBU is detailed later in this chapter].

By now we had received our complement of LANTIRN pods, so VF-211 jets were equipped with one on almost every flight over Iraq (except for those aircraft performing reconnaissance missions with TARPS, as they were not compatible with the latter store). The discussion was that if things heated up and we briefed to fly strike missions – which happened several times in response to Iraqi threats against U-2s conducting reconnaissance overflights of the country – we would quickly download the CBUs and upload GBU-12 and GBU-16 LGBs.

On every flight we could practice using the LANTIRN pod. Given the well-designed controls and candid ready room critique of post-mission video in the debrief, everyone's learning curve was steep. By the time the deployment started I was the commanding officer, so for me, the debriefs were a chance to display professionalism in accepting corrective comments. I tried not to make the same mistake twice. Soon, I was consistently operating the pod correctly, just like the JOs.

For many years VF-211 applied "Brutus" to its color jets whenever the unit spent Christmas on deployment. In December 1997, BuNo 161856 was given the treatment, the aircraft featuring a festive "Brutus" complete with a sack of presents. The squadron's checkerboard marking was also modified with the addition of seasonal green in place of white, while the horizontal red, white and blue striping was replaced by bauble-laden holly! The LANTIRN-equipped fighter was photographed being led back to the carrier at the end of an OSW patrol by an F/A-18C from VMFA-314. (Cdr Dave Baranek)

During work-ups, we had practiced a lot of roll-in deliveries [of Mk 80 series dumb bombs] from altitude and low-level approaches to pop-up deliveries. However, following the receipt of more LANTIRN pods and our observation of the strict operational rules over Iraq, we mostly planned level drops of LGBs from medium altitude. From my perspective, LANTIRN simplified the technical aspects of delivery and greatly increased the likelihood of a hit. There were plenty of other factors to keep the mission challenging, such as countering ever-present surface-to-air and air-to-air threats that would be expected in combat (the F-14's default OSW load-out also included an air-to-air package consisting of an AIM-54C, AIM-7M and AIM-9M), and co-ordinating with the large strike packages in-theater.

Complexity, it seemed, was everywhere, although I also found that everyone was up to the task. We performed regular practice strikes involving the USAF and US Marine Corps, as well as coalition air forces. These were large operations, where positioning and timing were important, and we could judge each other's performance. I was frankly amazed at how quietly and smoothly they ran. Our missions included dedicated air-to-air patrols of the no-fly zone, simulated strikes and TARPS. Some flights were routine, but several times we ramped up the excitement by "riding shotgun" for U-2s while they gathered intelligence from high above Iraq. Of course we didn't escort them in any traditional sense, but we provided a "response package" should Iraqi forces

"Nickel 101" featured the Christmas "Brutus" marking on both of its twin fins. BuNo 161856 was delivered new to VF-31 in July 1984, and it later served with VF-24 and VF-213 before joining VF-211. Another "Fighting Checkmates" jet transferred to VF-41, the Tomcat subsequently saw combat with the "Black Aces" in *Allied Force* and OEF prior to it being retired to AMARC in December 2001. (Cdr Dave Baranek)

The late afternoon sun gives a slight golden tint to two VF-211 F-14A Tomcats with their tailhooks down, minutes before landing aboard *Nimitz* to complete yet another OSW mission in December 1997. As was standard practice for the unit in the NAG at this time, both aircraft carry one AIM-9M, one AIM-54C and two Mk 20 CBUs. Unseen by the camera is an AIM-7M in the aft belly station and another AIM-9M on the starboard shoulder pylon. "Nickel 106" is BuNo 162600 and "Nickel" 107 is BuNo 162608. (Cdr Dave Baranek)

Three F-14As from VF-211 orbit in low holding above *Nimitz* while the carrier turns into the wind in preparation for recovering the aircraft to signal the end of an OSW mission over Iraq in December 1997. In the lead is "Nickel 115" (BuNo 161276), a TARPS jet that is also equipped with an AN/ALQ-167(V) "Bullwinkle" ECM jamming pod – a store always carried by a Tomcat configured for a TARPS missions. "Nickel 102" (BuNo 161274) is devoid of any CBUs because it has been the TARPS jet's dedicated fighter escort. Finally, "Nickel 104" (BuNo 161615) is carrying VF-211's typical OSW strike-fighter load of CBUs and air-to-air missiles. (Cdr Dave Baranek)

make any offensive moves. They did not make this mistake while *Nimitz* and CVW-9 were in-theater.

The Tomcat served as a strike fighter with the Navy until 2006, and the additional capability provided by the LANTIRN pod undoubtedly extended its life, as well as increasing its usefulness. If you're wondering whether Tomcat crews enthusiastically adopted the additional air-to-ground mission, just look at the squadron cruise videos from the final years of the jet's service in the fleet and you will see a LOT of bomb-dropping segments.

CBU and the Tomcat

As noted by Cdr Baranek, his OSW deployment as CO of VF-211 coincided with a time when F-14s were routinely undertaking patrols over Iraq armed with CBUs, rather than LGBs or Mk 80 series dumb bombs. By then, the Rockeye II CBU, in its various iterations, had been widely used by A-6, A-7 and F/A-18 units for many years. In service from 1968, Rockeye IIs were extensively dropped by US Marine Corps aircraft during *Desert Storm*, when no fewer than 27,987 canisters were expended by American combat aircraft. Primarily seen as an anti-tank weapon, the Rockeye II has proven to be highly effective against armored vehicles.

The Rockeye II family of CBUs employed by US Navy carrier-based aircraft consisted of the Mk 20, CBU-99 and CBU-100. All three contained 247 Mk 118 Mod 1 or Mod 2 anti-tank bomblets, each of which weighed 1.32lb (0.59kg) and featured a 0.4lb (0.18kg) shaped-charge warhead of high explosives that released a jet of superheated, pressurized gas capable of penetrating 7½in (19.5cm) of armor or 30in (76cm) of concrete. The bomblets for the Mk 20 were carried in a Mk 7 dispenser, while the CBU-99 and CBU-100 utilized the improved SUU-76 dispenser.

Being a conventional free-fall weapon that required no specialist guidance, the Rockeye II was seen as a suitable store for the F-14 in its new strike fighter role. The Naval Air Warfare Center's Aircraft Division, based at Patuxent River, carried out F-14 flight trials with Mk 7 training dispensers in the early 1990s, dropping a salvo of up to four canisters from BRU-21/22 racks fitted to the jet's centerline stores pallets. These culminated with fleet clearance for the weapon in late 1992, and VF-24 became the first operational Tomcat unit to drop CBUs (on December 8, 1992) when a number were expended during CVW-9's final pre-cruise work-up during a ten-day embark on board *Nimitz* while the latter participated in *FleetEx 93-1B* off the coast of southern California.

Lt Cdr Mikal Kissick was the RIO of the aircraft involved in the first drop evolution:

Our skipper in VF-24, CDR Ted "T-Bear" Carson, was hot for the unit to be the first to fully embrace the "Bombcat" mission. We loaded and carried Rockeye daily [for the duration of the exercise], even though we were not yet cleared to drop them. We did that so as to give the armorers valuable practice in loading CBUs, thus exercising the ship's ordnance capability. One sunny day we launched off the boat, and I was flying with LT Kevin "Space" Casey about 50 miles [80km] from the ship, "arcing and sparking," when we got a call on the squadron common frequency from the XO, CDR Rob "Squatty" Adamson, who was also a fanatic about "firsts." He told us, "You are cleared by NAVAIR to drop." We were flying Blue Water [open ocean] Ops, and I replied to "Squatty" that we had no targets available to bomb, just waves. He responded, "I don't give a fuck what you do. You better not come back with them [CBUs] on the rails."

Although the ubiquitous Mk 80 series GP iron bomb was typically employed by fighter crews learning the ground attack mission in the Tomcat community in the early 1990s, NAWC's Weapons Division also performed compatibility trials with other ordnance for use by fleet units. One such weapon was the Mk 20 CBU, which could contain Rockeye, CBU-59 APAM or CBU-78 Gator munitions. Equipped with four cameras on suitably modified pylons in order to record weapons' release for future analysis, this F-14A was photographed by a chase aircraft shortly after releasing four inert Mk 20s over the Bloodsworth Island Range in Chesapeake Bay. (US Navy)

So, we looked for a clear patch of ocean and delivered the ordnance as advertised. The drop was a bit of an anti-climax, as I am not even sure the CBUs fused. They looked cool, nevertheless, with the bomblets covering an area the size of a football field. I am sure they provided many fish with "metal death." "Squatty" already had the message promulgated, and the signal went out – 'VF-24 "Bombcats" drop Rockeyes, becoming the first fleet fighters to do so.' Those were the days.

COMNAVAIRPAC Tomcat units subsequently led the way with CBU employment during work-ups and when on cruise. Six months after VF-24's Rockeye II drop, VF-21 expended its first CBUs on June 5, 1993 while on deployment with CVW-5 embarked in *Independence*. Lt Danny Merrill was at the controls of the F-14A involved, with Lt Jeff Reed as his RIO. Merrill recalled:

Left: Sporting decidedly non-regulation headwear, NFO Lt Mikal Kissick mans VF-211's Alert 5 jet on the flightdeck of *Kitty Hawk* in 1985. Seven years later, on December 8, 1993, while serving with VF-24 embarked in *Nimitz*, he was the RIO of the first operational Tomcat to drop CBUs. (Lt Cdr Mikal Kissick)

Below: VF-24's "Renegade 207" (BuNo 161856, which can also be found earlier in this chapter as VF-211's "Nickel 101") smokes up the tires on its main wheels as the jet lands on board *Nimitz* on June 15, 1992 during a CVW-9 carrier qualification evolution off southern California. VF-24 was flying in the same area from the same flightdeck six months later when it dropped CBUs for the first time. (US Navy)

BOMBCAT CO

As a fighter guy, in the early 1990s I wasn't really too interested in dropping anything. The story as to why I ended up dropping the Rockeye that day was, and still is, quite interesting, and I'm very proud to have done so. My CO, CDR Stan "Steamer" O'Connor, was supposed to have dropped it, but his jet went down on the catapult. He wasn't happy about that for sure! We had had very little training in the air-to-ground mission other than dropping little blue bombs behind the boat. The rest is history.

All I remember about the mission is that there were no land targets to speak of, so I dropped on what we used to call a "killer tomato" – it looked like an oversized beach ball that was deployed by the carrier into the ocean at a specific latitude/longitude. I can recall that my HUD went blank that day, so we just did a low drop from 500ft at 500 knots [926km/h]. It was a success, but only because the damn thing came off the rails!

VF-14 earned the distinction of being the first FITWINGONE unit to drop CBUs when it expended munitions during a series of strikes on the Bloodsworth Island Range in Chesapeake Bay, Maryland, and the nearby Virginia Capes Operating Area W-386 off the Atlantic coast in July 1993.

Right: Naval Aviator Cdr Rob Adamson was XO of VF-24 when the unit made Tomcat history by dropping the first CBUs expended by a fleet squadron. According to Lt Cdr Kissick, Adamson (who sadly passed away in January 1999) was a "fanatic" about "firsts"! (US Navy)

Below: Trials with live CBUs were initially undertaken by VX-4 and then continued by VX-9 when the former was merged with VX-5 in September 1994. The new unit flew a handful of Tomcats from Point Mugu for a decade, with F-14B BuNo 162913, which had previously served with VX-4, undertaking a series of test drops with Mk 20 CBUs. Following several years with the test community, this aircraft returned to fleet service with VF-103 in August 1997. It was lost in a mid-air collision with another Tomcat from the unit during a training sortie over the Mediterranean on July 30, 1998, both aircraft being embarked in *Dwight D. Eisenhower* as part of CVW-17. (Ted Carlson)

Cdr Doug Denneny became familiar with CBUs at around this time, too, while serving as a RIO with VF-213 at Miramar, the unit progressing through the final stages of its pre-cruise work-ups prior to deploying with CVW-11 on board USS *Abraham Lincoln* (CVN-72) in June 1993:

> I dropped probably ten or 20 in the early 1990s while training. The only overland location US Navy units could drop CBUs on was in the R-2507 range in a small area called "Irish Wash" [also known to crews as the "rock pit"] in the Chocolate Mountain Aerial Gunnery Range in southern California. It was in a restricted area east of the Salton Sea and north of Yuma. It was the only approved overland range for the dropping of CBUs due to concerns with the high rate of unexploded bomblets that made the ordnance difficult to clean up. This was an area used primarily by Marine Air units operating with MAWTS [Marine Air Weapons and Tactics Squadron] One.
>
> We'd get into the JMEM and plan CBU missions pretty extensively. The canisters were released manually, and you had to plan them pretty well to get the right effects/coverage over the ground. Yes, you could fly rather low with CBUs, and you only had to get close to the target prior to their employment, but with wind and other effects, it was hard to get the right coverage. Basically, you wanted to cover a triangle of land over the ground, and to achieve this we'd drop two at a time.
>
> I recall that near the end of the CBU era we dropped a lot of them to keep from having to de-militarize unused stocks of weapons. In other words, it was really expensive to rip the CBUs apart and de-militarize all those submunitions. It was much cheaper to take them out and blow them up.

Cdr James Howe also dropped CBUs during his various squadron tours with VF-32, VF-101 and VF-2 in the 1990s:

> Every squadron had some allotment, and since Chocolate Mountain was the only place you could drop the CBUs, you had to work quickly to expend them all while you were at NAF El Centro on detachment – I also dropped them when I went through the US Marine Corps-led WTI course at [MCAS] Yuma. The dud rate for the sub-munitions was way too high to allow them to be dropped

VF-41's "Fast Eagle 101" (BuNo 160394), complete with recently applied *STRIKECAT* nose art, heads for the Chocolate Mountain Aerial Gunnery Range in southern California in August 1994. The aircraft is armed with a live Mk 20 CBU, which was duly expended on the R-2507 range – the only approved overland site in the USA on which CBUs can be dropped. (Capt Paul Filardi)

on our standard ranges. We did a lot of weaponeering to calculate the bomblet pattern, size and orientation. It was dropped in the same fashion as a standard unguided bomb, but against primarily soft targets – people, soft-skinned vehicles and light armor.

As explained by Cdr Baranek earlier in this chapter, for several years from the mid-1990s, CBUs were "part of the default OSW load-out." This was because Joint Task Force-South-West Asia (JTF-SWA), which oversaw all operations in-theater, felt that the Iraqi Army was likely to launch a ground offensive into neighboring Saudi Arabia and Kuwait should Saddam rail against the no-fly zone and punitive economic sanctions. The CAOC would in turn instigate a quick-response air-to-ground action undertaken by aircraft already patrolling southern Iraq, which included Tomcats armed with CBUs. They would be tasked with hitting missile and radar sites or similar soft targets such as troop convoys, armored vehicles and tanks. With the CBU being the best area weapon then in-theater, it gave F-14 crews the greatest chance of inflicting casualties on Iraqi ground forces.

Above: The single Mk 20 CBU attached to "Fast Eagle 101" at NAF El Centro, prior to it being dropped on the R-2507 range. As with all live, armed external weapons carried by US military aircraft, it is marked with yellow stripes. Three types of Rockeye II CBUs were cleared for use by the F-14, with the Mk 20 being the first. Its 247 anti-tank bomblets were carried in a Mk 7 dispenser, while the follow-on CBU-99 and CBU-100 utilized the improved SUU-76 dispenser. (Capt Paul Filardi)

Right: VF-2's Lt Cdr Dale Horan and squadron XO Cdr Scott Stewart consult a weapons manual while conducting a preflight check of a CBU-100 attached to their F-14D in the NAG during CVW-2's 1997 deployment embarked in *Constellation*. Thanks to the upgrading of its jets with Block D02 software in the winter of 1996–97, VF-2 was at last able to deliver air-to-ground ordnance such as this CBU-100 from its F-14Ds during the 1997 cruise. (US Navy)

With the advent of more LANTIRN pods from 1997, Rockeye IIs were quickly replaced by LGBs, specifically GBU-12s and GBU-16s, on F-14s committed to OSW. Not a single CBU was ever expended by Tomcats against targets in Iraq, and the high rate of bomblet failures associated with the weapon precluded its use by VF-14 and VF-41 in Operation *Allied Force* in 1999. Ultimately, only two CBU-99s would be dropped in anger from an F-14, and this unique event occurred on the night of October 19, 2001, in support of Special Operations Forces (SOF) targeting the Taliban during the early stages of Operation *Enduring Freedom* (OEF). VF-213 pilot Lt John Saccomando and his RIO Lt Cdr Michael Peterson crewed the jet involved, with the latter providing the following detailed account of the mission:

We were part of a mixed section of one F-14D FAC(A) and an escort F/A-18C, flown by Lt Cdr Rey Molina [of VFA-22], that was tasked with controlling air wing CAS assets in support of a SOF raid

US Navy Reserve unit VF-201 was the last of four "part-time" squadrons to be equipped with the F-14, flying the aircraft for 11 years from 1987 to 1998. Tasked with supporting the frontline force and assigned to CVWR-20, the unit was cleared to operate most weapons available to fleet units bar PGMs – there were insufficient TARPS pods available to issue them to the Reservists. VF-201's "Hunter 102" (BuNo 158627), armed with two Mk 20 CBUs, rolls along the taxiway at El Centro on April 18, 1998. The unit had flown in from its NAS Fort Worth Joint Reserve Base, Texas, home for a week-long strike training detachment that saw it drop various Mk 80 series iron bombs and CBUs. By mid-November of that year, VF-201 had lost all of its Tomcats in preparation for the unit's transition to the F/A-18A in January 1999. (Dan Stijovich)

on the home of [Taliban leader] Mullah Omar on the outskirts of Kandahar. He lived in a well-guarded compound shaped like a diamond and protected by high walls.

The SOF package consisted of several helicopters, including ground force transports and gunships, two AC-130 gunships in an orbit directly over the objective and a "CAS stack" of strike fighters offset to the side of the compound. The concept was to cover the insertion, operation and extraction of the SOF units with fire support inside the compound from helicopter gunships and AC-130s, and with the additional firepower of the air wing assets to address threats outside the compound. There were several FACs involved, including SOF controllers with the ground element, plus an overall airborne controller and our mixed section with the F-14 FAC(A), so radio discipline was essential.

Our launch, initial tanking and flight en route to the rendezvous point were uneventful. We were loaded with two CBU-99 Rockeye CBUs, and intended to buddy-lase for Hornets from our air wing, these aircraft being armed with LGBs and LMAVs [AGM-65E laser Maverick air-to-ground missiles]. As a FAC(A) crew, we were allowed to control the entire airspace over the target area due to our supposed higher levels of situational awareness and comms with all the "players" involved. We wanted to use stand-off weapons from the CAS stack as much as possible so that we could deconflict the air wing aircraft from the fires of the AC-130s directly overhead the compound.

At first things were pretty calm, as the various packages checked in on station and we waited for the helicopter assault force to arrive. During this lull in the action, we left the primary AC-130 FAC(A) on scene to cover the initial force insertion while we headed to the tanker to top off our fuel. We wanted to be able to stay on station, if needed, during the actual insertion period, so we planned our tanking up-front in the operation.

As we headed to the tanker, we monitored the primary controlling frequency so that we could keep a check on the progress of the operation. We could hear the initial kick off and arrival of the SOF troops as we refueled. By the time we had topped off and got back on station to join the fight, all hell had broken loose inside the compound.

The helicopters had dropped off the SOF elements that were now entering the compound and were engaged in several firefights with the guards. Tracers and gunfire were visible on NVGs both

Armorers from VF-154 secure two live Mk 20 CBUs to a pair of BRU-32 bomb racks attached to the forward weapons pallets of an F-14A chained down to the flightdeck of *Kitty Hawk*. CVW-5 had only recently embarked on a three-month-long WESTPAC deployment with the forward-deployed carrier when this photograph was taken in mid-March 2001. (US Navy)

inside and outside the compound and into the surrounding hills. You could hear the automatic weapons fire in the background as controllers made frantic requests for, and directed, fires.

A heavy-caliber AAA piece then opened up on helicopters orbiting the compound, and we called out the threat. Lt Saccomando quickly set up to put the target in the center of our HUD so as to acquire it with the LTS and direct an attack against it. The weapon we were targeting looked like a ZSU-23-4 23mm system based on its rate of fire – the "golden firehose." With our NVGs, we could easily see the tracer rounds being fired at the helicopters.

However, before we got the chance to set up an attack, one of the AC-130 gunships opened up on the AAA piece with its 40mm cannon and completely destroyed the gun – a good secondary

The crew of VF-213's "Black Lion 107" (F-14D BuNo 164344) keep a close eye on a pair of F/A-18Cs from VFA-94 as they take it in turns to top off their tanks from a Diego Garcia-based KC-10A assigned to the 32nd ARS/305th AMW in October 2001 during OEF. Congestion on the tanker was a common problem facing Tomcat crews returning from long-range strikes short on fuel. This photograph was taken during frontside refueling, however, when fuel was not so critical for the F-14. The VF-213 crew that expended the only CBUs ever dropped in anger from a Tomcat were flying as part of a mixed section with Hornets from CVW-11. (Lt Tony Toma)

explosion was also seen emanating from the target. Viewing this attack through the NVGs was one of the most impressive examples of destructive firepower that I have ever witnessed. The crew manning the AAA piece had probably only got off three bursts of fire in total prior to being taken out by the AC-130.

We heard several controllers calling for CAS inside the compound, and it was hard to get a word in on the radio. They were, understandably, very excited, but the FAC(A)s on station had to make certain that we got an accurate description of the target, and that friendly forces would not be caught up in the effects, before directing fires in close proximity to them. Emotional, knee-jerk responses to requests could have easily resulted in fratricide.

We eventually received a call to take out guard towers on the compound wall that was delivering automatic weapons fire into the compound itself. Setting up for an attack, we confirmed with the SOF controller on the ground which towers he wanted destroyed, then instructed our Hornet wingman to turn in toward the compound while we entered the laser code for his weapons into our LTS. We located the guard tower, designated it with laser energy and told Lt Cdr Molina to call "good spot" when his LMAV had acquired the target. The LMAV was the perfect weapon to take out the guard tower; not only did it contain a 300lb [136kg] shaped-charge warhead, it would safe up, climb and dud well down range if it lost laser energy during the time of flight. This feature was incredibly valuable when it came to protecting friendly forces.

VF-213's "Black Lion 103" (F-14D BuNo 163899) prepares to take on fuel from a KC-10 over Pakistan as the aircraft heads north to Afghanistan. Photographed in early November, the jet already boasts an impressive bomb tally beneath its cockpit. The Tomcat is armed with GBU-12s, as well as a pair of AIM-9Ms for self-defense. The author has yet to see any photographs of F-14s in OEF carrying CBUs. (USAF)

Lt Cdr Molina called "good spot," and we "cleared him hot" on the tower as we arced across the target, lasing out of the right side of our aircraft where the LTS pod was located. A moment later, the cockpit filled with light. All the aircraft involved in this operation were flying around with their lights off, and we were using altitude deconfliction to avoid a collision. Our wingman was passing right above us as we "cleared him hot," and the LMAV's rocket motor lit up our entire cockpit as it headed downrange towards the target. We continued to lase the guard tower until the LMAV scored a direct hit, to the cheers of the SOF forces on the radio. They then directed us to take out the remaining guard tower on the opposite side of the compound, which we destroyed using the same tactics.

While this was going on, there were several skirmishes occurring inside the compound and in the high terrain that surrounded the target area to the north. Helicopter gunships were also trying to suppress a large amount of fire being delivered into the compound from AAA weapons sited on the ridgeline to the north, but they were vulnerable to these heavy-caliber guns, so we were directed to

However, the Rockeyes we dropped that night were the only ones expended from an F-14 in OEF, as the unfortunate side effect of this weapon was that unexploded cluster munitions would be left behind due to the high dud rate suffered by the bomblets when dropped on soft soil. This became a problem if friendly troops planned to occupy the area at any point following CBU use, as the bomblets could be set off if they were disturbed.

SOF teams didn't like employing CBUs in close contact either, but it could be done. We would weaponeer for a one-third pattern overlap between each bomb, predicting the pattern length and width for employment considerations. The CBU-99 was a good psychological weapon to kill/harass enemy troops and keep them stationary for other killing weapons.

Seconds before we reached the release point, while scanning the target area through our LTS, we saw a helicopter conducting a strafing run as it headed down the same ridgeline. We immediately aborted our attack and pulled away. In all the confusion on the radio, the crew of the helicopter had missed the fact that they had a Tomcat bearing down on the ridgeline and had failed to clear the area. The rest of the operation seemed to go smoothly after this, with SOF forces eliminating the remaining pockets of resistance and then re-boarding the transport helicopters in preparation for their exfiltration from the area. There were a few targets external to the compound remaining when we handed over the operation to the next wave of controlling F-14 FAC(A)s from CVW-8, and they cleaned them up and then covered the SOF helicopter egress.

Before Lt Saccomando and I flew back to the carrier, we were able to head to a nearby predetermined alternate target area – a vehicle staging depot east of Kandahar – and expend our Rockeyes on a group of military trucks and APCs that were foolishly parked in close formation. The CBU-99s produced a spectacular pattern on the ground, as well as several secondary explosions.

Naval Aviators and NFOs from VF-213 pose for a photograph with LGBs in the foreground and "Black Lion 101" (BuNo 164603) as a backdrop shortly after *Carl Vinson* came off the line for the last time during OEF. The crew that dropped the two CBU-99s during the campaign can be seen here, namely Lt Shaun Saccomando (second from left) and Lt Cdr Michael Peterson (standing in the shadow of the jet at far right). (Lt Tony Toma)

drop a string of Rockeye on the ridgeline. We avoided the AC-130 line-of-fire and descended to set up our Rockeye run. As Lt Saccomando pressed in toward the ridgeline for a level lay-down delivery, I cued the LTS out in front of us to capture the effects of the combined 494 Mk 118 bomblets contained in our CBU-99s.

Each of the canisters was fitted with an FMU-140 dispenser proximity fuse [DPF], which could be set with an arming time, and a height-of-function [HOF] trigger that would detect the CBU-99's height above the target and then open the dispenser to produce an optimum cluster weapon footprint on the ground. F-14A/Bs did not generate sufficient ballistic information via their elderly weapons computers to effectively employ the FMU-140 DPF fitted to the CBU-99. The F-14D, however, had a more modern stores' management system that collated accurate ballistics and allowed the aircrew to enter the HOF for the FMU-140.

The CBU-99 was a deadly weapon against tanks and armored personnel carriers [APCs], and was equally as effective against troops if they were traveling in a soft vehicle such as a truck or SUV.

Lt Cdr Peterson and Lt Saccomando were also involved in separate CBU events towards the end of VF-213's commitment to OEF. On December 2, 2001, Saccomando and Lt Cdr Robert Garretson, paired up as a FAC(A)-qualified crew, provided target co-ordinates for two USAF F-16Cs armed with CBU-103s. These weapons featured Wind Corrected Munitions Dispenser (WCMD) guidance tail kits that turned the CBUs into PGMs. Saccomando recalled the mission in his OEF war diary as follows:

Worked with "Tiger 12" [SOF FAC] 20 miles north of Kandahar. He gave us control of two compounds [USAF fighter bombers]. Gave the "Alps" flight [two F-16s] T-cubed [Tomcat Tactical Targeting, abbreviated as T3] co-ordinates [mensurated co-ordinates, which is the degree of accuracy required to drop GPS-guided weapons in a CAS scenario – software in the LANTIRN pod could generate co-ordinates four digits right of the decimal point] for their CBU-103s. Took over as FAC(A) and cleared them hot for two direct hits on the compounds. Controlled wingman [in a second F-14D] for vehicle and two buildings. Dropped two [GBU-12] bombs ourselves for good hits as well.

Lt Cdr Michael Peterson was also part of the VF-213 crew that provided a B-52H from the 28th Air Expeditionary Wing with T3 co-ordinates to finish off a Taliban convoy in northern Afghanistan after the Tomcat had "bottled up the lead and trail vehicles" with LGBs. The USAF heavy bomber, armed with 16 CBU-103s fitted with WCMDs, was flying at more than 40,000ft when it dropped the PGMs. (USAF)

Lt Cdr Peterson, also crewed with a FAC(A)-qualified pilot, provided T3 co-ordinates for a B-52H armed with 16 CBU-103s and flying at more than 40,000ft over northern Afghanistan.

We hit a convoy [east of Mazar-i-Sharif and west of Kunduz] with LGBs and bottled up the lead and trail vehicles. At that point most of the people got out of their trucks and fled into the surrounding hills, and we called in a B-52 to clean up the rest of the vehicles. We provided co-ordinates generated by the LTS and they dropped WCMDs. The time of fall from where the B-52 was dropping was over a minute, and it was amazing to see weapons delivered from such an altitude produce accurate effects on target.

As a pseudo-reconnaissance asset, LANTIRN generated co-ordinates for any target located on the FLIR. The T3 software modification, which had been rolled out in pods assigned to frontline Tomcat units from late 1999, increased the accuracy of the co-ordinates produced to such a degree that LANTIRN could now be used in the prosecution of mobile targets for

Left: Leaflets giving the frequency of the coalition's Arabic-speaking radio station are carefully packed into the metal rings of a PDU-5/B aboard *Constellation* in March 2003. CVW-2's Hornet squadrons and VF-2 dropped PDU-5/Bs shortly before the start of hostilities in Iraq. (US Navy)

Right and below: This leaflet was one of several different designs packed in their thousands into PDU-5/Bs and then dropped over southern Iraq prior to OIF I commencing. It urged Iraqi soldiers to abandon their weapons and surrender, thus avoiding planned air strikes. (US Navy)

the first time. T3 also allowed TACAIR assets to generate their own co-ordinates suitable for GPS-guided weapon (JDAM, JSOW and CBU-103 WCMD) employment. The first combat use of this capability was in OEF, when Lt Cdr Peterson's T3 LTS-equipped Tomcat generated co-ordinates for the CBU-103-equipped B-52.

Although no CBUs were dropped from Tomcats during either OSW or OIF, a handful of Psychological Dispersal Unit (PDU) air-launched freefall leaflet dispensers were expended by VF-2 in mid-March 2003. Essentially a CBU-100/B with its submunition payload removed and a cardboard payload sleeve filled with leaflet rolls and spacers installed in its place,

Left: VFA-137's "Falcon 404" (F/A-18C BuNo 164709), flying from *Constellation* as part of CVW-2, heads for southern Iraq toting a single PDU-5/B beneath each wing in mid-March 2003. The PDUs were dropped from a height of 35,000ft over towns and cities in the no-fly zone. The yellow band on each PDU-5/B indicates the location of the charge that broke the unit open once the canister had dropped away from the aircraft. The blue band denotes that the canisters' contents were non-explosive. (VFA-137)

Below: These PDU-5/B leaflet dispensers were dropped by a B-52H from the USAF's 419th Flight Test Squadron over the Point Mugu Sea Test Range in July 2017. Stores identical to them were expended by F-14Ds from VF-2 immediately prior to OIF I commencing on March 21, 2003. (USAF)

the PDU-5/B had an external appearance similar to a Mk 20 Rockeye anti-tank CBU. Lt Cdr Peterson, who was now serving with VF-2, recalled:

> Just prior to the official start of OIF I, we flew a few missions dropping leaflet bombs that warned Iraqi forces of the coming attack. Some instructed them to park their vehicles in large square formations to indicate that they would not target US forces. Our TARPS imagery subsequently showed several such formations just prior to the kick-off of the ground offensive.
>
> Planning for a PDU-5 mission was very similar to when using Rockeye, as it had the same basic container. However, there was a greater emphasis placed on wind speed and direction so as to achieve the maximum spread with the leaflets, which were wrapped tightly like a wad of dollar bills inside the dispenser, where intended.

CVW-2's Hornet squadrons and VF-2 dropped a total of 29 PDU-5/Bs shortly before the start of hostilities in Iraq. The author has been unable to confirm whether the remaining two Tomcat units committed to OSW/OIF I operations in the NAG in early 2003 also expended PDU-5/Bs.

F-14 units continued to drop CBUs as part of their strike training right up to the end of the aircraft's fleet service. VF-11's "Ripper 101" (F-14B BuNo 162912) taxies out at El Centro armed with a pair of Mk 20 CBUs destined for the R-2507 range. This photograph was taken on March 21, 2005, two-and-a-half months after VF-11 had been transferred from CVW-7 to CVW-17 to maintain force posture due to the latter air wing's "surge" status. CVW-17's previous F-14-equipped unit, VF-103, had by then commenced its transition to the F/A-18F prior to future assignment to CVW-7. (Mark Munzel)

"Ripper 105" (F-14B BuNo 162700) departs El Centro in afterburner as it heads for the nearby US Marine Corps-operated Chocolate Mountain Aerial Gunnery Range during VF-11's March 2005 detachment to southern California. Exactly three weeks after this photograph was taken on March 21, the aircraft arrived at AMARC for storage. (Dan Stijovich)

CHAPTER 6

AIR-TO-GROUND TRAINING

The tuition of Tomcat crews in how best to use their air superiority fighter as a precision bomber evolved during the 1990s, particularly following the advent of the LANTIRN pod.

As previously noted in this volume, the F-14 community started to embrace the "Bombcat" role from 1990 onward as units tentatively explored the jet's air-to-ground capabilities at ranges on both coasts. VF-101, as the Oceana-based FRS on the East Coast for COMNAVAIRLANT, had led the way in conjunction with FITWINGONE, whose 1990 command history noted:

FITWINGONE initiated F-14 air-to-ground training. Milestones accomplished include:

(1) Convened initial Strike Warfare Team meeting, March 1990, to identify Plan of Action and Milestones to implement air-to-ground capability for the F-14 community.
(2) Revised squadron Required Operational Capability/Projected Operational Environment to reflect air-to-ground mission.
(3) Incorporated air-to-ground syllabus at the Fleet Readiness Squadron [VF-101].
(4) Revised training and readiness matrix to reflect air-to-ground mission.
(5) Conducted F-14 air-to-ground FRS training for two air wings, which included VF-33, -102, -142 and -143. Completed ground school for VF-41 and VF-84.
(6) Initiated development of Advanced Air-to-Ground training program.

A similar program had been instigated by VF-124 and COMFITAEWWINGPAC for COMNAVAIRPAC F-14 units based at Miramar. The FRSs integrated strike training into the five training phases taught by instructors as part of a building-block approach for student pilots and RIOs destined for fleet squadrons. The Strike phase followed familiarization (FAM), where students were taught systems, basic handling, switchology, emergency procedures and Weapons Control Systems, introducing basic intercept tactics, radar and communications use, ECM and air-to-air gunnery.

Instructors and students then moved to either El Centro or Fallon (VF-101 undertook its first detachment here in the fall of 1992) for the Strike phase, both locations enjoying better weather and larger weapons ranges than were available at Miramar and Oceana. These factors allowed the FRSs to teach low-level tactical flying and navigation, higher altitude bomb delivery techniques and SAM and AAA defense, often on instrumented ranges. The Strike phase grew in importance with the retirement of the A-6 in early 1997 and the introduction of the precision-targeting LANTIRN pod for the F-14. The Air-to-Air Tactics phase then followed at NAS Key West, Florida, and the year-long Tomcat FRS course culminated with the must-pass Carrier Qualification, undertaken on a vessel sailing off either the east or west coasts.

Some squadrons welcomed this change in mission emphasis, while others fought it. However, by the fall of 1991, the TASS (devised and implemented by SWATSLANT) had been established to replace the FRS bombing course that had initially introduced F-14 aircrew to the strike mission. This evolved into the AARP in 1992, and units equipped with A- and B-model jets were now deploying with an improving strike capability. VF-14 of FITWINGLANT had completed the AARP in November–December 1993, the unit's command history for that year noting that the program "consisted of two weeks of strikes utilizing ranges from Pennsylvania to North Carolina, allowing aircrew to hone their skills in strike planning, weaponeering, CAS and night attack."

Despite the fact that the Tomcat was now clearly being seen as a multi-role aircraft, some of those flying the jet still felt that it was a fighter first and foremost. One such individual was future WTI and SFTI Cdr Jim "Puck" Howe:

When I was a young student in VF-101 in the early 1990s, the Tomcat was just starting to develop the air-to-ground capability that had been inherent in the jet since its creation. Looking back, it's pretty funny, as we didn't really have any idea what we were doing. It wasn't exactly giving a loaded gun to a child, but it was close. Moreover, no one really seemed too serious or too happy about the F-14's potential

VF-124 "Gunfighters" (1993)

Above: The two Tomcat FRSs were big units during the heyday of the F-14 in the 1980s and early 1990s, as this group shot of VF-124's instructor cadre and maintenance and administration personnel in 1991 clearly shows. Typically, a Tomcat FRS needed to have about 50 qualified instructors (25 pilots and 25 RIOs), more than 50 aircraft and in excess of 500 maintainers to ensure that the fleet received its quota of qualified aircrew and maintainers. At any one time, VF-124 could have as many as 100 students on its books, with two-thirds of them being new Category One pilots and RIOs – the unit would typically commence a new class every month. Forming the backdrop to this photograph are F-14A BuNo 161620 and F-14D BuNo 163894 (one of the first four D-models assigned to VF-124), as well as a T-34C. VF-124 had commenced integrated strike training for students just months prior to this group shot being taken at Miramar in spring 1991. (via Cdr Doug Denneny)

Right: An instructor and his student walk out to the VF-101 flightline, packed with three different models of Tomcat. They are passing an F-14A (all A-models had 100–150-series modexes) towards the F-14B ramp (with 200-series modexes). Dubbed "Tomcat Central," VF-101 consisted of almost 50 jets and 1,200 personnel during the final decade of the F-14's US Navy service. At any one time in the late 1990s, when this photograph was taken, VF-101's 50+ instructors would be training five classes of 16 students each. In 1996, for example, the unit sent 76 pilots and 80 RIOs to the fleet. To achieve such numbers, VF-101 would usually generate 62 daily events including flights and lectures. Again, in 1996, it tallied 8,300 sorties totalling 11,000 flying hours, burning $11 million in fuel. Although the bulk of these flights took place at Oceana, VF-101 supported five training detachments to the West Coast (NAS Fallon and NAF El Centro) and three to Key West, each taking 150 personnel, 12 jets and an 18-wheeler articulated truck packed with suitcases. (US Navy)

to become a "strike fighter." But our leadership kept telling us that a single-role aircraft would not last long in a newly cost-conscious Navy. If we didn't figure out bomb dropping, and quickly, the Tomcat was going to fade unceremoniously into oblivion. Enter "flexibility."

Within five years the F-14 went from being a Cold War relic protecting the battle group from "Bear" and "Backfire" raids that were never going to happen, to being the LANTIRN-carrying, self-escorting strike fighter of choice. The Tomcat, with its two-seat cockpit, a big payload and more fuel, greater speed and greater bring-back than the F/A-18, had become the "Bombcat." Much to the chagrin of our Hornet brethren, the toughest missions were now doled out to Tomcat squadrons.

Cdr Doug "Boog" Denneny had been an early believer in the F-14 as a strike fighter. Part of the west coast fighter community at the time, he explained in Brad Elward's comprehensive volume, *TOPGUN – The Legacy*:

At TOPGUN in 1993–95 we, for the first time, introduced an actual bombing flight – using Mk 76 25lb training bombs – for a Tomcat Strike Fighter mission as part of the training syllabus. We achieved this despite many in the F-14 community protesting that "we have to have the fighter mission." Conversely, there was a large cadre of us that said we needed to do more self-escorted strike missions. We needed to do more missions where we would actually go out and drop Mk 76 bombs at the end of the mission, and have our total mission success not be completely predicated on the number of enemy fighters we simulated as shot down, but to also take into consideration "did we get to the target area, and did we deliver our ordnance?"

There was a difference of opinion within the TOPGUN staff, with a number of instructors stating that "we are the Navy Fighter Weapons School; our job is to go back to our roots in Vietnam, to be the best in combat maneuvering, the best in defeating the radar-guided missile-type threat and in defeating Russian MiGs. The bombing stuff – it's easy." And so, there was a lot of sentiment going through TOPGUN that argued, "anyone can roll in and drop a bomb and get it fairly close, particularly if it's laser guided.

VF-101 had the largest inventory of Tomcats in the US Navy following the disestablishment of west coast FRS VF-124 in September 1994. The "Grim Reapers" had initially been tasked by the CNO to act as model manager for the Tomcat strike fighter program, developing an air-to-ground syllabus for the F-14 and training replacement aircrews in strike warfare. Among the aircraft involved in this early "Bombcat" work was F-14A BuNo 161133, which served with VF-101 for 17 years from May 1980 until shortly before it was stricken in September 1997. Fitted with two empty BRU-42 ITERs on its centerline, this aircraft was one of a number of F-14s assigned to VF-101 that had their fins repainted in the markings worn by recently disestablished units – in this case, VF-1 "Wolfpack" (although the squadron was disestablished in 1993, not 1995!). (David F Brown)

AIR-TO-GROUND TRAINING

The tough stuff is air combat. So, let's not waste our time and energy in training guys to do a strike mission that's taken care of in other ways." I thought that attitude was just wrong.

Around that same time, VF-211 at Miramar proved it could drop a large amount of ordnance in a 24-hour period, showing off the Tomcat's sustainability in this role. The brainchild behind that was the unit CO, Cdr [later Admiral] James "Sandy" Winnefeld, and his Operations Officer, Lt Cdr Calvin "Goose" Craig. VF-211 dropped a lot of ordnance in training to show how one Tomcat squadron could put a large number of big bombs on target.

Clearly, the writing was on the wall for the F-14 as a single mission aircraft in the US Navy of the 1990s. TOPGUN eventually realized this, and Class 05-93 proved to be the final course prior to the reintroduction of an air-to-ground component to the syllabus, which had been absent since the F-4 Phantom II-aligned courses in the mid-1970s. In the late summer of 1993, TOPGUN instructors were working long hours to formulate and then incorporate seven new air-to-ground lectures (CAS, Target Area Tactics, AGM-88 HARM, AGM-65 Maverick, LGBs and infrared targeting, AGM-62 Walleye II and Weaponeering/Fusing – not all of them directly applied to F-14 crews) into the curriculum.

In order to overhaul the course in the shortest time possible, a single class was cancelled, giving staff eight to ten weeks to finalise their lectures. "That gave us the time to plan to change the course from five weeks of just pure fighter tactics to introducing air-to-ground and what I would call the strike fighter syllabus, where you were actually dropping bombs, fighting your way into the target to get an air-to-ground weapons release, and then fighting your way out," explained F/A-18 pilot Cdr John Clagett, who was TOPGUN's Training Officer during this period.

The change could not come soon enough. "The Tomcat crews at that point realised that they had just done *Desert Storm*, and they didn't have much to show for it because they were mission-limited," stated Clagett in Brad Elward's book. "But they were in the process of developing air-to-ground weapons capability, so in some quarters of the community they

"Gun Fighter 153" (F-14A BuNo 161866) was another of VF-101's Tomcats to be adorned with an insignia of a disestablished unit, in this case VF-114. The aircraft was photographed returning to El Centro in October 1995 with a single Mk 76 Mod 5 blue bomb still attached to the BRU-42 ITER on one of its centerline stores stations. BuNo 161866 subsequently became VF-154's last Tomcat CAG jet (from January 2002), seeing considerable action during OIF I. It was sent to AMARC in October 2003. (Dan Stijovich)

supported the school's decision to go down that path. Even so, when we rolled out the 'strike' syllabus, there was a large percentage of the rank-and-file Tomcat community who thought the Hornet guys were ruining the course.

"This is where skipper [Capt Richard K] Gallagher helped out. He fielded the phone calls from the various Tomcat skippers and above (CAGs and admirals) who were questioning what we were doing. As the skipper of TOPGUN and a Tomcat guy, Gallagher was ideal to tamp down the embers of dissent. He told them, 'Strike was going to happen in the course and their community.' The transition was smoother [as a result] because no real groundswell built up. In my opinion, it went smoothly because Gallagher and, in part, some very smart Tomcat leaders knew it would help their transition to becoming the 'big strike' fighter [of the air wing]. The shift to air-to-ground was a very big moment in time for TOPGUN and, frankly, in retrospect, it was the return to the format the school had starting out – the original syllabus had a portion devoted to air-to-ground."

Bombs Away

The late Cdr Neil "Waylon" Jennings, who had been a TOPGUN graduate in the days when the school focused exclusively on the air-to-air mission, served with VF-213 during this time of great change in Fighter Wing, US Pacific Fleet (FITWINGPAC). Very much a proponent of the Tomcat as a strike fighter, he had the chance to expend plenty of dumb bombs from the F-14 in the mid 1990s:

Unfortunately, my time in the "Bombcat" ended in 1996 when I transitioned to Hornets. Those were great days for me personally, but I ended my association with the Tomcat just as it was cresting on being pushed out on its first deployment as a genuine strike aircraft. I did see a couple of things though while still flying the F-14. I went through all the initial training with VF-213, when we shifted our role from purely fighter to the strike fighter role. It was fantastic, because we dropped LOTS of bombs. I think I might have dropped more bombs in my last year in the "Blacklions" than I did in my first two years in the Hornet. It was really interesting stuff, because we were on the leading edge of the "Bombcat" wave, along with all the other F-14 squadrons that were in the early part of their turnaround cycle.

Ironically, after eight or nine months of training in the Hornet in 1996, I was assigned to VFA-97, and I rejoined the same air wing that I had just come out of, CVW-11. I then deployed aboard [USS] *Carl Vinson* [CVN-70] with the "Warhawks" [in 1998–99] and was right back on cruise with the "Blacklions." The skipper of VF-213, Cdr Mark "Shaker" Adamshick, was gracious enough to invite me to join his ready room as a former "Lion," and I watched many video replay "roll 'ems" that deployment with my former squadron mates as they dropped ordnance during training missions.

The "Blacklions" did a great job that cruise with their air-to-ground skills, especially in the FAC(A) role, where the Tomcat excelled. While flying OSW missions with VFA-97, I'd always see them on my way to or from the tanker, as their birds could stay out longer than the F/A-18A I was flying. Although the Hornet was a more accurate and versatile bomber, having been built from the ground up for the role, the Tomcat also definitely contributed to OSW missions.

By 1995, the Tomcat units that had survived the post-Cold War cull had become quite proficient in the strike fighter role, as RIO Cdr Tom "Tumor" Twomey explained:

Although we had not yet received the LANTIRN pod in the fleet, we could still employ LGBs by "buddy" lasing with F/A-18s that were equipped with the Nite Hawk FLIR pod. These missions were a big learning curve for us, but we honed our skills and became very effective when operating in a mixed section with an F/A-18. The F-14 could haul a lot of "iron" to the target, and that increased the effectiveness of our air wing.

The BRU-42 ITER paired with the Mk 76 Mod 5 blue bomb were critical training aids for budding "Bombcat" crews from the early 1990s through to the jet's retirement in 2006. Two feet (0.6m) in length and weighing 25lb (11.33kg), the weapon has a charge in the nose that goes off when the bomb hits the ground. The point of impact is marked by white smoke from the charge. According to a former A-6 pilot, "Naval Aviators can spend a considerable amount of time on the practice range dropping Mk 76s, as they don't cost a lot of taxpayer dollars and don't blow the target to smithereens like a '500 pounder' might." The numbers employed by Tomcat units in a calendar year varied depending on what stage of their pre-cruise work-ups they were at. For example, VF-14 dropped 1,446 Mk 76s in the first four months of 2001 prior to heading out on its final Tomcat cruise. Designed to mimic the ballistics of Mk 80-series weapons, the Mk 76 is a simulator for the 500lb (226kg) Mk 82 GP bomb. (Danny Coremans)

AIR-TO-GROUND TRAINING

In 1996–97, the near simultaneous retirement of the A-6 and the supply of LANTIRN pods to operational units revolutionized the F-14's capabilities in the precision strike role and drastically altered the way Tomcat crews were trained to perform bombing missions. According to LANTIRN pioneer Lt Cdr Dave "Hey Joe" Parsons:

As the Intruder community began to dissolve, the infusion of A-6 "refugees" into the Tomcat community in the mid 1990s brought considerable expertise to the air-to-ground mission. Among the early arrivals from the Intruder world was Cdr John Snedeker, who was a highly regarded A-6 B/N tactician who had flown combat missions in *Desert Storm* off *Saratoga*. He was selected as the first F-14 XO/CO from the Intruder community, joining VF-14 in April 1995 after completing his Tomcat transition training with VF-101. Prior to reporting to his new squadron, John, who was a neighbor of mine, mentioned to me that he was worried about his lack of knowledge in ACM. I told him that his knowledge of air-to-ground would put him in great stead with the Tomcat community. And it did!

FITWINGLANT subsequently took over the Medium Attack Weapons School Atlantic, which had been the precision-attack center of excellence during the A-6 era, folding it into SWATSLANT. It was here that the subject matter expertise was nurtured and validated to ensure that each squadron was honed to carry out its assigned mission, with precision strike at the centerpiece. A succession of Tomcat RIOs took command of SWATSLANT during the 1990s to usher in the air-to-ground capability and equally important FAC(A) mission.

Cdr Twomey saw the effect the Intruder crews had on the Tomcat community at first hand while serving with VF-2 in 1995–96:

My squadron took in a few transitioned A-6 pilots and B/Ns, with the remaining aircrew going to other F-14 squadrons. Overall, the Intruder community was just decimated. Only a few lucky aircrew were offered transitions – pilots and B/Ns also went to other communities too. In fact, a lot of the pilots transitioned on to the F/A-18. Many of the A-6 pilots and B/Ns that received orders to Tomcat units went to VF-101 Det Miramar, which was the dedicated F-14D training unit. Here, they studied the Category 2 syllabus designed for Naval Aviators and NFOs who already had a lot of flying experience. They learned the F-14D systems and flew a few training hops in the simulator, before getting into the jet for flights with VF-101 Det Miramar. The course lasted four to six months, after which crews reported to their fleet squadrons.

The F-14D was the last version of the Tomcat to take on the bombing mission, so the units equipped with the aircraft [VF-2, VF-11 and VF-31 in 1995–96] benefited the most from the arrival of former A-6 crews because the majority of the pilots and RIOs

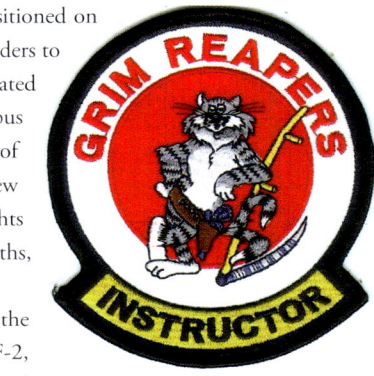

VF-101 "Grim Reapers" instructor patch (1997)

While the units equipped with F-14A/Bs had been cleared to drop live air-to-ground ordnance from 1991, the three squadrons flying F-14Ds were restricted from doing so until the late fall of 1995 due to software issues. This meant VF-2, VF-11 and VF-31 had to focus on the air-to-air and TARPS missions, and the latter two units worked on both elements with CVW-14 when the air wing sent the vast majority of its aerial assets to Roswell Industrial Air Center in April–May 1995 to participate in Exercise *Roving Sands 95*. This huge event was held biennially from 1989 through to 2001, when budget restrictions and real-world commitments resulted in its demise. As this panoramic view of the Roswell flightline shows, the US Navy aircraft (including F-14Ds of VF-11 and VF-31) operated in a joint environment with F-15Es from the 4th FW, which had flown in from Seymour Johnson AFB, North Carolina. (US Navy)

F-14D squadrons were the last to truly embrace the "Bombcat" mission. The arrival of the LANTIRN pod in the front line in 1996–97 coincided with an infusion of former A-6 pilots and B/Ns into the Tomcat community, specifically in units equipped with D-model jets. VF-2 welcomed a handful of ex-Intruder aircrew after they had completed the Category 2 syllabus (designed for Naval Aviators and NFOs with plenty of flying experience) with VF-101 Det Miramar. These individuals subsequently played an important part in the "Bounty Hunters'" 1997 WESTPAC/NAG cruise with CVW-2 onboard *Constellation*, which saw the unit give the LANTIRN pod its Pacific Fleet debut. (Cdr Tom Twomey)

already in these squadrons had very little experience of dropping bombs due to their fighter-only backgrounds. The ex-Intruder pilots and B/Ns worked well with the neophyte Tomcat bomber aircrews, teaching us the bombing business. We, in turn, taught them the fighter business, including ACM and the technical aspects of operating the jet's AN/APG-71 radar in an electronic jamming environment.

Once the A-6 pilots and B/Ns got a few hundred hours in the F-14D under their belts they were just as effective as any Tomcat-only fleet pilot and RIO. Initially, former Intruder pilots were paired with experienced F-14 RIOs, while the B/Ns flew with veteran F-14 pilots. This sped up the learning process and provided an additional safety factor as both flight time and experience built up. The Intruder aircrew were quick learners, as pilots and B/Ns that had been posted to the A-6 RAG when the jet was still part of the frontline force were generally at the top of their classes upon graduation from flight school. Delivering ordnance on time in a hostile environment required both skill and talent. The A-6 community had been very well respected in the air wing prior to its demise, and infusing "Bombcat" squadrons with this extensive experience greatly improved their mission effectiveness in the air-to-ground role.

F-14D Bomber

As touched upon by Cdr Twomey, the F-14D was the last Tomcat variant to fully embrace the strike fighter mission. Indeed, the aircraft was not cleared to drop live air-to-ground ordnance until the late fall of 1995 – more than four years after F-14A/Bs had started to do so. This was primarily because of long-running software issues with the interface between the aircraft's AN/APG-71 radar (unique to the D-model jet) and the mission tape that allowed the Tomcat to accurately employ air-to-ground weapons. Differences between the F-14D's wiring and that of the F-14A/B exacerbated the problem, with the latter aircraft featuring manual stores (including bombs) selection and the D-model being completely digital. Furthermore, with only 55 D-model jets being supplied to the US Navy, the majority of the funding for turning the Tomcat into a strike fighter was initially spent on the more numerous F-14A/B.

NAVAIR's test and evaluation personnel at NAS Point Mugu (Naval Weapons Test Squadron, F-14 Weapons System Support Activity and VX-9) and NAWS China Lake (Naval Air Warfare Center-Weapons Division) took several years to sort out the software bugs, flying endless configuration flights while working through validation and verification (V and V) checks. According to a source close to the Tomcat program who wished to remain anonymous, "the onerous part of [the F-14D software] V and V was the insistence by the west coast test community that it had to fly every possible [weapons] configuration, which was invariably time consuming and thereby inordinately expensive." By the mid 1990s, with defense budgets shrinking and funding for the F-14 being reduced, there was little money available to accelerate the V and V process.

Eventually, software Operational Flight Program (OFP) Tape D02 (based on an old A-model tape), which allowed F-14Ds to drop live ordnance, was cleared for installation from July 31, 1995. Tapes D03 and D03A would subsequently be released for the D-model Tomcat in 1997–98, with the former adding "GPS, improved air-to-ground capability and significant [cockpit] display enhancements," according to Cdr Dave Madsen, then military lead for the F-14 Integrated Program Team at NAS Point Mugu. The D03A tape "enhanced the F-14D's war-fighting capability through navigation and air-to-ground improvements." Updated software tapes were also developed for the F-14A/B.

The final hurdle stopping the D-model from initially fulfilling its full potential as a precision strike fighter when paired with a LANTIRN pod was the lack of a PTID in the rear cockpit. According to Cdr Doug Denneny:

> You could drop off the smaller MFDs [Multi-Function Displays, initially fitted in both cockpits of all F-14Ds] with LANTIRN, but it was not optimal. The resolution of the imagery on the

The front cockpit of the F-14D merged the old with the new in a "lashed-up, lump-it-on-top" kind of way according to those that flew it. Multi-function TIDs sat rather uncomfortably alongside old-fashioned "steam" gauges and dials, and systems integration between the front and back seats even in the D-model was virtually non-existent – just as it had been in the F-14A/B. Such "firewalling" meant that there were certain systems that only the pilot could control and others that only the RIO could operate. (Danny Coremans)

significantly larger PTID display, however, was exceptional compared to the image on the MFD or the older TID. This made the PTID a key tool in the effective employment of LGBs.

For many reasons, there were more F-14B/Ds needing PTIDs than there were PTIDs in the entire FITWINGONE inventory. The F-14Bs had to have a full complement of PTIDs [because they all had the D02 mission tape software], but the F-14Ds technically did not need them [early on because they lacked this upgrade at the time]. This resulted in all the B-model squadrons having PTIDs, while the D-model squadrons were left to share the few remaining PTIDs until the deploying unit was in the final stages of work-ups. Swapping out a TID and replacing it with a PTID was not a simple task, so the tendency was that once a D-model squadron got a PTID in a jet, it was pretty much going to stay in that aircraft and remain configured as a bombing platform during work-ups. PTIDs were hot swapped among deploying D-model squadrons for several years, with a unit only getting its full contingency immediately before heading out on cruise.

CVW-14 was the only air wing to boast two F-14D-equipped units in the 1990s, with VF-11 and VF-31 being assigned between 1992 and 1996. Their aircraft were updated with Tape D02 at Miramar during the late summer and fall of 1995, with both units dropping Mk 76 training bombs for the first time on August 14, followed by live Mk 80-series ordnance during CVW-14's detachment to Fallon in late September. As part of the fleet evaluation of Tape D02, VF-11 and VF-31 also dropped Mk 20 CBUs and TALD, thus allowing the D-model to match the F-14A/B in respect to the weaponry it could employ operationally.

This view of an F-14D rear cockpit clearly shows the size of the PTID screen (bottom). Above it is the Detail Data Display and to the right is the Multiple Display Indicator. The RIO's Hand Control Unit can be partially seen at the bottom of the photograph in front of the PTID. Another key addition to the D-model Tomcat's rear cockpit was the Sensor Slaving Panel mounted forward of the grab handle atop the instrument console. (Danny Coremans)

The original "fishbowl" TID in the rear cockpit of the F-14D was eventually replaced by the new 8 x 8in PTID, which presented both radar and LANTIRN data. The larger PTIDs were initially retrofitted into upgraded B-model jets (this photograph was taken in the back seat of an F-14B from VF-32) ahead of the then mission-restricted F-14D, as there was a shortage of them. "The Tomcat with LANTIRN was an awesome bomber," recalled Rear Admiral Jay "Spook" Yakeley (the first Tomcat acceptance test pilot at Grumman). "It was actually better than the F-15E, although my air force friends will beat me up for saying so. Really, there was no comparison, particularly when an F-14 was fitted with that huge [PTID] screen, which had much better resolution. With the Strike Eagle you could put the bomb on the building. With the Tomcat, you were putting the bomb into the third window from the left, from miles away." (Erik Hildebrandt)

AIR-TO-GROUND TRAINING

VF-11 pioneered the use of NVGs with the F-14 as early as the unit's WESTPAC onboard *Carl Vinson* in 1994. By the time this photograph was taken during the unit's next deployment two years later, most of VF-11's Tomcats had had "two redundant cockpit lighting systems" installed. One provided standard night lighting while the second was NVG compatible. All of the squadron's aircrew were fully qualified in the use of NVGs prior to embarking in *Carl Vinson* with the rest of CVW-14 and deploying in mid-May 1996. (US Navy)

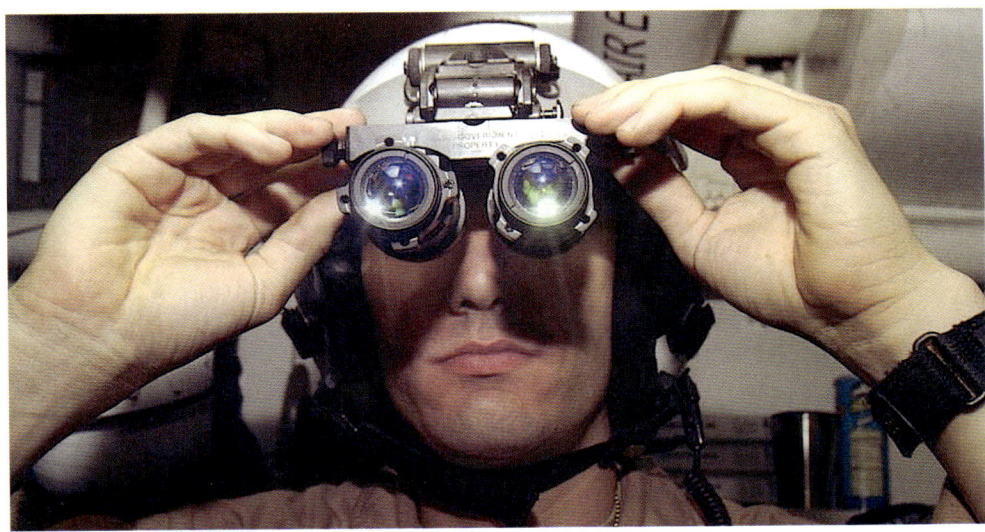

A pilot adjusts his MXU-810/U Mk IV Cats Eyes NVGs, manufactured by GEC Avionics Corporation, while suiting up prior to flying an OSW night mission. The NVGs are fitted with Generation III image-intensifier tubes, which could amplify light in the 600 to 900 nanometer range up to 30,000 times. F-14A Night Vision Imaging System testing with NVGs and compatible cockpit lighting was completed in December 1994, with F-14D testing concluding two months later. (US Navy)

On November 30, while embarked in *Carl Vinson* during COMPTUEX 96-1, VF-11 became the first F-14D fleet squadron to drop bombs (four Mk 76s) at night.

This sortie was the culmination of two years of work undertaken by the unit to turn the F-14D into a night-capable strike fighter, VF-11 having pioneered the use of NVGs with the Tomcat during its 1994 WESTPAC embarked in *Carl Vinson*. The following account of the squadron's work in this area appeared in the spring 1996 issue of *The Hook*:

> The addition of NVG capability allows the F-14 to expand its night operations to improve its effectiveness in strike warfare, CAS, TARPS, FAC(A) and CSAR. Normally, a new system is developed by test and evaluation organizations. In this case, COMNAVAIRPAC approved a plan whereby VF-11 modified an F-14D with the necessary NVG equipment. The Tomcat's cockpit was modified for NVGs by installing custom cockpit lighting lenses. In November 1994, the modified aircraft was ground tested at NAS Patuxent River and final modifications made. VX-9 at NAS Point Mugu conducted the operational flight tests and evaluations in February 1995 to conclude two years of effort by VF-11.
>
> The final aircraft design [used fleet-wide] is based on the original VF-11 design concept and incorporates two redundant cockpit lighting systems. One provides standard night lighting while the second is NVG compatible. The new system is slated to be installed in six squadron aircraft prior to VF-11's 1996 WESTPAC deployment. It will eventually be installed in all Navy Tomcats.
>
> Overcoming early challenges in the F-14D NVG concept, VF-11 set the standard for the Tomcat community in NVG tactics by developing standard operating procedures and a solid training program. Early in 1996 the "Red Rippers" qualified six aircrews to fly with NVGs, and the squadron will have the most aircrews fully qualified in their use prior to deploying.

The F-14D's software tape upgrades also allowed the aircraft to carry the LANTIRN pod, and VF-2 was the first D-model unit to deploy with this capability (six pods) when it undertook a WESTPAC cruise with CVW-2 embarked in USS *Constellation* (CV-64) in April–October 1997. Thanks to LANTIRN, VF-2 crews gave the F-14D its FAC(A) debut during the deployment. They also carried out basic Strike Co-ordination And Reconnaissance

F-14 NVIS (Night Vision Imaging System) patch (1995)

(SCAR, also known as "killer scout"), and crews qualified to fly both this mission and FAC(A) were also cleared to act as mission commanders for CSAR – all roles previously unavailable to F-14D units that lacked LANTIRN.

By the fall of 1997, VF-31 was the only D-model Tomcat unit still assigned to CVW-14, VF-11 having reverted to F-14Bs upon its posting to CVW-7 a year earlier. As part of its work-ups for a WESTPAC deployment on board USS *Abraham Lincoln* (CVN-72) in June 1998, VF-31 crews operated LANTIRN-equipped F-14Ds for the first time at Fallon during CVW-14's three-week strike integration training in October 1997. According to the winter 1997 issue of *The Hook*, the "'Tomcatters' employed LANTIRN on their initial strike mission. This feat opened many eyes and showed that the LANTIRN pod will be an extremely important tool on cruise. 'Felix' [VF-31] also led the way in FAC(A) and CSAR missions with their newly qualified FAC(A) and NVG-capable aircrews."

SWATSLANT staff had worked with VF-31 in the weeks leading up to the Fallon detachment as aircrew switched from the air-to-air emphasis of SFARP to the air-to-ground mission. Among the instructor cadre at SWATSLANT at this time was F-14 RIO Lt (later Capt) Randy "Abdul" Stearns. Fresh from attending TOPGUN at Fallon, he reported to the Oceana-based school as an SFTI. Despite being well versed in the Tomcat's strike capabilities by then, Stearns, like many of the F-14D aircrew he was instructing, was very much a novice when it came to the LANTIRN pod:

Prior to joining SWATSLANT, I had not had the chance to use LANTIRN, as there were only a small number of aircraft modified to operate it and a mere handful of pods available – and virtually all of them were forward deployed in the fleet. When I had finished my last deployment, with VF-143, in July 1996, we did not have any pods on cruise. Things only marginally improved during

Illustrating the F-14D's new-found precision strike capability, "Bullet 104" (BuNo 164351) descends on short final to the runway at Fallon in May 1998 carrying two BDU-59 LGTRs on its BRU-42 ITERs and a LANTIRN pod on the starboard shoulder pylon – both essential elements in LGB delivery training. VF-2 would expend 570 air-to-ground weapons during training in 1998, with the vast majority of them being BDU-59 LGTRs and Mk 76 Mod 5 blue bombs. Having served with VF-2 for almost five years, this aircraft was lost on August 9, 2001 when it crashed into the Bay of Bengal during a night-training flight from *Constellation*. No trace of the crew or wreckage from the jet were ever found. (Michael Grove via Mike Crutch)

By the time this photograph was taken at Oceana in July 2000, the F-14D was fast becoming the US Navy's ultimate precision bomber thanks to the provision of upgraded LANTIRN pods. These D-models, from VF-2 and VF-31, would soon be "armed" with the conical-finned full-scale Mk 83 1,000lb (454kg) practice bombs and BDU-59 LGTRs (on the white bomb trolley) seen in the left foreground. (Gert Kromhout)

my time as an instructor, with squadrons normally having only three to four pods to use throughout work-ups. This meant that LANTIRN was still pretty new to all the folks there, although it proved to be easy to use – aircrew dialed into the training syllabus pretty quickly.

SWATSLANT was home to the Subject Matter Experts [SMEs] on the LANTIRN more than TOPGUN at that time because we had more pod availability at Oceana – instructors and students got to train with it in the fleet on a daily basis. The biggest factor in our favor at SWATSLANT was that we had F-14 instructors who had real world experience with the LANTIRN, which TOPGUN lacked at the time. This meant that our students got the most from their lessons. From 1998, fleet units started sending LANTIRN-equipped aircraft for students to fly when training with SWATSLANT, and we also had a solitary pod that the school could use. TOPGUN had fully integrated the LANTIRN pod into its syllabus by then too.

Typically, when expending ordnance at Fallon, students would work over the B-17, B-19 and B-20 ranges. We also dropped a lot of LGTRs on the Dare County, Pinecastle and BT-11 (Pinet Island) bombing ranges on the East Coast. Again, the low numbers of pods in the fleet meant that each squadron had to really train its own crews when it came to operating LANTIRN. There were a lot of SWATSLANT instructors, including myself, that did not get access to a LANTIRN pod until we flew as an instructor in a fleet jet. Eventually, we had enough pods for every deployed squadron and for those back at home undergoing training.

SWATSLANT instructor patch (1995)

Each ten-aircraft fleet squadron usually had around seven pods due to their paucity in number. LANTIRN was not integrated into any of our simulators since it was a completely unexpected add-on to the aircraft that did not go through the normal NAVAIR procurement chain. The only trainer we had was a desktop computer with the hand-control unit attached to it. We would give aircrew one or two flights with the pod and some drops with a couple of LGTRs, after which they were normally good to go due to its ease of use. LANTIRN-capable squadron aircraft would also carry the pod on most missions, allowing crews to undertake on-the-spot training during overland flights or on cross-country hops.

The syllabus implemented by SWATSLANT was influenced by the lessons learned in combat by VF-41 in 1995 [Operation *Allied Force*] and VF-32 and VF-213 [Operation *Desert Fox*] in 1998 – these were the first occasions when Tomcat crews expended ordnance. In fact, most of what we taught about dropping ordnance came directly from lessons learned during these operations. The LANTIRN kneeboard package that was created for students by SWATSLANT was a direct result of all of these lessons, and it became the standard instructional "tool" for the pod going into OEF/OIF. We also had plenty of drops during OSW. By the time 9/11 happened in September 2001, the Tomcat community was very proficient in using the LANTIRN pod. Indeed, the lessons learned from 1995 through to 2001 culminated in the success enjoyed by F-14 squadrons in OEF and OIF.

Although SWATSLANT – which was consolidated within the Strike Fighter Weapons School, US Atlantic Fleet (SFWSL) on March 22, 2002, as part of the transition from the Tomcat to the Super Hornet – was like a "finishing school" for Naval Aviators and NFOs assigned to the F-14, it was in VF-101 that pilots and RIOs were initially introduced to the jet as a bomber. Among the instructors to serve with the FRS in 2002–03 was Lt Cdr (later Rear Admiral)

Above: Lt Randy "Abdul" Stearns (seen here as a commander) was an early arrival at SWATSLANT in late 1997, having recently attended TOPGUN. A qualified SFTI, he learned how to use the LANTIRN pod "on the job" with SWATSLANT at Oceana. (via Capt Randy Stearns)

Right: These F-14Bs from VF-103 were photographed during a BDU-59 LGTR training mission off Cape Hatteras, North Carolina, in 2001, LANTIRN-equipped "Victory 104" (BuNo 161442) buddy-lasing for "Victory 110" (BuNo 163221) prior to dropping its own weapon. According to the Naval Education and Training publication on aircraft ordnance, "the LGTR provides a low-cost training device permitting aircrews to realistically practice the employment of Paveway II LGBs. The BDU-59 duplicates the release envelope, terminal guidance and closely matches the time-of-flight characteristics of the GBU-10/12/16." (Rick Llinares)

John "Shorn" Saccomando, who came to the unit fresh from having seen combat in OEF with VF-213:

Pilots were separated immediately upon reaching the FRS, learning to fly either the F-14A, B or D. Once they reached the carrier qualification phase, F-14A/B pilots were again split up due to the types' different engines. The RIOs were separated into F-14A/B or F-14D streams upon their arrival at the FRS. Those training to crew the F-14A/B did not know which version of the jet they would be flying in the fleet until they were "patched" – given their unit assignments. The A/B-model jets both used the original AN/AWG-9 radar, so RIOs could go back and forth between the two types without any problem. The F-14D, however, had a completely different system in the AN/APG-71.

The strike course was only about 20 percent of the Tomcat syllabus, although this figure increased to about 35 percent when you included low-level route training. We did tactical formation training at an altitude of 500ft in preparation for the section maneuvering phase of the low-level "pop up" weapons delivery profile. The air-to-air syllabus was another 40–45 percent of the syllabus, while flying at the boat was the last 20 percent of the course. I don't know of any students that washed out due to the bombing syllabus. Ninety percent of pilots that failed did so because they could not hack carrier landings. I would estimate that about ten percent of every FRS class attrited because of problems at the boat. The only RIOs I ever saw leave the FRS departed for medical reasons.

The FRS syllabus was designed to teach the newly winged Naval Aviators to visually bomb using the advanced bombing symbology in the F-14. There were three methods that we taught – Constantly Computed Impact Point (CCIP), Computer Target (CTGT) and Manual. New F-14 pilots would get around six flights to practice pattern deliveries, before moving on to a more advanced part of the syllabus. Each pilot would typically expend six Mk 76 training rounds per mission, and their circular error of probability (CEP) would be calculated by eliminating the closest and furthest bombs from the bullseye, then averaging out the distances of the four remaining bombs.

The advanced strike syllabus included timed low-level routes with a planned Time on Target (ToT), followed by the graduate-level Self-Escort Strike (SES). During the latter, a section of F-14s would fight their way in on a timed route, execute an air-to-air intercept on bandits and then rejoin their route. Adjusting their timing to make their ToT, the section would then deliver their bombs, before fighting their way off-target again. We would hold the students to within +/- ten seconds of their ToT.

VF-103's color jets go in search of their target during a training exercise off the North Carolina coast in 2003. Both aircraft (BuNo 162918, closest to the camera, and BuNo 163217) carry dummy missiles and a single Mk 83 full-scale practice bomb. Once on deployment in the NAG on board *John F. Kennedy* in 2004, "Victory 100" dropped two LGBs and "Victory 103" four such weapons in support of the US Marine Corps offensive in Fallujah. Having been retired to AMARC in January 2005, just weeks after completing VF-103's final cruise with the F-14, both of these aircraft were eventually sold for scrap to HVF West in March 2008. (Rick Llinares)

Assigned to VF-101's commanding officer, "Gun Fighter 101" (F-14B BuNo 161287) rolls out along a taxiway at El Centro on a hot and hazy day in May 2002 at the start of a Strike phase bombing mission to one of the ranges at Yuma. The aircraft's impressive size contrasts markedly with the two Mk 76 Mod 5 blue bombs carried on its centerline BRU-42 ITERs. Students were taught low-level tactical flying and navigation, higher altitude bomb-delivery techniques and SAM and AAA defense, often on instrumented ranges, during their time on detachment at El Centro. (Dan Stijovich)

CCIP was the preferred method for dive deliveries. It provided a dynamic weapons fall line that extended from the HUD velocity vector to the Weapon Impact Point (WIP), which was depicted [on the HUD] with a starburst. When you rolled in, you dragged the line through your target until the WIP was on top of it, then pressed the pickle [bomb release] button. This method was extremely accurate, routinely producing hits within 25ft of the bullseye – more than close enough for a Mk 80, 82 or 83 iron bomb.

CTGT would see the pilot use designated co-ordinates on the ground to provide steering cues for level or dive deliveries. This method could be used for either straight and level or dive deliveries. For straight and level, you needed accurate co-ordinates, which would allow you to create a steering line in the HUD that had cues on it – upper and lower release marks – indicating when to release the ordnance. With this delivery method you would hold the pickle button while the upper and lower release marks converged, the bomb releasing when the solution was reached. CTGT could be pretty sporty to use in a dive, although when employing this method the pilot could "sweeten" the co-ordinates using his throttle cursor. Of course, you only had a second or two to slew the jet before the solution was met.

Manual was the least preferred method, and was just a dialed-in HUD reticle setting. This solution was static and was not wind-corrected. Nobody used this method unless it was as a last resort.

When conducting visual bombing, the RIO would back his pilot up on this roll in numbers and hits, keeping sight of other jets in the immediate area. Unlike the pilots, RIOs only completed one or two visual bombing hops at the FRS. Once they reached the fleet, their training on the LANTIRN was much more involved. Indeed, the RIO had exclusive control of the FLIR pod, while the pilot had release authority of the bombs.

Once pilots went to the fleet, they would further develop their air-to-ground skillset during SFARP, Air Wing Fallon and work-ups. Fleet squadrons had LANTIRN pods, but there were not enough for the FRS. Once pilots and RIOs got to the fleet, they would be trained to drop LGBs using the LANTIRN's FLIR capability. They would undertake SES missions, with LGTRs simulating LGBs, as part of a strike package. LGTRs had a seeker head that would guide on a laser code that the RIO had programed into the FLIR pod. Fallon provided a multitude of targets for RIOs to hone their FLIR skills against. Most pilots and RIOs would have the opportunity to drop five LGTRs and two live bombs during their work-up cycle prior to deployment.

Finally, students did not receive any TARPS training until they reached the fleet. Here, they were taught the various systems fitted within the camera pod, techniques for stand-off while photographing targets and setting up routes to get the best footage of all targets. The pod weighed about 1,850lb, and was a pain in the arse to land back aboard the ship with. If you got a little low or a little slow while landing with TARPS, you could count on pulling a one-wire!

VF-101's Strike phase was only about 20 percent of the Tomcat syllabus for student Naval Aviators and NFOs, although this increased to about 35 percent when low-level route training was included. By the time this photograph was taken from the back of a C-130J of the Rhode Island Air National Guard in June 2004, VF-101's once varied fleet of Tomcats, which numbered as many as 130 aircraft in the early 1990s, had been pared down to just the F-14D following the gradual retirement of the surviving A- and B-models in the fleet. Devoid of any external stores, BuNo 164601 is being flown by the Tomcat Demonstration Team crew, consisting of pilot Lt Jon Tangredi and RIO Lt Joe Ruzicka. Initially assigned to VF-124 in April 1992, this aircraft served with VF-101 Det Miramar, VF-31 and finally VF-101 again. (Erik Hildebrandt)

VF-101 color jet F-14D BuNo 163414 is put through its paces following an afterburner take-off from Oceana in September 2004. This jet was heavily involved in one of the final carrier qualification (CQ) periods for Tomcat students in March 2005 onboard *Theodore Roosevelt* while the carrier was sailing off the coast of Virginia. A veteran of Operation *Desert Fox* with VF-213, the aircraft was stricken by the US Navy shortly after completing the "TR" CQ. (Gert Kromhout)

VF-101 Class 04-04 student patch (2005)

VF-101 "Grim Reapers" deactivation patch (2005)

With the graduates of its final course (04-04) having been sent to VF-31 and VF-213 in early 2005, VF-101 was officially deactivated on September 30 that same year following more than 50 years of service, 29 of them with the F-14.

View From the Ground

Senior Chief Andy "Senior" Nelson, who was a US Navy SEAL for 27 years prior to becoming a JTAC (Joint Terminal Attack Controller, which was the terminology officially adopted towards the end of OEF to replace the more familiar FAC title), provides an illuminating insight into the Tomcat as a strike fighter from the end user's perspective:

Some observations from my time as a SEAL JTAC working CAS, and instructing at NSAWC [Naval Strike and Air Warfare Center] on Naval Special Warfare and on the JTAC Course of Instruction. One early spring, I was working with students from the JTAC course during an Air Wing det. On one CAS event, Capt Dave "Roy" Rogers, then NAS Fallon CO [1998–2001] and former N5

Despite having completed their final Tomcat deployment (with CVW-7) in July 2004, the crews of VF-11 continued to undertake training flights with their F-14Bs following the unit's transfer to CVW-17 due to its "surge" status. Armed with six Mk 76 Mod 5 blue bombs, "Ripper 107" (BuNo 161437) taxies out on a crisp spring morning at Oceana in March 2005. Originally built as an F-14A, this aircraft was the 15th Tomcat to be converted into a B-model. Flown to AMARC by VF-11 in April 2005, it was sold to HVF West for scrapping in March 2008. (Gert Kromhout)

[Director, Plans, Programs and Tactics in 1996–98] at NSAWC, asked to bring some visitors out to the range to observe the CAS event. When he arrived, I briefed the visitors on range safety and the schedule of events.

Shortly thereafter, a division of F-14s arrived on station armed with live Mk 82 and 83 bombs and GBU-16 LGBs. The F-14 squadron was hammering everything in sight. All bombs were on the correct target array, with ToTs under ten seconds. Capt Rogers (a former A-6 pilot) was standing next to me. I commented on how accurate the F-14s were with timing and bombing accuracy.

He replied, "For an organization that was designed solely for fleet air defense and air-to-air, they have wrapped their arms around the CAS mission and mastered it."

Compared to the Navy F/A-18s, the F-14s were the most reliable for showing up for CAS in questionable weather. The only exception to this were the US Marine Corps F/A-18s – Marine Air lives for CAS.

While briefing the CAS events during air wing training at Fallon, the standard or acceptable error for the ToTs was plus or minus ten seconds. I would challenge the aircrews to plus or minus five seconds on the first pass of each section. The bet was a six-pack of beer or the first round at the O club. I spent a lot of money paying off bets to the F-14 crews.

Navy F-14 FAC(A) aircrews were a welcome boon to special operations. A LANTIRN-equipped Tomcat with a FAC(A)-trained aircrew was a major force multiplier for small special operations forces. They could do it all, from controlling fixed- and rotary-wing air support to calling in sea- and shore-based artillery. The only other FAC(A) aircrews that could do that were those that flew US Marine Corps F/A-18Ds. However, the shortfall for them was having to use the Nite Hawk targeting pod. I would tell the Army Special Forces and SEAL operators to always request a Navy FAC(A) for support on combat operations.

One thing I always observed with the F-14 crews was that they were all big guys, usually standing more than 5ft 11in tall. Most looked like they played football for the Naval Academy. F/A-18 crews were generally 5ft 9in and under. The F-14s and the crews that flew them were the Dick Butkus [one of American football's most famous linebackers] of the strike fighters – big guys and big powerful planes ready to crush anything that came into their zone. They just looked scary when they were coming at you down low, and most F-14 flight crews looked like they could handle a good bar brawl. The F/A-18 was more of a stiletto of an aircraft, and its pilots acted like they should be wearing dockers and polo shirts, with cardigan sweaters tied around their necks. The latter description does not apply to US Marine Corps F/A-18 pilots, who are an entirely different breed.

VF-101 F-14D BuNo 163900 turns on to short final overhead Oceana in September 2004, the aircraft carrying a single Mk 76 Mod 5 blue bomb on its BRU-42 ITER. The aircraft is being flown by a student pilot, with an instructor in the rear seat – the latter is wearing the colors of VF-2 on his helmet. Initially assigned to VF-124 in March 1991, this aircraft later saw fleet service with VF-11 and VF-31 prior to joining VF-101 in 2004. It was retired early the following year. (Gert Kromhout)

VF-101 initially struggled to get its hands on LANTIRN pods due to their paucity in number. However, by late 2004, following the retirement of so many Tomcats, there were enough pods available to allow all the remaining "Grim Reapers'" jets to be permanently fitted with LANTIRN. Pod-equipped BuNo 163895 participated in one of the final CQ periods undertaken by VF-101, the unit working *Theodore Roosevelt*'s flightdeck as the carrier sailed off the Virginia coast in March 2005. (Gert Kromhout)

CHAPTER 7

OPERATION *SOUTHERN WATCH*

Aside from brief campaigns in the Balkans and Afghanistan, combat operations for F-14 pilots and RIOs took place almost exclusively over Iraq post-*Desert Storm*. In the wake of this conflict, a no-fly zone was created over southern Iraq, and for 12 long years Tomcat crews drilled holes in the sky policing these areas.

The first of these zones was established in the aftermath of *Desert Storm* in an effort to offer protection to the Kurdish population in northern Iraq from President Saddam Hussein's armed forces. Initially covering all Iraqi airspace north of the 36th parallel as part of Operation *Provide Comfort* in late 1991, the legality of this mission was mandated by UN Security Council Resolution 688. When the Shi'ite Muslims also began to suffer persecution in the south, a no-fly zone was created with UN backing as Operation *Southern Watch* (OSW) on August 26, 1992. Joint Task Force-Southwest Asia (JTF-SWA), consisting of units from the United States, Britain, France and Saudi Arabia, was established on the same date to oversee the day-to-day running of OSW.

As noted in chapter one, the F-14 was actively involved in the operation from the start, with the *Independence* battle group generating Naval Aviation's first OSW patrols over southern Iraq 24 hours after the no-fly zone was announced by President George H W Bush. CVW-5's VF-21 sent a division of aircraft into the area late in the afternoon of August 27, with one of the jets carrying both air-to-air missiles and two sand-filled Mk 83 blue bombs in what was the US Navy's first operational "Bombcat" sortie. More than six years would pass before Tomcats dropped live ordnance on targets in Iraq, however.

Like the operation in the north, which was officially titled Operation *Northern Watch* (ONW) on January 1, 1997, OSW saw US, British and French aircraft enforcing the Security Council mandate that prevented the Iraqis from flying military aircraft or helicopters below the 32nd parallel – this was increased to the 33rd parallel in September 1996.

The US Navy's principal contribution to OSW was the mighty carrier battle group, controlled by Fifth Fleet (formed in July 1995) as part of the unified CENTCOM, which oversaw operations in the region. Typically, an aircraft carrier would be on station in the NAG at all times, vessels spending around three to four months of a standard six-month deployment committed to OSW. Ships from both the Atlantic and Pacific fleets took it in turns to "stand the watch," sharing the policing duties in the no-fly zone with USAF and Royal Air Force (RAF) assets ashore at bases in Saudi Arabia, Kuwait, Bahrain and other allied countries in the region.

OSW's original brief was to deter the repression of the Shi'ite population in southern Iraq through the imposition of a no-fly zone, but it soon became obvious to the coalition that the Iraqi Army was more than capable of dealing with the disruptive elements in both the north and the south without having to involve the IrAF. Frustrated by its inability to defend the people it had encouraged to rise up and overthrow Saddam's regime in 1991, the US-led coalition subtly changed the emphasis of its ONW and OSW missions. This saw the systematic monitoring of Iraqi military activity in the area evolve from being a useful secondary mission tasking (after countering IrAF incursions into either zone) to the primary role of the crews conducting these sorties from the mid-1990s. By December 1998, the justification put forward by the US government for the continuation of both ONW and OSW was the protection of Iraq's neighbors from any potential aggression, and to ensure the admission, and safety, of UN weapons inspectors.

The Tomcat proved to be a key asset in OSW, although not because of its ability as a long-range fighter or, initially at least, its evolving strike fighter role. As had been the case in *Desert Storm*, the F-14's tried-and-tested TARPS capability provided JTF-SWA with the flexibility to monitor Iraqi military activity on a daily basis in good weather. Although the TARPS mission was seen as a necessary evil by most dyed-in-the-wool fighter crews, it nevertheless enabled the Tomcat community to make a concrete contribution to the daily enforcing of OSW. The TARPS missions also tended to be more eventful than the typically mundane and boring CAPs that were the "bread and butter" sorties of the F-14 units in the NAG in the years prior to the arrival of LANTIRN-equipped aircraft. Seasoned Tomcat crews

Opposite: Green-shirted maintainers from VF-24 and VF-211 work away on their F-14As in "fighter country" on *Nimitz*'s fantail in March 1993, oblivious to the close-maneuvering USS *La Salle* (AGF-3) – command ship for JTF Middle East. Prior to deploying with CVW-9, both VF-24 and VF-211 had become the first Miramar-based fighter squadrons to complete the AARP, whose content was virtually identical to the course taught to F/A-18 pilots. Just seven months after VF-21 had patrolled over southern Iraq in its F-14As "armed" with sand-filled blue bombs in late August 1992, VF-24 and VF-211 flew the first Tomcats committed to OSW, loaded with live Mk 80 series ordnance. (US Navy)

VF-211 (1993)

would be quick to point out that the absence of the IrAF in the no-fly zone for more than a decade was proof positive that the many thousands of CAP missions flown in that time achieved the desired result from JTF-SWA's point of view.

OSW Mission Anatomy

The actualities of a typical no-fly zone mission did not alter hugely throughout the duration of OSW, with most following a set pattern, as follows.

Thanks to the established routine of the operation, and the advent of secure e-mail communication between JTF-SWA's CAOC and the air wing aboard the carrier in the NAG, shipboard mission planners would usually get a rough outline of the Air Tasking Order (ATO) "frag" (tasking) about 72 hours before it was due to be flown. As each day passed, more information would be relayed to the ship to the point where, 24 hours before the package was due to launch, its participants had a detailed plan of where they were going and what they were doing, as well as the role being played by other supporting assets sortied from shore bases.

On the day of the mission, assigned crews (five Tomcats would be committed to the evolution, with four entering Iraqi airspace and the fifth launching as the airborne spare) started their OSW briefing about two-and-a-half hours prior to take-off. This was an air wing-wide meeting that was usually attended by all aircrew participating in the mission. The gathering lasted for 30–45 minutes, after which Tomcat crews would return to their own squadron ready room and conduct the division brief applicable to their part of the mission – this ran for about 15 minutes. Crews then broke up into sections to conduct individual briefs, where they would discuss things like in-flight emergencies and what to do during the sortie from a single aircraft standpoint. This process would effectively see the participating units go from "back row," to "mid-level" to "micro view."

One of the air wing's biggest advantages when compared with shore-based OSW assets was that all mission elements briefed together, face-to-face. Air wings would do this on a near-daily basis when on cruise, talking at length about various mission profiles and operational developments. This also allowed the US Navy to run bigger packages (typically consisting of F-14s, F/A-18s and EA-6Bs, with S-3B tankers and ES-3A EW and E-2C AEW aircraft remaining over Kuwait or the NAG in support of the primary strike aircraft) into Iraq. USAF groups, on the other hand, all briefed separately and then met up to support each other inbound to the Box, as the southern no-fly zone was dubbed by coalition aircrew.

CVW-7 and *George Washington* briefly undertook OSW missions from mid-October 1994 in support of Operation *Vigilant Warrior*, which was mounted following the movement of two Iraqi Republican Guard divisions towards the border with Kuwait. All of the aircraft seen in this photograph participated in missions during the carrier's spell in the NAG, although only the "pointy-nosed jets" (F-14Bs, F/A-18Cs and the A-6E and EA-6B) would have gone "feet dry" into southern Iraq. This deployment would be VF-142's last, with the unit being disestablished in April 1995. (US Navy)

Tomcat crews would go "feet on the deck" to their jets 45 minutes prior to launch, by which time the aircraft was fully fueled, all systems (bar the engines) were up and running thanks to the jet's auxiliary power unit and the pylon-mounted weapons, or TARPS, had been secured.

The jet was then pre-flighted at deck level for around 10–15 minutes, after which the RIO would climb aboard and commence his radar systems and avionics checks. With 30 minutes to run to launch, the carrier's Air Boss (who runs the flightdeck) would call "starts away," and the aircraft fired up. With everything functioning correctly, the five Tomcats would be unchained and marshalled in a pre-ordained order to one of the ship's four catapults for launching. Having successfully departed the carrier, the pilot would find the duty tanker using the aircraft's radar in air-to-air search mode, rather than "breaking comms" by talking on the primary control frequency known as "Strike" in order to get a steer from an AWACS controller. Having located the tanker, he would join the line of aircraft forming off its left wing, waiting his turn to cycle through and "top off" the Tomcat's capacious tanks. With the refueling complete, the pilot positioned the jet back in the formation, but this time off the tanker's right wing.

The air spare would return to the carrier at this point if the four primary F-14s had refueled successfully and all the mission-crucial systems – weapons, radio, radar homing and warning receivers and avionics – were functioning correctly.

Most groups were then divided into two mini packages once on station so as to cover any Periods of Vulnerability (known simply as vul time to aircrew). The OSW tasking was easier to perform when dealing with smaller divisions of aircraft, rather than sending in a huge formation of jets that tended to get in each other's way. The first group, including a section of Tomcats operating exclusively as fighters, fighter-bombers or in the TARPS role, would commence its vul time, leaving the second package on station for a short while longer prior to it also being committed. There was a brief period of overlap between the two packages as a result of this tactic being employed.

A coalition from France, Britain and the United States initially enforced the UN Security Council mandate that prevented IrAF aircraft flying over southern Iraq from August 1992. The French contribution to OSW usually took the form of Mirage 2000C fighters, with this example hailing from *Groupe de Chasse* 1/2 "Cigognes." Its pilot is waiting his turn to refuel from a KC-10A of Air Mobility Command's 22nd Air Refueling Wing in March 1995, the tanker's single hose being occupied by one of two F-14Ds from VF-2 receiving post-mission fuel over Kuwait after completing their time in the Box during an OSW mission from *Constellation*. All three fighters are carrying live air-to-air missiles, with the Mirage 2000C armed with Matra Super 530Fs and R550 Magic-IIs. (Ian Black)

Each package had a designated vul time in the Box according to the ATO, and each of these time slots had been meticulously worked out by JTF-SWA's CAOC. This organization, based in Saudi Arabia, was responsible for all no-fly zone mission planning, and it created a daily ATO for coalition participants (both naval and shore-based aviation assets).

It was standard operational procedure that if the package did not make it into southern Iraq during its set time slots, the jets would not be allowed "over the beach". Once in the Box, the aircraft pressed on along pre-planned routes until reaching their designated patrol station in south-eastern Iraq. While performing an OSW mission, crews remained in constant contact with one of four air traffic control agencies. One of these would be in a combat information center aboard a US Navy AEGIS-equipped *Ticonderoga*-class cruiser in the NAG, another in a purpose-built radar control facility in Kuwait near the Iraqi border, a third in the "big wing" tanker that was on station supporting the package and the fourth in an orbiting E-2C Hawkeye or E-3 Sentry AWACS. These controllers would keep all TACAIR elements in Iraq updated on what was happening in response to the mission.

Thanks to the Tomcat's immense range, it was not uncommon for F-14 crews to double cycle during OSW missions. Both jets would stay inside the Box for the entirety of the sortie, while the Hornet section it had ingressed with remained on station for as long as the jets' fuel permitted before being replaced by two more F/A-18s. The Tomcats would finally depart with the second Hornet section when the pilots of the latter aircraft declared that they needed fuel. This effectively meant that the F-14s had stayed on station for twice as long as the F/A-18s, and all on a single tank of fuel.

With the mission completed, the Tomcats would go "feet wet" back over the NAG and head along pre-planned routes south to a tanker. The latter would be either a USAF KC-10 or KC-135, an RAF VC10 or TriStar or two "organic" S-3B Vikings that had sortied with the strike package from the carrier and then stayed on station over the NAG, awaiting its return. Cycling through the refueling procedure once again, topping off their tanks to about 500lb (226kg) above what was needed to land back aboard the ship, the aircraft would overfly the carrier, proceed back into marshal and then wait their turn to

VF-2 conducted its debut F-14D cruise onboard *Constellation* between November 1994 and May 1995, with this being the first of only two such deployments that would see a single squadron embark 14 D-model jets. Part of CVW-2, the unit performed OSW missions for 73 days while in the NAG, flying DCA, TARPS and force defense sorties, as its F-14Ds had no air-to-ground capability at the time. "Bullet 111" (BuNo 159613), photographed on one of the unit's first OSW sorties on January 15, 1995, is DCA mission-configured with live AIM-9M, AIM-7M and AIM-54C missiles. This aircraft was still serving with VF-2 eight years later when it dropped LGBs, JDAM and iron bombs during OIF I. (US Navy)

"Bullet 107" (BuNo 163903) flies in close formation off the starboard wing of a Mirage 2000C (flown by RAF exchange pilot Flt Lt Ian Black) from *Groupe de Chasse* 1/2 "Cigognes" during a patrol in March 1995. Subsequently passed on to VF-213 in November 1997, this aircraft was one of two Tomcats from the unit involved in the unsuccessful IrAF MiG-25 engagement of January 5, 1999 that saw two AIM-54Cs "fired" at the fleeing Iraqi fighters in the no-fly zone. Following spells with VF-101 and VF-31, BuNo 163903 was reassigned to VF-213 once again in July 2002 and it duly dropped 19 bombs in OIF I while serving with the unit. (Ian Black)

recover. A typical OSW mission usually lasted around four hours, depending on whether a target was bombed during the course of the patrol.

OSW Highlights

Following a series of widespread coalition air strikes on targets in southern Iraq in January 1993 (with naval assets provided by CVW-15 embarked in *Kitty Hawk*), subsequent OSW operations in the Box passed off primarily without incident for the next three-and-a-half years. Throughout this period, F-14 units assigned the TARPS mission diligently photographed vast tracts of southern Iraq in order to monitor troop movements, locate AAA and SAM batteries (which were frequently mobile) and identify search radar sites.

The closest the Tomcat got to engaging Iraqi forces during this period came on September 3, 1996, when F-14Ds from VF-11 and VF-31, assigned to CVW-14 embarked in *Carl Vinson*, escorted two B-52Hs from Andersen AFB, Guam, tasked with hitting Baghdad power stations and communications facilities with 13 AGM-86C air-launched cruise missiles during Operation *Desert Strike*. No fewer than 43 RGM/UGM-109 Tomahawk cruise missiles were also fired from vessels within the *Carl Vinson* battle group, these weapons targeting air defense sites in southern Iraq. Although CVW-14 had initially been put on alert to carry out the attacks, the administration of President Bill Clinton decided to use cruise missiles instead.

Desert Strike has been authorized following the launching of the Iraqi Army's August 31 offensive (the largest of its type since 1991) against Kurdish forces in Erbil, in the north of the country. Fearing a similar attack on non-Sunni Muslims in the south of Iraq, President Clinton decided to act. The targets hit had been specifically chosen to attempt to further degrade the air defense infrastructure, as SAM and AAA sites, supported by a network of command, control and communication locations, had been increasingly challenging OSW missions. Following the cruise-missile strikes, TARPS-equipped F-14Ds from VF-31 flew bomb hit assessment (BHA) missions over southern Iraq. In the wake of *Desert Strike*, the northern boundary of the no-fly zone was also extended to the 33rd parallel, giving patrolling aircraft more territory to cover.

The *Carl Vinson* battle group was relieved by the *Theodore Roosevelt* battle group in late November 1996, with CVW-3, embarked in CVN-71, seamlessly taking over the OSW tasking from CVW-14. VF-32, whose five TARPS-capable F-14As were equipped with the first (four) TARPS-DI pods in fleet service, immediately generated imagery of key sites in southern Iraq. The revised pod had had its KS-87 "wet" film camera replaced with a digital equivalent that could transmit imagery via a UHF radio link. One of the pilots who routinely flew TARPS-DI missions with VF-32 in 1996–97 was Lt Joe Barnes, who recalled at the time:

> The TARPS-DI tasking involves flying a detailed route to get real-time target area information that we can then transmit back to the Battle Group Commander [Rear Admiral Gregory G Johnson] onboard *Theodore Roosevelt*. This data is then used in future strikes. Iraq has become fairly adept at moving its high-value missile systems, and it is therefore vitally important for the coalition to continue to locate these threats via such

Leaving his RIO and a squadron-mate in conversation on the flightdeck of *Carl Vinson* in September 1996, former A-6 pilot Lt Rick LaBranche climbs up the side of VF-11's "Ripper 107" (BuNo 159603) and prepares to stow his flight bag in the front cockpit. Both this aircraft and the F-14D from VF-31 chained down alongside it are configured for the DCA mission in the Box, being armed with AIM-9Ms, AIM-7Ms and AIM-54Cs. VF-11 and VF-31 escorted two B-52Hs from Andersen AFB, targeting Baghdad power stations and communications facilities during Operation *Desert Strike* on September 3, 1996. (US Navy)

OPERATION SOUTHERN WATCH

Aircraft from CVW-1 commence a launch cycle from *George Washington* to signal the start of an OSW mission in late November 1997. Earlier that month, the carrier and its battle group had been ordered from the eastern Mediterranean to the NAG by President Bill Clinton to support UN efforts aimed at compelling Saddam's regime to allow independent inspection of known WMD sites. Typically, two or three S-3B tankers (this jet is from VS-32) would launch ahead of the strike aircraft heading into the Box so as to provide them with frontside fuel. Tomcats would depart next, and here two F-14Bs from VF-102 are being attached to the launch shuttles of both bow catapults. Hornets duly followed, with all three examples in this photograph being F/A-18Cs from VMFA-251. Finally, an EA-6B (this aircraft is from VAQ-137) and an E-2C (from VAW-123) would complete the launch cycle. (US Navy)

reconnaissance missions. On these flights, it is not uncommon to see surface-to-air gunfire from several areas in southern Iraq, although the fire normally take the form of an unguided barrage, and is therefore of no real threat to any coalition aircraft. It seems simply to be a way for the Iraqi Army to appease their Brass!

Having escorted both strike and reconnaissance aircraft in the NAG since 1991, Tomcat crews were finally given the chance to drop bombs on targets in Iraq during the early hours of December 16, 1998, with the launching of Operation *Desert Fox*. A four-day aerial offensive ostensibly aimed at curbing Iraq's ability to produce Weapons of Mass Destruction (WMD), this campaign was also triggered by Saddam's ongoing unwillingness to co-operate with UN inspections of known weapons sites. Many observers believed that the primary aim of *Desert Fox* was to attack the Iraqi leadership in a series of decapitation strikes. To this end, a presidential palace just south of Baghdad was hit, as were buildings that housed the Special Security Organization and the Special Republican Guard.

In the vanguard of these precision strikes on the first three nights of the operation were the F-14Bs of VF-32, flying from *Enterprise*. Part of a 33-aircraft force launched by CVW-3, the Tomcats headed into Iraq in the wake of concentrated Tomahawk missile attacks. The opening attacks of *Desert Fox* were exclusively flown by US Navy carrier-based aircraft, as an unnamed Tomcat strike lead from VF-32 explained to a reporter from *Stars and Stripes* magazine:

> The first night was all Navy, no Air Force – not even their tankers – or Brits. It was designed for a single cycle so as to achieve the element of surprise. Our Tomcats were loaded with two 1,000lb [454kg] GBU-16 LGBs, and our target was within Baghdad city limits. Tomcats were assigned most of the hard targets because of the aircraft's LTS capability – collateral damage was unacceptable. We found our targets and "schwacked" them. To watch those buildings go away through the LTS cockpit display was impressive. We were opposed by ballistic-launched SAMs and plenty of AAA.

Opposite: Until the advent of the LANTIRN pod and PGMs, the F-14's most important mission in OSW was as a tactical photo-reconnaissance platform when configured with TARPS. Weighing a hefty 1,760lb (798kg), the pod remained in use with the fleet until late 2004, and in its final years of service there were three main types fielded. They were the legacy "wet" film pod, TARPS-DI, which used digital cameras that allowed the shots to be viewed in the cockpit and sent back to the carrier or other Link-16-capable aircraft over encrypted UHF, and TARPS-CD (completely digital), fitted with digital cameras that auto-sent the imagery when within range of a receiving station. This VF-102 F-14B (BuNo 163221), photographed on an OSW mission in early 1998, is also carrying an AN/ALQ-167(V) "Bullwinkle" ECM jamming pod forward of the TARPS store, as well as AIM-7M and AIM-9M missiles. (US Navy)

Right: When VF-102 made its OSW deployment onboard *George Washington* in 1997–98, all of its jets wore full-color markings despite US Navy regulations limiting such decoration to two aircraft per unit. This was the last time a Tomcat squadron went on cruise with its full complement of aircraft featuring colorful insignia. (US Navy)

As with all Tomcat squadrons since *Desert Storm*, VF-102 were restricted to primarily flying DCA and TARPS missions in 1998–99, despite its Tomcats having LANTIRN. In order to retain proficiency in the air-to-ground mission while on cruise, VF-102 regularly used the Camp Udairi bombing range in Kuwait – indeed, BuNos 162920 (foreground) and 161435 are seen inbound to the range on February 11, 1998, each jet being armed with a solitary BDU-59 LGTR. (US Navy)

The outstanding performance of VF-32's LANTIRN-equipped F-14Bs on the first night of *Desert Fox* was largely ignored, or misreported, in the mainstream press at the time, although defense analyst Michael Dennis was fulsome in his praise of the aircraft and its crews in his "Potomac Currents" column in the spring 1999 issue of *The Hook*:

> The night *Desert Fox* started, the reporters in Baghdad and the so-called experts in Washington studios all-knowingly explained that the explosions heard in the distance were those of Tomahawk cruise missiles, since the only manned aircraft allowed over the city limits were the USAF's F-117 and the B-2. Unfortunately, none of the USAF aircraft were available that night.
>
> Also that night, an NBC correspondent was unable to understand why some of the Iraqi AAA was shooting higher than low-level, since the Tomahawk missiles flew low. More than one network reported seeing twin streaks of flame in the sky, accelerating and gaining altitude. This was sagely explained as a Tomahawk having been hit and sent out of control. When national imagery was released of both Baghdad and other targets, showing extremely accurate bomb hits that had caused great damage, one again received sagacious "if you can understand what I know" explanations of the improved accuracy of the Tomahawk. This interpretation was uncontested then and since either by the Department of Defense Public Affairs or by the Chief of Naval Information.

Above: Very occasionally Tomcats would be forced to land ashore due to technical problems or poor weather in the NAG while operating from carriers in-theater. VF-11's fully armed "Ripper 211" (F-14B BuNo 162919) spent a short time on the ramp at Ahmed al-Jaber Air Base in Kuwait on March 19, 1998, parked on the transient ramp near the wrecked hardened aircraft shelters that had been bombed in *Desert Storm* some seven years earlier. VF-11 was part of CVW-7 at the time, embarked in *John C. Stennis* for the vessel's around the world cruise. (USAF)

Right: VF-154's "Knight 102" (F-14A BuNo 161620) approaches a USAF tanker with its refueling probe extended during an escort mission for a TARPS Tomcat from the same unit in the Box on April 11, 1998. Assigned to CVW-5, the squadron was participating in *Independence*'s final deployment prior to the vessel's decommissioning in September of that same year. VF-154 was the only Tomcat unit at that time to have all 11 of its jets fully capable of carrying a LANTIRN pod or TARPS-DI. Ironically, this particular aircraft appears to be carrying neither store. (USAF)

As some of you will know, however, VF-32 did go "downtown." The explosions heard in the distance by reporters were, in fact, 1,000lb GBU-16s delivered by Navy aircraft. The Iraqi gunners were not as dumb as the television experts – the "twin streaks of flame" were probably Tomcat afterburners! Nevertheless, the Washington pundits searched for Tomcats everywhere and concluded they would be unable to destroy Saddam's WMD. In great part this was due to US reliance on carrier-based aviation, which could not possibly (they said) bring to bear the intensity of a land-based strike force. The strikes went on with little individual publicity until two B-1Bs dropped 500lb (226kg) dumb bombs on a Republican Guard target. These, of course, did damage sufficient for the operation to be terminated.

By Christmas, it gradually emerged that the timing of *Desert Fox* had nothing to do with WMD, but everything to do with an attempted coup by the 3rd Republican Guard Corps based around

Tomcat units committed to OSW would also occasionally carry out Operation *Sea Dragon* maritime surveillance sorties during the course of their patrols in the NAG, the jet's two-man crew, its long range and the LANTIRN pod making the F-14 well suited to such missions. This well-weathered F-14D (BuNo 163904) from VF-31, embarked in *Abraham Lincoln* with CVW-14 in August 1998, is armed with an AIM-54C and a rarely seen Mk 7 CBU that contains Mk 20 Rockeye bombs – an unusual combination for an F-14 that proved popular during OSW. BuNo 163904 subsequently participated in OIF I with VF-31 in 2003. (Lt Jim Muse)

Enterprise, with CVW-3 embarked, sails at speed in the NAG in December 1998 on the eve of Operation *Desert Fox*. The vessel had relieved *Dwight D. Eisenhower* (with CVW-17 embarked) in-theater on November 23. CVW-3 had had three weeks to get into the swing of OSW operations before it was called upon to carry out strikes originally planned by CVW-17 for Operation *Desert Thunder*, which came within 20 minutes of execution on November 11. (US Navy)

Basra. The strikes were not targeting WMD, but instead were designed to keep the rest of the Republican Guard occupied to give the coup a chance. In the event, the coup failed and the operation ended.

Following the first night over Baghdad, the F-14s joined in mixed formations with F/A-18Cs and British Tornados. The F-14 payload appears to have been either two 1,000lb (454kg) GBU-16s or 2,000lb (907kg) GBU-10s or, against hardened targets, two 2,400lb (1,088kg) GBU-24 penetrators. The targets were mainly Republican Guard installations, often to the north of Baghdad. The F-14s additionally flew fighter sweeps and reconnaissance. Each crew flew five or six missions in the first four nights of the operation.

Although senior Naval Aviators and NFOs undertook the early strikes during *Desert Fox*, the high mission tempo maintained by VF-32 meant that all the pilots and RIOs in the unit eventually saw combat during the four-day operation. Among the junior pilots sent into action was Lt Bryan McRoberts:

I joined VF-32 in October 1998, one month prior to deployment. Like several other FRS classmates, we arrived in time to make 18 weeks of the usual six months of work-ups. What I lacked in experience I made up for in enthusiasm. A month into our deployment, we assumed that we would have a few more days of OSW flying before *Carl Vinson* arrived and we headed back through "the Ditch" for some time in the Med.

Expecting a full day of diplomatic meetings between the UN and the Iraqis on December 15, we were sat in the ready room waiting for things to kick off when the CO [Cdr Gary D Galloway] stood up and said that we were going in. It was going to be primarily a CVW-3 effort, and the first strike was launching at 0200hrs [on December 16]. Listening eagerly as the CO called off the names of the crews involved, I knew mine would not be on the list. Everyone watched from the ready room as the first strike launched. We waited anxiously for their return, wondering if they had hit their targets. We were also unsure about what kind of resistance the Iraqis would offer. One by one, each aircraft returned. Everyone split their time between planning, watching CNN and getting the inside story from the returning crews. I got very little sleep that night.

For the next two days, I got up and checked the flight schedule, hoping to see my name in print. I felt as if I was sitting on the bench, silently screaming "Put me in, Coach!" By the third day, just when it looked like the new guys were not going, the CO made the huge decision to pick me. Here I was still trying to get my sea legs, and the skipper was going to let me take a $40 million aircraft, loaded with live ordnance, over Iraq and into combat.

I launched at 0300hrs. On only my second flight wearing NVGs, I headed to the rendezvous stack for the tanker, looking for the other 14 aircraft in my strike. It was by far the largest evolution I had ever participated in. Luckily, when it came time to refuel, I got in the basket on my first attempt and didn't hold anybody up. We pressed on time. Looking across the horizon, I could see the flashes of other strikes hitting their targets, and the AAA tracers trying to chase them down. It took a few minutes to register, but I finally realized that those tracers were coming from my target.

We were still 11–12 minutes out when things seemed to calm down over the target, but I knew what was coming.

My RIO acquired the target early. Now all that was left to do was to pickle the two 1,000lb [454kg] bombs, avoid being shot down by AAA or a SAM, go back to the tanker for some back-side gas and then head to the ship for a night trap. No problem! Concentrating on staying in position while watching for the telltale glow and plume of a SAM launch, we made our way into the target area and dropped our ordnance. Unaware at this point if we had hit the target (we later found out that we did), I was focused on the AAA that was now flying around, undoubtedly meant for me. I couldn't help thinking that those guys had some nerve to be actually shooting at us!

Egressing from the target, seeing the tracers still flying up into the now-empty sky, we proceeded to the tanker, got a squirt of gas and headed to the ship as the sun was coming up. Holding in the marshal stack, still more awake than I had ever been, I tried to forget what I had just done and concentrated on what I still had to do – trap. Thank God for that "Combat OK." It wasn't the prettiest pass, but I trapped first time.

Desert Fox ended the next day. I realize that the opportunity to actually use the skills we are trained for come rarely, and I am grateful that I was given the chance to prove that I was capable of employing my skills.

On the eve of *Desert Fox*, "Mag Rats" from *Enterprise*'s G-3 Division attach a seeker head to a bomb body, forming a 2,400lb (1,088kg) GBU-24 "bunker buster" in the process. F-14Bs would drop 26 of these penetrator LGBs, with a further four being expended by F/A-18Cs, during the course of the 70-hour campaign. (US Navy)

Red shirted "ordies" from VF-32 prepare to load 2,000lb (907kg) GBU-10s (foreground) and 1,000lb (454kg) GBU-16s (partially gray bombs to the rear) on to F-14Bs chained down to the fantail of *Enterprise*. A number of the weapons have had messages scrawled on them in chalk by the armorers of the G-3 Division that assembled the PGMs below deck. (US Navy)

OPERATION SOUTHERN WATCH

Aside from the Tomcat making its combat debut with the LANTIRN pod in *Desert Fox*, female Naval Aviators and NFOs also saw action for the first time during the operation. Of the seven that expended ordnance on targets in Iraq, three were assigned to VF-32. Lt(jg) Andrea Quy, who was the RIO for Cdr Will Cooney during a strike on December 18, recalled her experiences of *Desert Fox*:

Late in the afternoon of December 15, we were notified of the execute order, which included both manned aircraft and Tomahawk strikes on Iraq. Although no one really believed we would actually go, we all went into strike mode. I remember feeling envious as the crews came back from their strikes and more launched throughout the night. Everyone was so excited about bringing back the BHA tapes of their targets blowing up. All of us JOs wanted the chance to prove we could do it too, and we got the opportunity on the third night.

I wasn't concerned at the time about them shooting at us – I was more worried about screwing up my one opportunity to drop live ordnance in combat. Manning up with my pilot in our F-14B was no different from any normal launch. As we entered Iraq, we could see explosions and AAA more than 100 miles away in Baghdad, for we had launched another round of Tomahawks that night.

NVGs are a wonderful thing. I was head-down from the initial point [IP] to the target, as the RIO usually is, looking for the target. After finding it, and waiting for the bomb release cues, I looked outside for possible SAM launches. They were shooting at us! I guess they must have been pretty mad, because red and white tracers from AAA were everywhere. After that, it wasn't difficult to want to drop bombs and get out of town quick. I really didn't think of the danger involved until we had dropped the bombs, at which point I realized that we were several hundred miles from good guy country. Most importantly, we ended up with an OK 3 wire on the pinky morning recovery at 0630hrs.

CVW-3 (1998)

CVW-3 (1998)

An armorer from VF-213 lines up a GBU-24A/B penetrator LGB on its trolley beneath an F-14D's port BRU-32 bomb rack prior to attaching the weapon to the jet on the flightdeck of *Carl Vinson* on December 19, 1998. This "bunker buster" was one of eight such bombs dropped by the unit on the last night of *Desert Fox* in an attack that gave the F-14D its combat debut. (US Navy)

On the last night of *Desert Fox* (19–20 December), CVW-3 jets were joined by aircraft from CVW-11, embarked in *Carl Vinson* – the latter had only entered the NAG eight hours earlier. At the heart of the CVW-11 strike force were four LANTIRN-equipped F-14Ds of VF-213 each armed with two GBU-24A/B 2,400lb (1,088kg) penetrator LGBs. Giving the D-model jet its combat debut, the unit targeted IrAF hangars at two airfields near Baghdad, recording a 100 percent success rate. VF-213 also lased targets for the trio of Hornet units (VFA-22, VFA-94 and VFA-97) assigned to the air wing. Aside from the airfields, CVW-11 aircraft (including Tomcats) also attacked nearly 50 targets at six Iraqi military sites in the south of the country in 14 strikes, expending 20 precision-guided and 60 laser-guided munitions.

Tomcat RIO and former A-6 B/N Capt Tom Hagen, CAG of CVW-3 during *Desert Fox*, is debriefed by air-wing intelligence personnel onboard *Enterprise* after participating in the first strike of the four-day campaign in an F-14B from VF-32. A live CNN feed providing details of his night's work is being shown on the television screen behind him. (US Navy)

By the time *Desert Fox* had come to an end, CVW-3 and CVW-11 had undertaken 25 strikes, totaling more than 400 sorties. VF-32 had completed 38 sorties, dropping 111,054lb (50,373kg) of ordnance consisting of 16 GBU-10s, 16 GBU-16s and no fewer than 26 GBU-24 penetrator LGBs. The latter proved to be the PGM of choice against hardened aircraft shelters, HQ bunkers and command and control buildings. Not all of the Tomcats sortied were carrying bombs, however, as both VF-32 and VF-213 also conducted a series of escort CAPs for USAF B-1Bs committed to the operation from day two of the campaign.

F-14 RIO and former A-6 B/N Capt Tom Hagen was the CAG of CVW-3 during *Desert Fox*:

> The four-day offensive was the culmination of my career because it showed what an aircraft carrier and carrier battle group can do independent of host nation support or a long trail of support from the States. We are an enabler unlike anything else the military owns, and we were clearly the right force for the job because of the desire for surprise.
>
> When you are in a leadership position, whether as the CO or XO of a squadron or as an air wing commander, you should have a conscience about whether or not you have prepared and trained your people well. We have had a lot of assistance getting prepared from type wings, NSAWC at Fallon and weapons schools back at the home bases, but when you put all the elements together on a deployment and are challenged as we were by the magnitude and intensity of *Desert Fox*, you are always going to be curious, maybe anxious, about whether or not you can deliver the goods.
>
> I wanted to make sure that we sustained the same energy level we took into the first night, that we didn't fall into any complacency traps based on any one particular reaction from the Iraqis, and that we gave our aircrews and maintenance people as good a shot at success as we possibly could. I think we did those things very well. The results speak for themselves.

Strapped into an F-14B from VF-32, Hagen had flown in the first strike of *Desert Fox*, followed by another mission on the third night of the campaign. A veteran of *Desert Storm* with VA-75, he found flying into Iraq in the back seat of a Tomcat a little less nerve-racking than when sat in the right seat of an Intruder in 1991:

> I was not quite as anxious as I had been in *Desert Storm*. There is still some healthy nervousness that helps you produce when the chips are down, as they are during a combat mission. I felt that same sense of, "I have to do everything right, because every trip I make in here successfully is one less that someone else will have to make." Depending on where you were for certain targets in *Desert Fox*, it was equal to or less intense than *Desert Storm*. Overall, the resistance was less, but any one missile is lethal. There is a certain amount of risk that goes with flying over a foreign country when you're

This photograph of the Taji fabrication plant, located 18 miles (29km) north of Baghdad, was taken just hours after the opening strikes of *Desert Fox*. This was one of a number of images released by the DoD to the press that revealed just how precise the attacks had been. Taji, home to Iraq's indigenous long-range missile program, had been hit by PGMs dropped from F-14Bs – Tomcats from VF-32 also took this post-strike imagery. (US Navy)

This BHA photograph generated by VF-32 shows four direct hits by LGBs on hangars at Al Sahra airfield, near Tikrit in northern Iraq. Again, these strikes were the work of F-14Bs from the unit. The photograph was shown to the media at a Pentagon press briefing on December 18, 1998 – just hours after the target had been hit. (US Navy)

engaged in combat. The aircrews handled it magnificently and focused their attention on getting to the target, putting weapons on it and then egressing.

We had very rigorous debriefs so that we were aware of what was going on out there, and equally rigorous briefs to make sure people knew what we were still up against. In spite of our early successes, there were still a lot of unknowns over Iraq. With each successive night, there was greater preparation by the Iraqis, greater indications that they were dispersing, and a greater likelihood that they could construct sites for their integrated air defense systems.

I was confident in the strike leads assigned because I had thoroughly scrubbed every plan we flew in there. I knew that we were asking for things that were very doable. We were trained, the airplanes and their systems worked and the aircrews displayed exceptional airmanship and discipline in carrying out their missions.

Having experienced at first hand the multinational and multi-service "mid-air madness" of *Desert Storm*, Hagen insisted on thorough communications among the various coalition strike elements once USAF and RAF jets joined *Desert Fox* on the second night of the operation:

Through e-mail notes, face-to-face liaison and phone calls, bombers, fighters and tankers all ended up on the same page in *Desert Fox*. This was in stark contrast to *Desert Storm*. Because of the magnitude of the force structure we had in 1991, just working with our own forces, let alone against the formidable Iraqi forces, was a greater task in many respects than anything that came up during *Desert Fox*. This time, we were almost singularly responsible for the success or failure of the tactical side of the house, and our assets were key to the protection they [USAF and the RAF] needed to get over the target area. Talking with each other made the integration go well.

"Coondawg" in *Desert Fox*

VF-32's XO during *Desert Fox* was Naval Aviator Cdr Will "Coondawg" Cooney, who provided the following account of the operation specifically for this book:

I had just completed F-14 refresher training after being plugged into VF-101 as the XO for ten months, which included carrier qualifications onboard *John F. Kennedy* in early December. Prior to that tour I had spent two years on the Joint Staff in the Pentagon after leaving VF-32 as a department head in March 1996. I rejoined VF-32 onboard *Enterprise* in the NAG along with Lt Cdr Scott "Whiskey" Calvert on December 10, 1998.

The following day CAG called a meeting with all of CVW-3's COs and XOs to inform us that *Desert Fox* might be on the horizon. Capt Hagen handed me two strike folders to plan for the missions on nights two and four. CVW-17, embarked in *Dwight D. Eisenhower*, had performed the initial strike plans for *Desert Fox* [and come within 20 minutes of launching Operation *Desert Thunder* – the precursor to *Desert Fox* – on November 11, only for Saddam to start co-operating with UN inspectors again] that were turned over to us when CVW-3 and *Enterprise* arrived in the NAG on November 23.

Now I hadn't led more than two jets around in almost three years, yet CAG had decided to make me a strike leader. Furthermore, I hadn't flown a jet equipped with the LANTIRN pod, nor had I flown with NVGs – the only training I had received with the latter was in the NVG "lab" at SWATSLANT prior to joining *Enterprise*. Obviously, I wasn't feeling up to speed, but I inherited a great strike planning team and the "Gypsy" [VF-32] RIOs I flew with were outstanding. With more than 3,000 hours in the Tomcat, I just had to get familiar with two new systems – how hard could that be?

Night One of *Desert Fox*, I flew on the first event with Lt "Ichy" Acheson, who was the squadron NVG instructor. We were parked aft of the island facing towards the port side. It was approaching midnight, and we didn't know if the operation was going to be scrubbed at the 11th hour, as had happened to the boys in CVW-17 on "Ike." Any doubts we had were soon dispelled when, off the port side, I saw numerous Tomahawk Land Attack Missiles [TLAMs] being launched from the small boys [surface ships] in our battle group – it was an impressive sight to see. That was soon followed by the Air Boss calling over the 5MC [Main Circuit – public address system specifically for the flightdeck] to start the go aircraft. The flightdeck immediately swarmed with life getting the go jets started and catapults manned.

Before and after photographs of Republican Guard barracks buildings in Baghdad and Abu Ghurayb following precision strikes by CVW-3 (and specifically VF-32) on December 16, 1998 – the opening night of *Desert Fox*. This imagery was taken by an F-14B TARPS jet from VF-32. (US Navy)

The plane captain of "Gypsy 102" (F-14B BuNo 161426) gives the jet's windscreen a final wipe over before declaring the aircraft "up" for its next sortie. During the four days of *Desert Fox*, VF-32 dropped 111,054lb (50,373kg) of ordnance – 16 GBU-10s, 16 GBU-16s and no fewer than 26 GBU-24 penetrator LGBs. Like "Gypsy 102," most the unit's ten F-14Bs carried bomb tallies beneath their cockpits. (US Navy)

The first night was all Navy, with no USAF or Brits supporting us. It was designed as a single cycle with no tanking to increase the element of surprise. Launch went as planned and the strike package rendezvoused overhead the ship. Now came FAM [lesson] One of NVG training. I had the goggles mounted on my helmet as we passed 5,000ft, climbing up. I initially had a "helmet fire" while trying to get used to looking through a soda straw (seriously, that's what looking through NVGs resembled) and attempting to figure out who was who. Thankfully, strike elements were separated by latitude overhead the carrier.

The Tomcats were carrying 1,000lb [454kg] GBU-16s, and our targets were just inland near the coast of Iraq. My wingman joined on me initially (thank God), but during the ingress my jet started having some navigation and LANTIRN degrades. I passed the lead to my wingman, and that's when the "fun" began, having never flown as a wingman wearing NVGs. Now, along with the helmet fire and "killing snakes in the cockpit," I was having to try and fly formation for the first time with NVGs.

I kept looking through the NVGs and under them, trying to keep visual on my lead. I repeatedly asked him to turn his external formation lights on and off so I could see him. Why my poor RIO just didn't eject us over water, I'll never understand! On the first pass through the target area, my lead destroyed his target, but our LANTIRN failed and we didn't drop our weapons. We circled around and tried a buddy lase drop, but our jet's issue continued and we were unable to get our LGBs off. We headed back to the ship for an uneventful night trap, after which I downed the jet for navigation/LANTIRN issues, and the fact that I had turned the stick and throttle grips into diamonds from squeezing them so hard learning to fly with NVGs.

On night two, I was the strike lead, with Lt "Meat" Overbaugh as my RIO. This time we were using USAF tankers, and RAF Tornado GR 1s also participated. We were heading further north to make life miserable for the Republican Guard. The strike package consisted of three Tomcats each carrying two 2,000lb [907kg] GBU-10s, eight Hornets carrying either GBU-16s or GBU-10s and a HARM-equipped SEAD package, with the Tornado GR 1s bringing up the rear. The GBU-10s effectively removed any building from the desert floor – pretty impressive to

Although there were no "big wing" USAF and RAF tankers available to support the first night strikes by CVW-3 during *Desert Fox*, US Navy strike packages made use of KC-10s, KC-135s and VC10s during the latter stages of the campaign. Here, two KC-10As from the 60th Air Mobility Wing are on station to provide frontside fuel for a pair of thirsty F/A-18Cs from VFA-105 and VF-32's "Gypsy 132" (BuNo 163409). Sections of Hornets and Tomcats worked closely in *Desert Fox*, with crews flying LANTIRN-equipped F-14Bs not only designating their own targets for LGBs but also buddy-lasing for F/A-18C pilots – a first for the F-14 community and a precursor for operations to come with the Tomcat in the Balkans, Afghanistan and, once again, Iraq. (Capt Will Cooney)

watch when you were not being distracted by AAA. NVG FAM Two was completed with a little less stress.

On night three, I flew a double header, launching on the first event with Lt Cdr "Snap" Courtney as my RIO. We had a pair of 2,400lb [1,088kg] GBU-24 penetrators strapped to the belly of our jet. *Desert Fox* was the first time the Navy had used the "bunker-busting" weapons for hard-to-find targets in combat. "Snap" did an awesome job locating our target, and being a good "stick monkey," I pickled when he told me to. We trapped three hours later, NVG FAM Three completed.

One hour later, I was briefing for my second sortie of the night with Lt "Traps" Quy, one of our female RIOs. This strike had us carrying two GBU-10s again as we looked to shake up the Republican Guard. We successfully engaged our target and headed back to trap onboard as the sun was just peeking above the water. I managed to get about three hours of sleep before getting up to complete my strike planning for night four – my second mission as a strike lead. I would be flying with "Meat" Overbaugh again, and we were carrying two more GBU-24s destined for yet more bunkers. This strike consisted of three Tomcats, eight Hornets, four Tornado GR 1s, four F-16CGs and the HARM/SEAD package of Prowlers and Hornets.

By night four, the Iraqis knew we were coming. Our targets were up north again, and they were filling the sky with AAA. I could easily see it through the NVGs while we were still way south on the ingress. We eventually dropped both GBU-24s simultaneously into the same hole in what we referred to as "consecutive miracles." With the mission a success, we headed back to the carrier to complete what was my third strike in 24 hours. I was now also NVG and LANTIRN qualified. I'm not sure how many pilots could say they qualified during combat missions. I owed most of that success to the great folks I flew with.

Carrying both a GBU-12 and air-to-air missiles, VF-32's "Gypsy 132" (BuNo 163409) takes on fuel from a KC-10's port underwing refueling pod shortly after *Desert Fox*. The aircraft boasts an impressive bomb tally beneath its cockpit. The last time this aircraft had such a tally applied was in late 1990 when it was marked with nine symbols and the nickname *Camel Smoker* (see page 16), plus water-soluble desert camouflage, while assigned to VF-24. Following its service with VF-32, the aircraft subsequently saw action with VF-11 in both OEF in 2002 and OIF III in 2004. (Capt Will Cooney)

Above: VF-32's CAG jet (BuNo 162916) was adorned with a bomb tally that reflected the unit's efforts as a whole during *Desert Fox*. Above the "scoreboard" was a unique sword and campaign ribbon marking seen on this aircraft only. The titling remained on the F-14B well into 1999. (David F Brown)

Right: VF-32 personnel come together beneath *Enterprise*'s island for a group photograph with "Gypsy 100" in the Atlantic during the final days of the carrier's 1998–99 deployment. Squadron XO Cdr Will Cooney and CO Cdr Gary Galloway are standing in the front row. (via Capt Will Cooney)

VF-32 performed tremendously during *Desert Fox*. All the aircrew made it over the beach at least once, which was quite impressive as there had been some major hurdles for the unit to get over. Only seven of its 14 pilots had gone through Air Wing Fallon training with VF-32, two senior pilots had arrived one week prior to *Desert Fox* and the squadron had only two LANTIRN pods from Fallon until its arrival in the NAG, when four more were cross-decked over from "Ike." But none of that stopped VF-32 from dropping more than 111,000lb [50,300kg] of ordnance in four nights while achieving a 100 percent sortie completion rate.

OSW Post-*Desert Fox*

Although *Desert Fox* lasted just four days, its consequences were felt right up until OIF in March 2003. Proclaiming a victory after UN weapons inspectors had left Iraq on the eve of the bombing campaign, and stating that he no longer recognized the legitimacy of the no-fly zones, Saddam brazenly challenged patrolling ONW and OSW aircraft by moving mobile SAM batteries and AAA weapons into the exclusion zones. Both were used in the coming months, and Iraqi combat aircraft also started to push regularly into the Box.

The adoption of this more aggressive stance by the IrAF almost resulted in a US Navy Tomcat claiming its first Phoenix missile kill when, on January 5, 1999, two F-14Ds from VF-213 launched two AIM-54Cs at MiG-25s that had penetrated the no-fly zone. The Iraqi jets had already turned back north and were making a high-speed run for home by the time the Tomcats launched their missiles at very long range. Neither weapon hit their intended targets.

Iraq's open defiance to OSW meant that coalition aircraft patrolling in the Box were now regularly locked up by fire-control radar and engaged by AAA and unguided SAMs on a near-daily basis. In the post-*Desert Fox* world, these violations provoked a swift, but measured, response from JTF-SWA. Typically, such missions were devised within the CAOC-approved pre-planned retaliatory strike framework, and they soon became known as Response Options (ROs). The latter allowed no-fly zone enforcers to react to threats or incursions in a co-ordinated manner through the execution of agreed ROs against pre-determined targets such as SAM and AAA sites and command and control nodes. The first such attack by ten aircraft from

"Black Lion 111" (BuNo 161159) idles on the flightdeck of *Carl Vinson* while awaiting the call to taxi on to one of the carrier's waist catapults for a dusk launch in the NAG in March 1999. By then, VF-213 had participated in *Desert Fox*, prosecuted a number of ROs (including the destruction of anti-ship missiles) and unsuccessfully engaged IrAF MiG-25s in what proved to be an action-packed deployment. One month short of seven years after this photograph was taken, BuNo 161159 – again serving with VF-213 – would participate in the last US Navy Tomcat combat sortie on the night of February 7–8, 2006. By the time of its retirement to the National Museum of Naval Aviation at NAS Pensacola, Florida, in April 2006, BuNo 161159 was the Tomcat with highest amount of flight time, totaling 7,719 hours. (Tony Holmes)

CVW-11 post-*Desert Fox* took place on January 23. Regular strikes would subsequently occur for the remainder of the air wing's time in the NAG.

VF-213 prosecuted a number of ROs, including the destruction of newly-placed CSSC-3 "Seersucker" anti-ship cruise missiles that the Iraqis had moved on to the Al Faw peninsula in an effort to threaten shipping transiting Kuwaiti waters. Two F-14Ds, accompanied by a pair of F/A-18s from CVW-11, used LGBs to destroy the weapons on February 2. A first cruise Naval Aviator, who wished to remain anonymous when interviewed by the author onboard *Carl Vinson* in the NAG in March 1999, gave the following account of VF-213's time committed to OSW in 1998–99:

> The dual cockpit of the Tomcat is an asset for the missions that we are performing on OSW, particularly when we are flying a jet equipped with the LANTIRN pod. We have an unfair advantage with the latter because we have two guys to split the workload, using a targeting system that produces superb imagery on our cockpit displays. Hornet pilots have their work cut out for them using the Nite Hawk pod, which generates imagery that is nowhere near as clear, while still flying form and having to find the target. In the Tomcat, the pilot gets on with flying the mission while the RIO concentrates exclusively on targeting.
>
> One of the deployment highlights for me has been getting to use the F-14D's AN/APG-71 radar operationally. The interface between the system and the Tomcat's unique AIM-54C missile is truly awesome. Despite a recent uptick in IrAF incursions into the no-fly zone, the Iraqi fighters never directly challenge us. The pilots of these aircraft are well aware of the performance envelope of the Phoenix missile, which means no aerial threat wants to come anywhere near us. It is almost like carrying a nuclear weapon in respect to its deterrence value.
>
> I logged a little time in the F-14A at the RAG, where I used the old AN/AWG-9 radar fitted in the earlier Tomcat models. The guys in the fleet that are still using it really have to work hard to

VF-213 (1998–99)

Tailhook down and wings swept forward, "Black Lion 103" (BuNo 163899) descends in a gentle left turn towards *Carl Vinson* following an OSW patrol. The aircraft is very lightly armed and lacks a LANTIRN pod, indicating that it may have been an air spare for the primary strike package that pushed into the Box. *Sea Dragon* sweeps were routinely carried out by aircraft that were ultimately not required for OSW missions. (VF-213)

Six of VF-213's 11 F-14Ds are chained down to the flightdeck at the stern of *Carl Vinson* between launch and recovery cycles while the carrier sails at modest speed through the calm waters of the NAG. The vessel was on station with Fifth Fleet from December 19, 1998 through to March 22, 1999, when it was relieved by *Enterprise* and CVW-3. (Tony Holmes)

get the best out of the radar – it would scare me to have to do half the things I have done on this deployment with the old radar. I often get ribbed by the more senior members of my squadron for being a "D-baby," as I have really known nothing else but the Tomcat "Delta" since I joined the F-14 community. Those guys, on the other hand, have all been through the "school of hard knocks" with the AN/AWG-9 on previous WESTPACs.

Aside from its superb radar and superior targeting pod, the Tomcat's outstanding endurance is also a major plus factor for the jet. Just today, we flew an OSW mission that saw us stay inside the Box for the entirety of the sortie. We had ingressed with two Hornets, who stayed as long as their fuel permitted, then were replaced by a second pair, and we left when the replacement jets declared they needed more fuel. In essence, we had remained on station for twice as long as the Hornets had on a single tank of fuel. We can also carry more ordnance than the F/A-18, and have a greater "bring back" capacity when trapping aboard at the end of the mission.

I wouldn't mind staying here indefinitely in order to continue flying the OSW missions that I've been undertaking for the last three months, as this is what I've spent all those years working up to do. For example, the mission I flew into Iraq today was exactly what I had been trained to do. To have had this experience on my first deployment is exceptional. My skipper [Cdr Mark H Adamshick] has been in the Tomcat community for more than 20 years, yet this is the first time he has dropped live bombs on an enemy.

Everybody is flying at least once a day at the moment, sometimes twice, so we are all getting plenty of hours in the cockpit. I have been told by my more experienced squadron mates that OSW missions are not what they used to be [since *Desert Fox*] – continuous patrols for days on end. Now, we fly a short, focused mission as part of a strike package, hit a target, and come out. That said, mission-planning is considerably more complex than in the old days.

Despite the high sortie tempo, generally, our 11 jets [one of which served as a parts source "hangar queen"] have held up well during the deployment. Nevertheless, obtaining spares has been the biggest problem facing the squadron. There simply aren't enough parts in the supply chain for the F-14D, particularly when it comes to the AN/APG-71 radar and its related cockpit displays and avionics. We have no problem dealing with general airframe maladies that may affect the wing sweep mechanism or fuel system, for example, as bits for these are readily available throughout the Tomcat community. Sadly, this is not the case for F-14D-specific systems.

We have also had modest G restrictions placed on our aircraft, which have been brought in principally to lengthen the jets' service lives, rather than restrict our operational flying – I am still going to pull 20G if it means avoiding being shot down by a SAM or an air-to-air missile!

With the uprated avionics, bigger engines and LANTIRN, the F-14D is effectively the naval equivalent of the F-15E Strike Eagle – this WESTPAC has proven that. Yet despite the fact we have embraced the strike fighter role, and really enjoyed the LANTIRN-heavy missions that we have performed out here, after a while it is refreshing to get back flying the F-14 as a fighter, pure and simple. Loaded up with air-to-air missiles only, and tasked exclusively with defending the guys that are going to drop the bombs, the life of a fighter pilot in a Tomcat – particularly a D-model – is still one to be envied!

By the time *Carl Vinson* had been relieved in the NAG by *Enterprise*, whose CVW-3 had been conducting Operation *Deliberate Force* missions over Bosnia-Herzegovina since mid-January

VF-213's officer cadre and senior chiefs pose in front of "Black Lion 101" (BuNo 163893) in April 1999 during *Carl Vinson*'s long journey east after completing its NAG deployment. The unit, led by Cdr Todd Miller (front row, sixth from left), completed 115 missions totaling more than 615 combat hours during OSW. It dropped 20 LGBs and supported a further 11 PGM strikes by F/A-18s from CVW-11. (VF-213)

F-14 "BOMBCAT": THE US NAVY'S ULTIMATE PRECISION BOMBER

1999, VF-213 had flown 70 air-to-air and air-to-ground missions (230 combat sorties) and 45 TARPS missions (with more than 580 targets "imaged") totaling in excess of 615 combat hours during OSW. The unit had dropped 20 LGBs on targets in Iraq, and also supported another 11 strikes by F/A-18s from CVW-11 while on deployment.

After almost a month of OSW missions, during which time F-14Bs from VF-32 would again undertake strikes on Iraqi air defense targets below the 33rd parallel, *Enterprise* was relieved by *Kitty Hawk* on April 17. The latter vessel was at the start of its first WESTPAC following its posting to Yokosuka, Japan, having replaced *Independence* as the US Navy's forward deployed carrier in July 1998. CVW-5 was embarked in *Kitty Hawk*, and the air wing was returning to the NAG less than ten months after completing its previous OSW assignment because *Theodore Roosevelt* and CVW-8 had been retained in the Mediterranean as part of the US Navy's

Right: VF-154 (1998)

Below: Armed with AIM-7Ms and AIM-9Ms, well-weathered "Knight 103" (F-14A BuNo 161293) comes under tension prior to launching from one of *Kitty Hawk*'s two waist catapults at the start of an OSW mission in June 1999. Nicknamed *Mi Evy*, the aircraft is also adorned with combined Battle "E"/Safety "S"/"Clifton"/Golden Wrench award markings and a Phoenix missile silhouette beneath the pilot's cockpit. All of VF-154's Tomcats had received the MMCAP, DFCS and FTI upgrades prior to CVW-5 and *Kitty Hawk* being sent on their unplanned NAG deployment in April 1999. (Tony Holmes)

commitment to Operation *Allied Force* (which commenced on March 24, 1999) in Kosovo. The carrier and its air wing would not relieve *Kitty Hawk* and CVW-5 until July 15.

VF-154, which had become the air wing's solitary Tomcat squadron following the disestablishment of VF-21 in January 1996, was equipped with ten F-14As. Although the majority of these aircraft were close to 20 years old, they were amongst the most capable A-models then in fleet service. All had been wired for TARPS-DI and LANTIRN in January 1998 at the start of CVW-5's final WESTPAC embarked in *Independence*. Indeed, as the forward-deployed squadron, VF-154 was the only Tomcat unit at that time to be permanently equipped with its own LANTIRN pods, leaving Oceana-based squadrons to share the remaining 24 among themselves.

By the time VF-154 unexpectedly returned to the NAG in mid-April 1999, the unit's Tomcats had received yet more upgrades as part of the Multi-Mission Capability Avionics Program (MMCAP). This saw the AN/ALR-67 Radar Warning Receiver (RWR) system fitted, aircraft made compatible with the LAU-138 missile rail/BOL chaff dispenser (capable of dispensing 160 chaff or IR packets per rail), installation of a PTID in the rear cockpit, modification of the venerable analog 5400 mission computer with digital capabilities from the F-14D's AN/AYK-14 system and the addition of twin Military Standard 1553B databases.

Furthermore, VF-154 became the first unit to deploy with a full complement of F-14s that had been fitted with the Digital Flight Control System (DFCS) developed by GEC Marconi (and based on equipment installed in the Eurofighter Typhoon). DFCS replaced the analog flight control system, which had been found wanting in the high angle of attack regime and when the jet was in landing configuration. DFCS greatly increased departure resistance and provided improved recovery capabilities when the Tomcat did depart controlled flight.

It also made the aircraft appreciably easier to handle on approach to landing. All Tomcats had received the DFCS upgrade by the late summer of 2000.

The final upgrade for VF-154 saw the unit's F-14s gain the Fast Tactical Imagery (FTI) capability, which, according to Capt Rick McHarg, commanding officer of CVW-5, "allowed Tomcat crews to transmit digital imagery from a LANTIRN pod back to the ship over a range of up to 120 miles in near real time via the aircraft's UHF radio link. This provided the battle group commander with BHA in just a matter of minutes [following a strike]."

During 91 days on station in the NAG, VF-154 flew 397 sorties in support of OSW and enjoyed a 91 percent mission completion rate. More than 9,000lb of ordnance was dropped during three separate RO strikes in Iraq, with 67 percent of all assigned targets being destroyed. LGBs were again the weapon of choice thanks to VF-154's thorough familiarity with the LANTIRN pod. "As an ex-A-6 B/N, when it comes to finding the target in a LANTIRN-equipped F-14 and then hitting it with an LGB, the Tomcat is a superior platform thanks to the pod being GPS-driven," explained Capt McHarg.

Fellow NFO Lt(jg) Rich Hill, very much at the other end of the experience scale from his CAG, concurred. "Despite the age of our jets, the squadron has done some excellent work on this cruise. We have consistently dropped LGBs right in the 'pickle barrel' when called upon

Mi Evy returns to *Kitty Hawk* at the end of an uneventful OSW mission in June 1999. This sortie was one of 397 flown by VF-154 during its 91 days on station with Fifth Fleet in the NAG, the unit dropping more than 9,000lb (4,082kg) of LGBs in three separate RO strikes during that time. BuNo 161293 subsequently saw action in OIF I again as "Knight 103," ending the campaign with 51 bomb symbols beneath its cockpit. (Tony Holmes)

to do so, proving the value of LANTIRN. The addition of this pod has made the F-14 the best bombing platform in the Navy."

Theodore Roosevelt and CVW-8 arrived in the NAG in mid-July following their commitment to *Allied Force* (see next chapter), the air wing's VF-14 and VF-41 having completed approximately 15 percent of the NATO sorties flown and destroyed 30 percent of the targets hit during the operation. Robert Wilcox, author of *Black Aces High,* which detailed VF-41's role in *Allied Force*, gave the following summary of the OSW missions flown by both Tomcat squadrons:

> After Palma, the *Roosevelt* sped to the Persian Gulf, where the "Black Aces" helped enforce UN no-fly zone rules over Iraq. It wasn't a war; it was more like a border patrol or police action. But it was still against the hateful Saddam Hussein, and the first time a modern carrier had fought in two combat zones during a single cruise. The Gulf had its perils. The heat was ghastly – sometimes 120°F on the flightdeck, and worse below, if that can be imagined. Although some ordnance was expended, mostly the squadron just went through the motions of bombing, performing all associated tasks except for the actual drop. There had to be a "triggering event" for that to be authorized. They used the runs as practice.
>
> Several times the "Black Aces" were involved in air wing "rope-a-dope" traps to lure Iraqi MiGs into flying violations [over the no-fly zone]. The object was to have grounds to shoot them down. But the wily plans never bore fruit.
>
> The "Black Aces," in Iraq, were honed and deadly – a fighter squadron at its peak. They had coalesced in Kosovo. On the rare occasions when strikes were authorized, direct hits became the norm.

Sister-squadron VF-14 also played its part in CVW-8's OSW commitment, with a single Tomcat being involved in the air wing's first strike in southern Iraq on July 17, 1999. According to the unit's Lt Jim Stanley, "Lt Keebler McDowell and LCDR Snapper Bull destroyed their target, making them the envy of all the aircrews in the ready room as the first 'Tophatters' to successfully hit targets in two separate theaters of operation in the same deployment." The squadron's command report for 1999 provides further details of VF-14's operational highlights while in the NAG:

> Having developed innovative and exceptionally effective FAC(A) tactics over Kosovo, the unit employed them in southern Iraq and subsequently exported them to other battle groups. It perfected mixed section employment tactics using the F-14 LANTIRN targeting system and the F/A-18 laser Maverick. VF-14 recorded a perfect 100 percent success rate while guiding 35 weapons. These tactics were passed down and employed extensively by *Constellation* [CVW-2] and *Kennedy* [CVW-1] battle groups, which have recorded similar results.
>
> VF-14 led the one and only successful FAC(A) strike performed by Air Wing EIGHT in Iraq during Operation *Southern Watch*. VF-14 FAC(A) aircrew guided four laser Mavericks to direct hits, demonstrating to Commander Joint Task Force Southwest Asia (CJTF-SWA) the usefulness of this previously pioneered tactic. Based on the success of this strike, CJTF-SWA directed this tactic as the standard for employing weapons in high collateral damage areas.
>
> Deputy JTF-SWA J3 acknowledged that the one dedicated FAC(A) mission led by VF-14 over Iraq was the finest strike CJTF-SWA had seen during his tenure in Riyadh, Saudi Arabia, and it was in turn briefed up the Chain of Command to the Joint Staff.
>
> Due to a detailed LANTIRN Targeting System and delivery techniques training course within VF-14, the squadron enjoyed an increasing ordnance delivery success rate during its deployment. The unit's GBU-24 success rate was an unprecedented 100 percent in Operation *Southern Watch*.
>
> VF-14 was the first squadron to successfully employ Fleet Tactical Imagery operationally during combat. Following a strike in Iraq, VF-14 captured Bomb Hit Assessment images and transmitted them back to USS *Theodore Roosevelt*. The transmission was re-transmitted to JTF-SWA and was received and verified before the originating aircraft returned overhead. This near real-time imagery proved the viability of reporting target destruction and threat assessment for follow-on packages.
>
> VF-14 was also the first squadron to operationally test and utilize the Tomcat Tactical Targeting (T3) LANTIRN software. This led to the recent clearance to use designated target co-ordinates via LANTIRN for Global Positioning System (GPS) guided munitions, allowing for real-time targeting of high threat mobile tactical targets with GPS-guided stand-off munitions previously reserved for fixed targets.

As noted in chapter three, the T3 upgrade to the LANTIRN system (which reached VF-14 after *Allied Force* had concluded but in time for OSW) was crucially important to the Tomcat community in the aircraft's final years of service with the US Navy. According to Lt Cdr Dave "Hey Joe" Parsons:

> Another parallel development critical to the broadening of the "Bombcat" role was the necessity for crews to observe ROE when it came to verifying targets. LANTIRN became an important tool to satisfy ROE because of its unique ability to produce precision co-ordinates which the USAF LTS could not do. The implementation of T3 allowed Tomcat crews to generate GPS targeting co-ordinates for JDAM and satisfy onerous ROE [this was a first for a TACAIR platform, as although some strategic USAF aircraft like the B-2 had GATS (GPS-Aided Targeting System), there were no tactical USAF assets that could do this at the time].
>
> Initially, digital frame grabs via LANTIRN Video Transmitter Receiver System [LVTRS] allowed Tomcat crews to obtain clearance to drop by sending images back to the ship for approval. Later, during OEF and OIF, they used LVTRS to exchange images with SOF elements on the ground.

LVTRS eventually gave way to the Remotely Operated Video Enhanced Receiver [ROVER] in December 2005, which had even greater capabilities.

When the *Theodore Roosevelt* battle group departed the NAG on August 26, 1999, its place in-theater was taken by the *Constellation* battle group. CVW-2 was embarked in the veteran carrier, with VF-2 undertaking its third WESTPAC with the F-14D. Like the Tomcats of VF-14 and VF-41, the D-model jets assigned to the unit had the software installed in their LANTIRN pods that gave them FTI capability – a first for a fleet squadron equipped with the F-14D. VF-2 was keen to prove its worth before arriving on station in the NAG, as the following report by the squadron in the winter 1999 edition of *The Hook* confirms:

In a demonstration of the FTI's effectiveness, CAPT Danny "Grinch" Knutson, ComCVW-2, led a long-range strike with four Tomcats against *Kitty Hawk* as it returned from the Gulf. Flying more than 800 miles towards the *Kitty Hawk* Battle Group, "Grinch" and his RIO, LCDR Greg "Chundar" Eden, acquired the carrier with LANTIRN and simulated a bombing strike against it.

Once the simulated bomb was released, a recorded image of the attack was forwarded to another F-14D and relayed onward [via FTI] to *Constellation*. After printing the image and presenting it to the battle group commander, it was copied and forwarded to *Kitty Hawk*. All this took place while Capt Knutson was still 600 miles away from Connie. Once the transmission was received, CAG Knutson gave *Kitty Hawk*'s crew a victory fly-by and then headed home to *Constellation*.

Thanks to the FTI software, an F-14 with a suitably modified LANTIRN pod became a highly effective reconnaissance asset. Crews now had the ability to take images from the pod and send them back in real time to the carrier, other LANTIRN-equipped Tomcats and ground units with laptops and radios that had had the FTI software installed. As VF-14 had shown, BHA imagery could be sent to the carrier while aircraft were still returning from a mission in the Box, greatly shrinking the targeting cycle time for the CAOC.

Constellation and CVW-2 commenced OSW patrols on August 29, and 11 days later, on September 9, following significant opposition to recent missions, the air wing launched Operation *Gun Smoke*. Some 35 of 39 AAA and SAM sites targeted for destruction by VF-2 in the Box were eliminated in a series of precision strikes that saw the largest expenditure of

Eight of VF-2's ten F-14Ds sit on the ramp at Oceana in early June 1999, just days away from flying cross-country to North Island and embarking in *Constellation* for CVW-2's WESTPAC/NAG deployment. Heading the line-up is VF-2's "Bullet 100" (BuNo 163901), which features the names of the unit CO and XO rather than the CAG and his deputy on the starboard side of its canopy rail. (Takashi Hashimoto)

The crew of VF-2's "Bullet 104" (F-14D BuNo 164351) prepare to strap into their aircraft during Operation *Gun Smoke* on September 9, 1999, their jet armed with two GBU-12s. This 24-hour OSW offensive saw CVW-2, embarked in *Constellation*, expend the most ordnance in combat in a single day since *Desert Storm*. VF-2 led the air wing by destroying 35 of the 39 targets it was assigned – S-60 (57mm) and KS-19 (100mm) AAA pieces and SAM sites – around Basra. (US Navy)

Carrying two GBU-12s, "Bullet 103" (F-14D BuNo 163900) is cleared for launch by the catapult officer during *Gun Smoke*. A second Tomcat from VF-2 can be seen idling behind the raised jet blast deflector, awaiting its turn to ride the waist catapult. BuNo 163900, flying as "Bullet 104," dropped 40 bombs in OIF I with VF-2. (US Navy)

ordnance in a single day since *Desert Storm*. VF-2 made effective use of the FAC(A) tactics recently implemented by VF-14, as Cdr Larry Burt, CO of CVW-2's F/A-18C-equipped VFA-137, recalled: "VF-2 crews would attack an AAA site with their own LGBs, then bring the Hornets in with LMAVs. The F-14 guys would find another gun, call in the Hornet, join up on the run-in and lase the target with their LANTIRN pods. This combination proved deadly, with one F-14 crew destroying ten guns on one mission alone – four with their LGBs and six with LMAVs."

The F-14Ds of VF-2 were critically important to the success of this campaign, and aside from dropping LGBs and lasing targets for CVW-2's trio of F/A-18C units, a single Tomcat also fired an AIM-54C at an IrAF MiG-23 – it is believed that the fighter outran the missile, which was fired at long range.

During the first nine months of 1999, US and British aircraft had flown 10,000 OSW sorties and dropped more than 1,000 bombs on 400 targets. Having added to this total, VF-2 departed the NAG with *Constellation* on November 7. Its place as the resident Tomcat unit was taken by VF-102, assigned to CVW-1 embarked in *John F. Kennedy*.

Millennium Deployment

Immediately prior to commencing OSW operations on November 1, CVW-1 had participated in Exercise *Bright Star 99*. A biennial multinational event (which included aircraft from seven nations) held in Egypt, the exercise saw allied nations in the region operating side-by-side with CENTCOM units (US Navy, US Marine Corps, USAF, US Army and SOF). CVW-1's involvement in *Bright Star 99* was tailored towards its up-and-coming operations in the NAG, with VF-102 and the air wing's trio of F/A-18C units (VFA-82, VFA-86 and VMFA-251) dropping both live LGBs and inert bombs on a range in Egypt and on the deserted island of Avgo Nisi, north of Crete.

Once in the NAG, CVW-1 aircrew were briefed by their counterparts in CVW-2, who explained that their aircraft were experiencing opposition from Iraqi AAA in the no-fly zone on virtually every patrol into the Box. Sure enough, this proved to be the case on CVW-1's first OSW mission, with an RO strike being immediately authorized after Tomcats and Hornets were fired upon. The offending AAA site was neutralized with LGBs.

VF-102's CO on this deployment was veteran Tomcat pilot, and former TOPGUN instructor, Cdr Mark Molidor, who provided a detailed snapshot of OSW operations in the

VF-102 joined CVW-1 embarked in *John F. Kennedy* on September 17, 1999 for what was dubbed the "millennium deployment." Prior to heading to the NAG, the carrier docked in the Port of Valletta, Malta, on October 7 for a brief spell of R&R. Some of CVW-1's Hornets and Tomcats had flown off the ship four days earlier, however, in order to participate in Exercise *Frisian Flag* at Leeuwarden Air Base, in the Netherlands. Among the aircraft left behind by VF-102 was a spotless, but drab for a CAG jet, "Diamondback 100" (F-14B BuNo 162700). (Joe Ciliberti)

Some of the 705 fixed and rotary winged aircraft to participate in Exercise *Bright Star 99* form up behind a B-1B in the traditional flypast of the pyramids at Giza. CVW-1's contribution to the formation was three F/A-18Cs and a solitary F-14B (from VF-102). Hornets and Tomcats dropped both live LGBs and inert bombs on a range in Egypt during the course of the 16-day exercise, which ran from October 15–31. (USAF)

new millennium when interviewed by the author in February 2000 during VF-102's spell in the NAG:

We have the oldest aircraft here on the carrier. Indeed, two of my jets are more than 20 years old. This means our Tomcats take a lot more maintenance to remain airworthy than any other aircraft type in CVW-1. This is primarily because of their age, and the fact that they have been upgraded so often that there are many more systems within the airframe of a Tomcat than any other aircraft on the flightdeck. We are the reconnaissance platform, the fighters and the bombers for the air wing, and all these responsibilities require dedicated systems within the jet.

The flipside of being oldest is that a lot of the design criteria that went into the F-14 in the 1960s has made the jet superior to other air wing assets in several important areas. For example, we have far more gas than anybody else, and we can go faster than anybody else. These are important "luxuries" exclusive to Tomcat operators. These pluses do come at a cost, however, in terms of maintenance. At the moment, we are averaging 65 to 70 maintenance hours per one flight hour, whereas a typical Hornet squadron usually averages between 23 to 25 maintenance hours per one flight hour.

This increased workload means that we have more personnel to share the maintenance burden. We have 245 troops aboard the ship at present, compared with 175 for each of the three Hornet units within CVW-1. Some 85 percent of VF-102's personnel are maintainers. The high levels of maintenance associated with the Tomcat in its twilight years has been the main factor in the Navy accelerating the service entry of the Super Hornet.

Generally, we will aim to have seven jets up on the flightdeck conducting ops while assigned to OSW. The maintenance of the F-14 tends to go in cycles. We had a spell not too long ago where we had four of our aircraft down for prolonged periods of time through maintenance issues, whereas at the moment we seem to be at the top of the cycle, with everything running smoothly.

I have a very experienced maintenance team staffed by key people who I recruited specifically to come and work for me some two years ago. Having such individuals within the unit makes a huge difference to our operability. You tell them before they join the unit that they are going to have to work hard thanks to the Tomcat's present maintenance record. However, this means the 180 days that we are out on cruise will pass far quicker due to the high work rate. Time can pass incredibly slowly on the boat when you are sitting around waiting for something to break.

This has been my busiest deployment to date, and I am on my fifth cruise. As a "JO" [junior officer], you would generally get 80 to 90 traps and fly 150 to 170 hours per cruise. On this deployment everyone will easily get more than 100 traps and close to 200 hours in the jet. We are flying well above the norm at the moment. The sorties themselves are pretty good too thanks to the variety of missions we get tasked to perform. It is not all Defensive Counter Air [DCA] or interdiction. One day we will do a TARPS mission, the next we will drop LGBs and lase for the Hornets and 24 hours after that we will join USAF Eagles on a DCA protection CAP mission. All this variety keeps the young guys really motivated and pumped up in the jet. Each crew will have flown several DCA/CAP, TARPS and bombing interdiction missions over the beach as part of CVW-1's OSW commitment by the end of this deployment, which is a great experience for the first cruise "nugget" Naval Aviators and NFOs.

Back in September 1999, the squadron deployed with five brand new pilots and RIOs, and you have to slowly work these fellows up in-theater. Initially, you have them fly wing on an experienced crew for four or five missions so that they can get to see the operational environment in which we are operating here in support of OSW. Once they have observed how the missions work, and we are comfortable with their progress over the beach, we will let them lead four or five missions. We always sortie in sections of two jets or divisions of four or six, just like the Hornet squadrons in CVW-1. We will continue to closely watch the new crews' progress in the lead slots, and critique their performance. We have to be right on top of their development, as these five crews will be the leaders for VF-102's next combat deployment in 18 months' time.

Due to our small fleet of just nine jets, we have to closely co-ordinate our ability to fulfill CVW-1's taskings with Strike Ops aboard *Kennedy*. We cannot physically launch a four-jet division and then put up four more in the next cycle. We can do a four-jet launch and then follow this up with two more aircraft, holding a single Tomcat as a spare ready on the flightdeck in case one of the primary F-14s goes down with a technical fault. The first four will land soon after the second two have launched, and the former will then be readied for the next cycle. By operating in this way, we

Cdr Mark Molidor led VF-102 throughout the unit's 1999–2000 deployment with CVW-1. A TOPGUN graduate and multi-cruise veteran, Molidor had the distinction of flying Jordan's King Abdullah II (an accomplished helicopter pilot) in the back seat of his Tomcat on November 4 during Exercise *Black Shark* in the Red Sea. His Royal Highness experienced both a launch and a recovery during the flight from *John F. Kennedy*. (US Navy)

With King Abdullah II of Jordan strapped into the rear cockpit, Cdr Molidor takes on fuel from an S-3B assigned to VS-32 during a brief flight from *John F. Kennedy* (visible in the background) while the carrier sails in the Red Sea. (US Navy)

have flown 20 sorties today, and we will do 24 tomorrow. This is exactly the kind of tempo we have maintained throughout our time on OSW.

Usually, each serviceable Tomcat will fly between three and five sorties per day while on OSW. This compares favorably with a typical USAF fighter squadron, whose jets rarely do more than one flight per day, and never more than two. We fly our aircraft very hard. If you are up in the crow's nest looking aft at the fantail, you will see a lot of maintenance personnel constantly around our jets between launches. They will be checking out external panels, feeling over the wing sweep gloves, testing certain hydraulic lines that have been known to cause issues in the past and changing out faulty avionics boxes.

Our maintainers are quite adept at rectifying serviceability issues with our veteran jets, with many of them having worked on Tomcats for a number of years. They have experience of fixing most gripes. Where we get real problems, however, is when a crew comes back to the ready room and explains what is happening with a jet, and the senior personnel in our maintenance department respond with, "Man, we haven't seen that in a long time!" They then have to dig into the manuals and various NATOPS [Naval Air Training and Operating Procedures Standardization] publications to figure out what the problem is, and how to solve it.

If it is a hydraulic or fuel problem we can generally fix it pretty quickly. The F-14 is much like a car in that respect. If a hose breaks, you get your technicians to fix it. We have very good airframers and hydraulic specialists who can fix these snags quickly. And they need to be good at their jobs as we are essentially operating a very sophisticated 20-year-old car!

Despite being the best in the business, our maintenance folks were defeated by a major issue with "Diamondback 102" just two months into the deployment. While performing a routine inspection, they found a large crack in one of the mounts that holds the engine up at the back of the airframe. We had to do a one-time fly off and send it home. The crack has been caused by the bigger General Electric F110 turbofan engines fitted in the F-14B, which are heavier than the original Pratt & Whitney TF30s and create a lot more torque due to their higher levels of thrust. Despite these mounts having already been significantly beefed up, we are still discovering cracks,

Cdr Molidor prepares to join the landing pattern overhead *John F. Kennedy* in "Diamondback 103" (F-14B BuNo 163227) on November 4, having swept the jet's wings back to 68 degrees and extended the tailhook. Shortly after this photograph was taken, with the Tomcat flying at 350 knots (648km/h), he would pass down the starboard side of the carrier at 800ft. His subsequent break into the pattern would see the jet bank to port at between 45–60 degrees – a maneuver that undoubtedly left an impression on Jordan's King Abdullah II. (US Navy)

so more strengthening is now being incorporated into the airframe. These inspections are routinely carried out on all F-14Bs to catch such cracks before they pose a hazard to the crews' safety.

According to Cdr Molidor, VF-102's success on this deployment was due to his unit, and his bosses in CVW-1, being fully aware of the strengths and weaknesses of the F-14B, and employing the jet in such a way that crews maximized the former and minimized the latter:

The Tomcat is not a truly swing-mission aircraft in the same way as the Hornet. We know our mission before we launch, and we stick to it. Having said that, we can drop bombs and then fly as a fighter. If it comes to the air-to-air or air-to-ground mission, any crew in VF-102 can handle both right now. However, within those crews we have certain guys who are specialists at particular missions. We have a cadre of TARPS experts, for example, and they can tell you about every wire in the pod, and what it does. We need our specialists in each of these disciplines, as they can then teach the rest of the crews in their particular area of expertise.

I would hand pick these specialist crewmen if, for example, the squadron was tasked with carrying out a critically important TARPS mission. I would make sure my TARPS specialist was in the backseat of my jet, as I would then know that the mission would be successfully completed first-time round. If the scenario called for precision bombing, I would make sure that my best air-to-ground man was again in the lead jet.

Having flown both the fighter-only and the multi-role versions of the Tomcat, it has been a pleasure to be able to finally come out here and perform the missions that we have been training to do for so long at Fallon, China Lake, El Centro and Vieques. The boys are constantly training to fly these missions when we are back at home, allowing them to come to the ship fully prepared for the deployment. Usually, however, when you go on cruise you sit around and do very little real mission flying, which is a huge disappointment. I was on the Operation *El Dorado Canyon* strike on Libya in 1986, and up until this cruise I had done very little real combat flying in the intervening 14 years. When we came here this time and got to drop bombs as part of the OSW mission, it was a fitting reward for all the hard work put in by the unit in the 18 months of training leading up to this deployment.

We have dropped 18 bombs to date, and of those, 16 have been direct hits on their intended targets. The key to our success has been the LANTIRN pod system. There is nothing presently in the US military to touch it. We took the USAF pod and, for once, the Navy did the smart thing and acquired the upgraded version. This meant that we literally had a better pod than the equivalent equipment then employed by the Air Force. You attach it to an aircraft that goes faster farther and with a lot more gas, and you have a great precision bombing platform.

When it comes to visual bombing, the Hornet kills us, being dead-on accurate every time. We are fairly accurate, but you really don't want to be dropping bombs on targets visually below 20,000ft nowadays, particularly in the AAA- and SAM-rich environment of southern Iraq. In respect to LGB drops, however, there is nothing that can touch the F-14. As a result of the LANTIRN pod, we also do a lot of "spiking" of targets for the Hornet because it has the older Nite Hawk FLIR fitted as standard. This is very much a hands-on and head inside the cockpit system – unenviable traits when over hostile territory. The Hornet guys regularly have trouble seeing targets with the FLIR if conditions are not perfect, but our LANTIRN system works well in most weather conditions and at night, and the light attack squadrons are often blown away by our imagery upon their return to the ship.

On February 27, *John F. Kennedy* departed the NAG and headed for home, its place in-theater being taken by USS *John C. Stennis* (CVN-74) the following day. The latter vessel had CVW-9 embarked, which included VF-211 equipped with ten FTI-capable F-14As. The air wing commenced OSW missions on March 2, and ROs would again be approved after the patrolling aircraft were engaged by AAA and locked up by SAM radar sites. Lt Nathan Ballou of VF-211 reported on the unit's operations over southern Iraq in the fall 2000 issue of *The Hook*:

Over two-and-a-half months, the squadron flew 166 missions totaling more than 380 hours, dropped 16 LGBs on military targets and exposed more than 25,000ft of film [on TARPS sorties]. VF-211 also flew DCA missions enforcing the no-fly zone. The Iraqis must have learned that the "Checkmates" had recently earned the Boola-Boola award for air-to-air missile-firing prowess, for there was not a single violation of the zone while VF-211 was standing the watch.

By the time the author photographed "Diamondback 105" (BuNo 162692) in the NAG in February 2000, the aircraft had dropped three GBU-12s on Iraqi targets during missions in the Box. Aircraft from CVW-1 undertaking OSW sorties were targeted by AAA virtually every time they patrolled the no-fly zone during the "millennium deployment." This aircraft later saw action with the unit in OEF in 2001–02 before participating in VF-32's final Tomcat cruise in 2004–05, where it again flew in Iraqi skies as part of OIF III. (Tony Holmes)

Although VF-211 dropped 16 LGBs and provided FAC(A) target guidance and buddy lasing for CVW-9's three Hornet units during the course of 166 OSW missions from *John C. Stennis* in two-and-a-half months in 2000, Tomcat crews still had to train hard while on deployment in order to remain operationally ready. Carrying a solitary BDU-59 LGTR on a BRU-42 ITER, "Nickel 101" (F-14A BuNo 162696) heads for the Camp Udairi bombing range in Kuwait so that its crew can conduct some LGB strike training. (US Navy)

Aside from dropping its own ordnance, VF-211 also provided FAC(A) target guidance, buddy lasing and BHA for the three F/A-18C units (VFA-146, VFA-147 and VMFA-314) within CVW-9. Capt Mark Christenson, assigned to VMFA-314, described his squadron's OSW experience as follows:

> While the missions over Iraq did not come as frequently as in past years, the flights proved more exciting as each visit to Iraq had the potential to become a rapid-response air strike. On several occasions, VMFA-314 pilots, along with other CVW-9 crews, responded to conduct real-time targeting of Iraqi military targets as directed by JTF-SWA. Using precision weapons, direct hits were later confirmed [often by TARPS-equipped F-14s] after Black Knight pilots dropped their ordnance, and they were credited with the destruction of several Iraqi surface-to-air defense sites.

John C. Stennis left the NAG on May 22, two days after *Dwight D. Eisenhower* arrived in-theater following an extended spell in the Adriatic supporting NATO peacekeepers in Kosovo (Operation *Joint Guardian*) and the NATO stabilization force in Bosnia-Herzegovina (Operation *Joint Forge*). CVW-7, which included F-14B-equipped VF-11 and VF-143, would spend two months undertaking OSW patrols. As alluded to by Capt Christenson, the Iraqi response to these missions had reduced in intensity by mid-year. Nevertheless, CVW-7 would fly 530 sorties totaling 1,550 hours through to July 26, when *Dwight D. Eisenhower* departed the NAG. The highlight of the deployment for the Tomcat units came on June 17, when jets from both squadrons became the first aircraft in *Dwight D. Eisenhower*'s then 21-year history to drop weapons in combat – LGBs were used to knock out Iraqi air-defense targets – after launching from the carrier.

VF-103 (2000)

Effectively mirroring the *Dwight D. Eisenhower*/CVW-7 deployment, USS *George Washington* (CVN-73) and CVW-17 took up station in the NAG on July 22 after briefly supporting *Joint Guardian*/*Joint Forge*. VF-103, flying F-14Bs, would complete more than 200 combat missions through to September 24, when *George Washington* left the theater.

VF-11's "Ripper 200" (F-14B BuNo 161433) is prepared for a night OSW mission onboard *Dwight D. Eisenhower* on June 22, 2000. Five days earlier, Tomcats from both VF-11 and sister-squadron VF-143 had become the first aircraft in the carrier's then 21-year history to drop weapons in combat – LGBs were used to knock out Iraqi air defense targets – after launching from the vessel. (US Navy)

During the first nine months of 1999, US and British aircraft flew 10,000 OSW sorties and dropped more than 1,000 bombs on 400 targets in Iraq. This level of action was sustained into the new millennium, and between March 2000 and March 2001, coalition aircraft were engaged more than 500 times by SAMs and AAA while flying a further 10,000 sorties into Iraqi airspace. Among the units to spar with Iraqi ground-to-air defenses during the latter half of 2000 was VF-103 "Jolly Rogers," embarked in *George Washington* as part of CVW-17. Manned up and with their F110 engines idling, these F-14Bs are chained down to the stern of the carrier ready to launch on their next OSW patrol on October 21, 2000. Note the bomb tally (two GBU-12s and a GBU-16) on the nose of "Victory 103" (F-14B BuNo 161435). (US Navy)

OPERATION SOUTHERN WATCH

CVW-17 had seen the first combat of the deployment (and the first from the flightdeck of *George Washington*) on August 11, with VF-103 joining F/A-18C-equipped VFA-34, VFA-81 and VFA-83 in an RO strike. Four more attacks by Tomcats and Hornets would follow up to September 14, with VF-103 having dropped 12 LGBs for 11 direct hits by the time the carrier departed the NAG.

CVW-14, embarked in USS *Abraham Lincoln* (CVN-72), commenced flying OSW patrols on September 26, with F-14D-equipped VF-31 dropping LGBs on six occasions between October 20 and December 27 – F/A-18Cs (from VFA-25, VFA-113 and VFA-115) also expended ordnance on these missions. The carrier and its air wing departed the NAG on December 29.

USS *Harry S. Truman* (CVN-75), with CVW-3 embarked, launched its first combat missions on January 3, 2001. The first bombs of the deployment to be dropped in anger were expended by F-14Bs from VF-32 and F/A-18Cs from VFA-105 and VMFA-312 on January 20 when Iraqi AAA sites were destroyed. These strikes came after coalition aircraft had been fired on 60 times in the first three weeks of 2001, US and British strike aircraft responding with bombs on 38 occasions.

The most comprehensive of the RO attacks (indeed, the biggest OSW strike since *Gun Smoke*) occurred on February 16 when CVW-3 targeted five command, control and communications sites. Again, VF-32 found itself in the vanguard of the one-day war, dropping LGBs, lasing for fellow Hornet strikers, running TARPS missions and conducting DCA sweeps in the Box. In the wake of this action, the unit's CO, Cdr Will Cooney, acknowledged the F-14's capabilities when it came to enforcing the OSW mandate: "The Tomcat's distinct size and power made it an intimidating foe to any enemy. With the big motors in the F-14B/D, its speed and power were very impressive. Coupled with size, large ordnance load and long legs, the Tomcat could really reach out and touch the bad guys in OSW."

VF-32 had dropped more than 16,000 tons of LGBs, and provided laser guidance for the F/A-18Cs of VFA-37, VFA-105 and VMFA-312, by the time *Harry S. Truman* departed the Persian Gulf on May 2 after being relieved by *Constellation* (with CVW-2 embarked).

Two F-14Bs from VF-32 fly in close formation with a pair of F/A-18Cs from VFA-37 during CVW-3's Fallon deployment in the late summer of 2000. For the air wing's trio of Hornet units and VF-32, the aim of the two-week exercise in Nevada was to prepare them for deployment and the potential operations they were likely to face. The key elements focused on were aerial refueling, long-distance transits to and from combat operations and shows of force, as well as the employment of Mk 76 Mod 5 blue bombs and BDU-59 LGTRs. Dozens of simulated strafing runs over hostile targets were also made on Fallon's live ranges. (US Navy)

Above: "Gypsy 101" (BuNo 161860) is marshalled forward after recovering back onboard *Harry S. Truman* in the NAG in January 2001. Note how the jet's LANTIRN pod has been painted in the same shade of gray as the Tomcat, but with the addition of Strike Eagle gray "tentacles." The US Navy permitted F-14 units to paint their targeting pods from late 2000. On February 16, 2001, just a few weeks after this photograph was taken, VF-32 found itself in the vanguard of the one-day war against five Iraqi command, control and communications sites, dropping LGBs, lasing for fellow Hornet strikers, running TARPS missions and conducting DCA sweeps in the Box. BuNo 161860 also served as "Gypsy 101" during VF-32's OIF I and III deployments. (Gert Kromhout)

Left: Armorers gather under the port shoulder pylon of "Gypsy 112" (BuNo 162922) prior to attaching two Mk 83 GP bombs to its forward centerline BRU-32 racks. The unguided nature of these weapons, combined with the presence of a blue AIM-9M drill round, suggests that the crew of this F-14B will be undertaking a training mission in the NAG from *Harry S. Truman*. (Gert Kromhout)

OPERATION SOUTHERN WATCH

Vice Admiral Charles Moore, Commander, Fifth Fleet, visited the newly arrived carrier just 24 hours after it had entered the NAG, and he spoke to the crew during an early evening edition of "Connie Currents," broadcast over the ship's 1MC (Main Circuit) public address system:

> No battle group in recent memory has done a better job than Connie. This ship has built a reputation over the last 40 years that can't be matched. She's an outstanding ship in every respect.
>
> All the training and preparations you did during work-ups are now going to bear fruit for the US while the Connie Battle Group is here representing the country on the front lines of this critical region. We're going to contain the Iraqis so they can't destabilize this region. We will do that by enforcing the no-fly zone so their forces can't move south and threaten Kuwait. There are forces from 18 nations that have joined us over the years in enforcing these sanctions. We've managed to keep Iraq in a box, and that's how we want to keep it.
>
> Aircraft from this ship represent 60 percent of the total strike capability we have in the theater. This carrier battle group is not just a "big player," it's "the player" in our effort to ensure the Iraqi threat doesn't destabilize the region.

While the pilot (Lt Kerry Kuykendall) gives her plane captain the thumb's up, the RIO prepares to enter the rear cockpit of "Gypsy 100" (BuNo 162916) prior to flying an OSW mission in March 2001. A single LGB silhouette is carried beneath the front cockpit, this aircraft having been involved in the February 16 operation of the previous month. Like Kuykendall, this aircraft had participated in *Desert Fox* in December 1998. It would also see combat in VF-32's OIF I and III deployments. (Gert Kromhout)

Constellation's CO, Capt Jamie Kelly, concurred with Vice Admiral Moore, stating, "This is where our training pays off. Every member of the crew should know that when they're doing their jobs they're contributing directly to our mission, which is putting planes in the air to enforce the no-fly zone and, when necessary, putting bombs on targets." Few of the latter would subsequently be hit, however, as CVW-2's VF-2 noted in its command history for 2001: "Over the hot Arabian Gulf summer, the Bullets [VF-2's callsign] logged more than 700 sorties in support of OSW. The political climate was challenging [as the ROE became increasingly restrictive], and few missions were allowed to drop ordnance [just four LGBs were expended in total]. While this was frustrating, the Bullets took advantage of the flights to hone the edge of their war-fighting skills, practicing tactics and weapons delivery techniques when not actually engaged over Iraq."

With its flightdeck dominated by F/A-18Cs from CVW-2, *Constellation* prepares for its next recovery cycle of aircraft returning from the Box after enforcing the no-fly zone over southern Iraq in June 2001. Five of VF-2's ten F-14Ds can be seen at the stern of the vessel, which means four more will be aloft on operations and one will be struck down in the hangar deck undergoing routine maintenance. (Tony Holmes)

The pilot of "Bullet 100" (BuNo 163894) is given the signal for launch by the catapult officer, thus denoting the start of a training mission from *Constellation* in June 2001. Although VF-2 completed more than 700 OSW sorties during the carrier's three months in the NAG, it dropped just four LGBs in anger as the ROE in-theater became increasingly restrictive. (US Navy)

"Black Aces" Return to OSW

On August 4, *Constellation* was relieved in the NAG by *Enterprise* and CVW-8, which included the F-14As of VF-14 and VF-41. The CO of the latter unit was Cdr Brian "Donger" Gawne, who provided the author with the following insight into VF-41's pre-cruise work-ups and subsequent operations in the Box:

> Before we left on cruise, we knew that we would be dropping ordnance in combat, as every carrier since *Desert Fox* had seen action in OSW. VF-14 and VF-41 had won all the awards following their exploits during the 1999 *Allied Force* deployment, with the "Tophatters" picking up the Clifton award as the top fighter squadron and the "Black Aces" becoming the first Tomcat unit to receive the McCluskey award as the top strike fighter squadron in the Navy. Both units were manned by some battle-seasoned veterans who had honed their skills in combat in Kosovo and during OSW. These Naval Aviators and NFOs gave both units the impetus to do well during the 2001 deployment.
>
> The 1999 veterans in VF-41 were adamant that they were going to train like they had fought on their previous cruise, so they were merciless when it came to getting new arrivals in the unit – including myself, initially as XO and then as CO – up to speed in flying the mission. We had no LANTIRN mission trainers available aside from a half-arsed computer simulator, which meant that crews familiarized themselves with how to use the pod in-house during training flights involving other Naval Aviators and NFOs who had employed the system in actual combat. We essentially learned how to use the LANTIRN pod by flying with it.
>
> VF-41 had proven that the F-14 with the LANTIRN pod was the FAC(A)/SCAR platform of choice during *Allied Force*. Clearing the targeting pod for use by the Tomcat revolutionized the way the jet was flown in combat in the final years of its life. LANTIRN had allowed the F-14 "bomb truck" to evolve into a mission-critical platform that could deliver PGMs, as well as providing FAC(A) and bomb targeting for all TACAIR assets.
>
> We knew how important FAC(A) crews had been during the 1999 cruise, so we wanted to get as many Naval Aviators and NFOs as possible qualified to fly this mission before deployment. The LANTIRN pod was the critical tool for FAC(A), so we needed to get hold of as many of these as we could early on in our work-ups. There really weren't many in the system in 2000, as at that point the Navy still had 11 frontline Tomcat units which were all clamoring to get LANTIRN pods on to their jets. You may have been left with one after completing your cruise and returning to Oceana,

Tanks topped off, VF-14's "Camelot 207" (BuNo 161292) drops away from a KC-135R of the 100th Air Refueling Wing during a training mission in the eastern Mediterranean in July 2001. Early the following month, VF-14, as part of CVW-8 embarked in *Enterprise*, would commence OSW patrols in the NAG. Already a veteran of *Allied Force*, BuNo 161292 would experience further combat in OEF during its 2001 deployment and, following a transfer to VF-154, in OIF I in 2003. (Tony Holmes)

with the rest being sent to other squadrons that were just about to deploy. You might then have received two more by the time the unit was sent to Air Wing Fallon, with the last of the six – which was the usual pod complement when deployed – arriving just prior to the squadron going on cruise.

We had no technical publications such as a NATOPS for the LANTIRN, which was essentially an Air Force pod for the F-15E that the Navy had bought off-the-shelf. The maintenance department was equally in the dark when it came to servicing and repairing the system, and therefore relied heavily on tech reps from Lockheed Martin.

VF-41 had suffered prior to its last cruise by receiving mission-ready aircraft just weeks before going on deployment. It was still having modifications – including the fitting of DFCS – done to the jets in the final days prior to leaving on cruise [on March 26, 1999]. Ideally, you needed to get your F-14s before heading to Air Wing Fallon, as it would take six months to groom them to the standard you required to get the mission done.

By 2001, the Tomcats we were operating were close to 30 years old, and they needed fettling like a classic muscle car of the same era in order for them to attain their peak performance. There were numerous modifications that had to be installed during the turnaround period between cruises, and all the bugs that inevitably cropped up in the jets following long spells on the ground needed to be worked out by the maintenance folks. Further compounding the problem when it came to aircraft maintainability, in the Navy, you would always transfer out your best jet [if it was required by a unit already on deployment] and accept the shittiest one in return!

[Cdr James J] "Dog" Bauser, who was CO of the unit during the work-ups, told me that "unless they are going to give us the Klingon cloaking device, I am not accepting any new jets after Fallon. We will deploy with what we have, otherwise our maintenance folks are going to struggle to get them modified and then mechanically sorted out before we cruise." He succeeded in getting our full complement of 11 aircraft prior to us tackling SFARP.

We pushed to have five LANTIRN pods and all of our NFOs in the squadron before attending Air Wing Fallon. A study after Kosovo had shown that an NFO's success rate when it came to hitting a target with an LGB using the LANTIRN pod was substantially higher for those individuals who had joined VF-14 and VF-41 pre-Fallon than for those who arrived on the unit after the air wing work-up. NFOs that arrived post-Fallon would still be learning how to use the pod effectively during the early stages of the cruise, thus limiting their combat effectiveness.

Even with all of our jets and aircrew in place well in advance of the deployment commencing, the squadron's experience with using LGBs was limited by the fact that VF-41 was allocated just three PGMs to drop during our interim training cycle. We therefore fell back on LGTRs, and, even then, each crew got to drop just two or three of them. We also managed to drop a number of dumb bombs in training, but other than exercising the jet's ordnance system, these were of no value when it came to learning how to accurately deliver LGBs. It was better to go out with no bombs and simply practice with the laser. The shortage of LGBs also meant that maintenance crews and armorers received little training in arming our jets with such weapons.

Due to the limited number of PGMs made available to VF-41, we had to rely on our mission recorders to tape our training runs on pre-determined targets. We would devise a route to and from the target, plan our fuel burn, pick our mission way points, devise the optimum attack angles, fly the sortie and then come back to Oceana and thoroughly scrutinize the tapes. Every time you missed the "target" on one of these flights, the ex-Kosovo guys would agonize over what you might have done wrong – we were routinely asked "did you lase too soon, or did you lase too far?" They became very good at doing the "CSI" of finding out why we didn't "hit" the target.

In addition to having only a limited opportunity to drop LGBs pre-cruise, we did virtually no strafing on the ranges. There was a question mark at the time over the reliability of the rounds then being used by Tomcat units in the fleet, and we also struggled to find a range where we could fire our guns. The Atlantic Fleet Weapons Training Area on Vieques Island, off the east coast of Puerto Rico, had always been our last stop prior to heading east on deployment, but it was no longer available

Three F-14As and two F/A-18Cs sit chained down to the deck over *Enterprise*'s bow catapult two in August 2001. CVW-8 was in the middle of its five-week commitment to OSW when this photograph was taken. Climbing away from the carrier in phase five afterburner at the start of yet another lengthy mission over southern Iraq is VF-41's "Fast Eagle 104" (BuNo 158630), which would subsequently drop 24 LGBs during the opening phase of OEF. (US Navy)

to us [following the range's temporary closure in April 1999 after a mishap]. We ended up getting some strafing passes in at the ship's wake as we sailed across the Atlantic, although our supply of ammunition at the time was limited.

Fully aware that we were going to perform OSW missions during the second half of our deployment, the CAG staff of CVW-8 made sure that we made up for the limited amount of bomb dropping we had done during our work-ups by hitting seven different ranges in Europe and North Africa. In the three months prior to us heading into the NAG, we sent detachments to Tunisia, Israel and Corsica, and participated in the Joint Maritime Course war game off Scotland. By the time we exited the Suez Canal and headed south into the Red Sea at the end of July 2001, we were as ready as any squadron could possibly be for OSW. In my 20 years in Naval Aviation, I had never previously experienced so much training in such a short space of time as we had done in the three months leading up to this point in the cruise.

We thought that all of this training was to get us ready for Iraq, but little did we realize that OSW was to get us really ready for OEF. We got a good turnover from VF-2 embarked in *Constellation*, then spent all of August in the NAG flying over Iraq. We knew from VF-41's previous OSW cruise in 1999 that it was going to be hot in theater at this time of year – the temperature on the flightdeck was typically 135°F. Our deck crews would walk into the ready room looking like they had just

"Big wing" tanker support was critical to the success of OSW during the ten-and-a-half years that missions were flown into southern Iraq, with USAF KC-135s and KC-10s and RAF VC10s and TriStars flying countless sorties primarily from bases in Qatar and Bahrain. All armed with air-to-air missiles to perform the DCA role, F-14As from VF-14 and VF-41 and F/A-18Cs from VFA-87 take it in turns to receive mid-mission fuel from a 2nd Aerial Refueling Squadron/305th Air Mobility Wing KC-10A during an OSW patrol in early September 2001. (Cdr Brian Gawne)

stepped out of the shower. Despite these harsh conditions, which continued when we moved east to the Arabian Sea for OEF, none of my sailors complained. When you are dropping live weapons, the whole ship takes on a different persona. If a young sailor knows there's a meaningful reason why he is working so hard, he will give you 18 hours a day of hard graft.

We dropped three GBU-12s and four GBU-16s on targets in southern Iraq during our short time manning the OSW station [sister-squadron VF-14 expended four GBU-12s and four GBU-16s]. Only five of the 11 strikes we flew into Iraq were executed. Our targets were what we dubbed the "Frankenstein" SAM sites, which the Iraqis had modified so as to make them difficult to detect – the SA-2/3s fired from these sites had been fitted with different types of seeker heads, and the launchers would be regularly moved around. I dropped my first bomb in anger during one of these mission, with CAG [Capt David J] Mercer as my pilot on this strike.

A lot of the missions we were flying were generated in response to Iraqi AAA being shot at us during routine patrols. We would then be cleared to retaliate. Collateral damage was a prime concern throughout OSW, and we were not allowed to drop ordnance unless there was a Predator unmanned aerial vehicle [UAV] on station that could film the results of our attack. We required film of the drop so that we had plausible deniability to refute any collateral damage claims made by Saddam's regime in the Arab press.

Time-sensitive Target [TST] strikes were the order of the day when it came to dealing with AAA and SAM batteries in OSW, as we needed to be able to hit them as soon as they shot at us, or shortly thereafter. Of course, we also needed UAV and EW support in order to satisfy the recently enhanced CAOC strike criteria. Our first strike was against an antenna relay tower, and although the first LGB hit the target, the two that followed had their laser signals interrupted by the debris cloud from the detonation of the first weapon. Once back aboard the ship, I remember agonizing for several hours as I explained to the senior officers on *Enterprise* how we had slung a couple of LGBs down range because they had lost their laser guidance. There was much wailing and gnashing of teeth, but that was typical of the mindset that we had created in the decade that OSW had been running.

We executed our fifth, and last, strike on September 9, turned over the OSW tasking to CVW-11 [embarked in *Carl Vinson*, which was then still heading westward across the Indian Ocean – it would never make it to the Persian Gulf] and sailed south through the Strait of Hormuz. It felt like Christmas as we left the NAG behind. We had had a very successful cruise, we had not lost anybody, we had had some great port calls and we had Cape Town and Rio de Janeiro to look forward to.

On September 11, we welcomed the first no-fly day for more than a month, and most of us enjoyed a rare sleep in. We were also making the most of the cooler temperatures. Having done some work in my stateroom, I finally walked into the ready room just

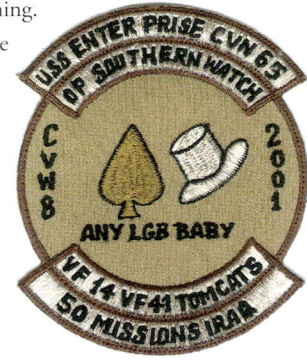

VF-14 and VF-41 (2001)

before 1500hrs, and was immediately told that airliners had flown into the World Trade Center. We then watched the events unfold live, as the ship had a terrific satellite link to the US. We were all in a state of shock, and I eventually called a squadron meeting and told everyone that we weren't going home. We all felt the ship turn around, and we knew that we still had work to do on the cruise. We knew within days that we would be bombing Afghanistan.

OSW Pause

The steady escalation of the conflict in southern Iraq was only brought to a halt, albeit temporarily, by the devastating attacks on the World Trade Center, in New York City, and the Pentagon, in Washington, D.C., on September 11, 2001. The subsequent declaration of the War on Terror by President George W Bush saw US carrier battle groups under Fifth Fleet control removed from their OSW station and pushed further east into the Arabian Sea and Indian Ocean in order to support OEF in Afghanistan (*see* chapter nine for details). With the bulk of the tactical air power in this conflict provided by carrier aircraft flying arduous four- to eight-hour missions over land-locked Afghanistan, OSW no-fly zone operations by the US Navy were drastically scaled back. This allowed the Iraqis to move more air defense weaponry below the 33rd parallel.

By the spring of 2002, the Taliban regime had been removed from power in Afghanistan and the US government's focus of attention returned once again to its old foe in the region, President Saddam Hussein. Proof of this came with the arrival of *George Washington* and its embarked CVW-17 in the NAG in late August 2002 following seven weeks of flying OEF missions over Afghanistan. The vessel's subsequent assignment to OSW on September 1 marked the first time a carrier battle group had performed this mission in almost a year. Eight days later, the air wing (including VF-103, equipped with F-14Bs) conducted its first RO strikes – with LGBs – after being repeatedly fired on by AAA and SAMs during patrols over southern Iraq.

The adoption of ROs as the primary means of enforcing the OSW mission evolved to match the coalition's desire to ensure the safety of its crews flying over Iraqi territory. Initially, the near-immediate air strike response to surface-to-air fires (SAFIREs) involving AAA or SAMs that had been the norm after *Desert Fox* was replaced by delayed, punitive strikes that were usually flown the same day as the no-fly zone violation took place. This RO evolved following the September 11, 2001 attacks into an even more considered approach, whereby the coalition introduced the policy of attacking any Iraqi military target in the southern no-fly zone. It did not even have to be the one that prompted the RO in the first place. This, in turn, led to the adoption of pre-planned RO methodology in the final months of OSW as the battlefield for OIF was prepped.

After almost two months conducting OEF missions over Afghanistan, CVW-17's spell in the NAG was to last just three weeks. In that time the air wing flew 170 missions and dropped bombs on five occasions during RO strikes. Aside from expending a handful of bombs, VF-103

VF-103 was the first Tomcat unit to benefit from the more aggressive RO policy instigated in OSW in 2002 post-OEF, dropping five GBU-12s in RO strikes during its 21-day spell in the NAG. However, poor LGB work in the Box in September 2002 by a CAG staff RIO flying with the unit almost resulted in the severing of an Iraqi oil pipeline north of Basra, and the CAOC immediately banned any further bombing by F-14s in-theater. Here, VF-103's CAG jet (BuNo 162918, christened *SAN ANTONIO ROSE*) returns to *George Washington* with its two GBU-12s still very much in place shortly before the bombing mishap. Assigned to VF-103 in December 1999, the aircraft became "Victory 100" six months later and subsequently completed three combat tours (2000, 2002 and 2004) with the squadron. (Capt Dana Potts)

also tail-chased an IrAF MiG-25 "Foxbat" north for more than 150 miles after the interceptor had boldly penetrated the no-fly zone.

George Washington was relieved of its OEF responsibilities by *Abraham Lincoln*, with CVW-14 embarked, on September 11, 2002, the latter carrier and its air wing being in the early stages of a marathon ten-and-a-half-month cruise. The vessel was then transferred to the NAG in late October, commencing OSW patrols on the 29th after a relatively quiet time over Afghanistan. CVW-14 featured three types of strike fighter on this deployment, with the F-14Ds of VF-31 and the F/A-18Cs of VFA-25 and VFA-113 being joined by the F/A-18E Super Hornets of VFA-115 as the new type made its operational debut. The aircraft routinely flew mixed sorties into southern Iraq through to December 6, but only a handful of PGMs were dropped on targets.

This was all about to change, however, with the Bush administration's focus on presenting the case for war against Iraq due to the latter country's alleged development and stockpiling of WMD. "Links" between Saddam's regime and Osama Bin Laden's al-Qaeda network were also played up, and the end result of all this talk was the decision by US Secretary of Defense Donald Rumsfeld to step up the level of response to Iraqi threats to US and British aircraft conducting OSW missions from September.

The unannounced escalation was part of Operation *Southern Focus*, which was not a publicly declared offensive. Its aim was to counter Iraqi attacks on coalition aircraft, which were seeing as many as 15 SAMs (usually unguided) being fired at them per day during the final months of OSW – there were 651 attacks between June 2002 and 19 March 2003. *Southern Focus* concentrated on AAA and SAM batteries, early warning radar sites and command and control facilities – the fiber-optic cable network associated with the latter, which kept Baghdad in communication with southern Iraq, was repeatedly hit. Effectively preparing the battlefield for the coming ground campaign, *Southern Focus* came to an end with the commencement of OIF.

As the sun sets in the west, an F-14B from VF-103 heads north over the NAG before crossing the Iraqi coast and commencing a night OSW patrol. After seven weeks of its aircraft flying OEF missions over Afghanistan, *George Washington*'s assignment to OSW on September 1, 2002 marked the first time a carrier battle group had performed this mission in almost a year. (Capt Dana Potts)

A key weapon in the revised RO for OSW was JDAM, which became CAOC's "bomb of choice" post-OEF thanks to it being wholly autonomous after release, unlike laser-guided or electro-optical munitions whose accuracy can be affected by bad weather or poor targeting solutions. A clinically accurate weapon against fixed targets, which proliferated in OSW, JDAM bombs then available to the US Navy (GBU-31 and GBU-32) were effectively standard Mk 83 1,000lb (454kg) and Mk 84/BLU-109 (penetrator) 2,000lb (907kg) unguided munitions fitted with a GPS guidance control unit (GCU), mid-body ventral strakes and a tail unit with steerable control fins. The GBU-38, based on the 500lb (226kg) Mk 82, was cleared for use by the F-14B/D post-OIF I.

Developed by PGM pioneer Boeing in the mid to late 1990s, JDAM differs from other GPS-guided weapons (AGM-130 and EGBU-15) in that it guides completely autonomously after being released – it cannot be steered or fed updated targeting data once dropped.

The "baseline" JDAM is considered to be a "near precision" weapon, the bomb's GCU relying on a three-axis INS and a GPS receiver to provide its pre-planned or in-flight targeting capability. The INS is a back-up system should the GPS lose satellite reception or be jammed. With GPS guidance at its heart, JDAM can only be employed by an aircraft fitted with an on-board GPS system so that GPS-computed co-ordinates can be downloaded to the weapon for both the target itself and the weapon release point. That way the jet's onboard INS remains as accurate as possible while the bomb is acquiring a GPS signal after being released over the target. This means that the jet has to have a MIL-STD 1760 databus and compatible pylon wiring in order to program the bomb's aim point, intended trajectory shape and impact geometry.

Achieving initial operational capability in 1997, JDAM made its combat debut during the NATO-led bombing campaign (Operation *Allied Force*) in Serbia and Kosovo in 1999. It was then employed during OSW, primarily by US Navy F/A-18s, until the weapon really began

Having come out of the Box for mid-cycle fuel, the crew of VF-31's "Bandwagon 101" (BuNo 164600) await their turn on the hose while the pilot of an F/A-18C from VFA-25 tops off his tanks from an RAF TriStar of No 216 Sqn in early November 2002. The latter jet was flying from Muharraq, Bahrain, where the RAF usually had at least one TriStar or VC10 permanently deployed supporting coalition jets venturing into Iraq. "Big wing" tankers were in great demand in both OSW and OIF, and the TriStar was a particular favorite of US Navy TACAIR pilots. "We would take fuel off anyone who had it available in-theater – USAF, US Navy, RAF or RAAF," CVW-2 staff officer and F-14 pilot Lt Cdr Dave Grogan explained. "I personally preferred to tank from a British L-1011 TriStar rather than a USAF KC-135 or KC-10, as the basket in the Lockheed jet was a lot easier on our probes than its USAF equivalent. The RAF jet lighting system was also much easier to work with at night." (Lt Cdr Jim Muse)

to capture headlines during OEF thanks to the exploits of Hornet units operating from the various carriers assigned to the conflict. JDAM finally made its combat debut with the F-14B in March 2002, again in OEF, and on the eve of OIF I with the F-14D – the A-model Tomcat could not employ JDAM because the jet lacked a digital databus.

Aside from its stunning accuracy in OEF, the weapon also proved popular with crews because it could be dropped in level flight from high altitude, thus allowing aircraft to stay well above any SAM or AAA threats. Depending on the height and speed of the delivery platform, JDAM can be released up to 15 miles away from its target in ideal conditions.

Following several embarrassing targeting failures of LGBs dropped by Tomcats in OSW in the late summer of 2002, the CAOC began to favor the employment of JDAM weaponry almost exclusively. This continued up until the final battlefield prepping in early March 2003. Restricted to LGBs only, the F-14D units committed to the last months of OSW felt more than a little frustrated at being cut out of the action due to their lack of JDAM compatibility. There was also a feeling within the Tomcat community that the USAF-dominated CAOC was using this as an excuse to keep the choice RO missions for land-based units preparing for OIF. One senior NFO serving with VF-2, which commenced OSW missions with CVW-2 from *Constellation* on December 19, 2002 after the carrier had relieved *Abraham Lincoln*, recalled:

We felt that the Air Force-led CAOC was saving the best targets for USAF assets, few of whom had JDAM capability in any case. Navy crews believed there was some consternation within the Air Force that a carrier-based aircraft such as the F-14 could conduct deep strike missions into western Iraq against high-value targets that the USAF was determined to save for itself, and its aviators. This left us Tomcat crews feeling like the bastard children of the OSW coalition.

I experienced this at first hand right from the start of our time on station in the NAG. On December 26, CVW-2's Hornets conducted a big strike with JDAM near An Nasiriyah in retaliation for the shooting down of a Predator UAV four days earlier. We really wished we could have dropped but there was no trust in-theater for Tomcats with LGBs – JDAM was now the name of the game. This was very frustrating because the CAOC would allow USAF aircraft to drop LGBs but they wouldn't let us.

We were pretty sure this was because of an incident involving a previous F-14B squadron. An inexperienced aircrew's errors with an LGB during an OSW mission almost resulted in an oil pipeline being hit miles from the target. That soured the CAOC's view of the F-14 as an LGB-employing aircraft. The F-14D with its LANTIRN and PTID was an exceptional LGB delivery platform, but all it took was one issue with a jet in another air wing to ruin future LGB employment opportunities for the entire F-14 community.

The CAOC was the final decider of what jet would be allowed to drop during an OSW retaliatory strike. Once a coalition aircraft was shot at by Iraqi forces, the CAOC made the decision as to what target would be struck with a JDAM. The CAOC did allow other platforms such as the Strike Eagle to employ LGBs during OSW, so there was frustration in the VF-2 ready room with not being able to drop such weapons. That mindset, of course, would change in OIF, when the CAOC didn't have

Its LGBs still on their racks, "Bullet 111" (BuNo 159613) returns to *Constellation* at the end of yet another fruitless OSW patrol in January 2003. A month later, on February 28, this aircraft became the first F-14D to drop a JDAM operationally. Although the LGB-equipped Tomcat had not been the preferred CAOC weapon of choice at the time, this would soon change once OIF I commenced in March 2003. Delivered to the US Navy as an F-14A in late 1975, BuNo 159613 was remanufactured as a D-model 15 years later and subsequently served with VF-11 and VF-2. The latter unit retired it to AMARC in June 2003, and the veteran fighter was sold to HVF West for scrapping in July 2009. (VF-2)

control over every mission and LGBs were used to a great extent off VF-2's F-14s in the later weeks of the campaign.

The fallout from the wayward LGB was a painful lesson for the Tomcat community, and we were very upset at our comrades from Oceana putting a "turd in the punchbowl" for everyone that followed. It was very, very frustrating to be over targets and not be able to drop our LGBs.

A typically fruitless OSW sortie for VF-2 at the time went as follows. Almost always flying at night, and using NVGs, we would launch with our weapons (usually two LGBs), refuel from USAF tankers, check in with myriad control agencies, do a roll call at the beginning of a vul window and then press out and perform either our Strike Familiarization, TARPS or DCA missions. We would then head back to southern Kuwait and get mid-cycle gas.

Sometimes an RO was called away. Routinely, the AWACS controllers would read out the call-signs of the guys they wanted to be "droppers" and then switch them up to another radio frequency over which they would relay specific targeting information. The unwanted jets – usually F-14s from VF-2 – were then told to return to the carrier. Not being given a mission to perform was like being picked last in football. The Hornets were doing great work with JDAM, while we were left to act as high-speed cheerleaders.

During this period, I had even been the overall strike lead for the CVW-2 jets in the Box when the USAF controllers came on the radio and called my guys away to do their thing. We would simply turn around and go home, later slapping the Hornet guys on the back when we saw them at the mission debriefing in CVIC [carrier intelligence center]. Although it was nice to see "my strike"

Two months of intensive flying synonymous with operations in the NAG had inflicted a heavy toll on VF-2's CAG bird, F-14D BuNo 163894, by the time it was photographed being attached to waist catapult three onboard *Constellation* in late January 2003. The aircraft's refueling probe door was removed to prevent it from being damaged while tanking from USAF KC-135s and KC-10s. (Iwan Bögels)

equipped with PTIDs for those drops was limited since the squadron didn't get its full complement of PTID-modified F-14Ds until right before cruise. From senior leadership levels, it became unnecessary to take the risk of pushing for the Tomcats to drop during OSW when a 1,000lb JDAM off a Hornet would meet the CAOC's needs.

Having been ruled out of the precision strike role, VF-2 reverted to flying CAP for much of its time prior to OIF. This was a role that the F-14D was particularly suited to, as Cdr Denneny explained:

During OSW CAP missions, the F-14D crews had a substantial range, speed and time on station advantage over every other coalition fighter platform, and it was the only aircraft, along with the F-15C, that covered the most northern no-fly zone regions close to Baghdad. Flying an aircraft with Link-16 JTIDS [Joint Tactical Information Distribution System] and the improved IRST [Infrared Search and Track], VF-2 crews had amazing situational awareness of who was airborne around them. And they could track them at a considerable distance with the jet's AN/APG-71 in medium-PRF [pulse-repetition frequency] mode.

At night, F-14D crews flew their jets right along the most northern part of the OSW no-fly zone – close enough to see the lights of Baghdad on a clear night and very clearly on their NVGs. The aircraft's radar was powerful. RIOs were able to track MiGs during OSW flying out of airfields around Baghdad. The rub was that those MiGs would come south, but only when the Tomcats and Eagles weren't on those CAP stations. During one of the times [December 23, 2002] when there wasn't a manned northern CAP station, an IrAF MiG-25 came south into the no-fly zone and shot down a US [MQ-1] Predator drone. When this event forced the CAOC to put more fighters on CAPs to deter additional shootdowns, Tomcat crews quickly realized that they were protecting an unmanned asset with a manned one.

D04 to the Fore

The crucial piece of equipment required by VF-2 to permit it to employ JDAM was the D04 mission tape, which allowed the F-14D to communicate with GPS-guided ordnance. Cdr Denneny explained that the unit had lobbied for D04 (the F-14B-equivalent, OFP321A, had been installed in B-model Tomcats for almost a year by then) to be made available prior to CVW-2 deploying:

When VF-2 set sail for the NAG on November 2, we were well prepared for potential combat operations. Our ten F-14Ds were in excellent condition. Our morale was high, and we were pretty confident that our timing was going to be good for what we considered was a foregone conclusion that we would invade Iraq.

"Ordies" from VF-2 attach a pair of GBU-12 LGBs to the forward centerline BRU-32 racks of an F-14D on the flightdeck of *Constellation* on March 5, 2003. Tomcats armed with LGBs were not trusted by the CAOC when it came to undertaking ROs against Iraqi targets during the final months of OSW, JDAM having by then become the favored PGM. LGBs would, however, be expended in their hundreds by all five F-14 units committed to OIF I. (US Navy)

get the job done, especially after the hours and hours that I had put into the mission with pre-flight planning, it would have been far more rewarding had we been "cleared hot" to attack the target too. This lack of trust of the Tomcat and its LGBs motivated the VF-2 aircrew to push NAVAIR to release new software that would allow the F-14Ds to drop JDAM.

VF-2's XO for much of the unit's 2002–03 cruise was Tomcat RIO Cdr Doug Denneny, and he was frustrated that the squadron lacked JDAM upon its arrival in the NAG and was unable to employ LGBs in-theater until the final weeks of OSW:

The VF-2 crews had earned for themselves an excellent reputation for hitting their targets at Fallon during CVW-2's air wing work-ups, using LGTRs and inert LGBs. Few of our aircrew had dropped live LGBs in training, let alone in combat. Additionally, the time spent training in jets

The biggest disappointment prior to deployment was that our aircraft were not yet configured with the new computer software referred to as D04. The latter gave the F-14D many excellent upgraded capabilities, including compatibility with the 2,000lb [907kg] GBU-31V(2) JDAM. We pushed very hard during work-ups to have the testing of D04 completed in time for the mission tape to be installed in our aircraft. Unfortunately, the test community and NAVAIR were around two years behind schedule on D04, and during the summer of 2002, while we were in the final stages of work-ups, they found further problems with the tape that caused still more delays. Our chain of command and my CO [Cdr Andrew Whitson] and I were not interested in taking an immature tape on cruise, so we left without what we considered to be an incredibly important war-fighting tool.

Once we arrived in the NAG, we quickly realized just how important JDAM had become to the overall prosecution of OSW. We could also see that our participation in the early phases of any fully blown conflict would be jeopardized without D04. VF-2 began pushing for the tape's accelerated release, despite the reluctance of certain senior leaders in the deployed air wings, and throughout the Navy as a whole, to have the tape installed in our F-14Ds in time for combat.

As in most bureaucracies, the more senior you get, the more risk averse you become – this was definitely the case with D04. Unquestionably, their reticence was grounded in many sea stories that told of the perils of bringing immature systems into the operator's hands, and then having the operators hurt themselves, or someone else, due to anomalies in a software load. However, we were convinced to back the tape by ex-SFTIs Lt Cdrs Keith Kimberly and Michael Peterson, both now serving with VF-2, who had been intimately involved with the development of D04. They knew more about the tape than just about anyone else in the Navy, and they knew that the F-14D units needed it.

Thanks to some backroom lobbying by Tomcat proponents in the Pentagon, the senior leadership in the Navy was informed that the remaining tape testing could be rapidly completed through the immediate release of funding allocated for future expenditure on D04 in the months to come. Upon receiving this news, the Vice Chief of Naval Operations directed NAVAIR to spend the money immediately and get the tape out to the fleet. VX-9 and the F-14 software support team at NAWC China Lake worked hard to complete the final bomb drops in record time – all flight testing had been completed within 30 hours of the request being received! A rapid-action response team was then assembled from personnel within NAVAIR, VX-9 and NAWC, and they were sent to the three deployed carriers operating F-14Ds to install the D04 tapes into the 30 aircraft flying from carrier decks in the NAG and the Mediterranean.

By the time the D04 rapid-action team arrived in-theater, four of the five carriers to be committed to OIF were on station in the NAG or in the eastern Mediterranean. And it was in the latter region that the ten Tomcats of VF-213, serving with CVW-8 aboard *Theodore Roosevelt*, became the first F-14Ds to be upgraded with D04 from February 12, 2003. The rapid-action team then headed south-east to the NAG and *Constellation*.

While VF-2 started its accelerated work-up with JDAM, the D04 team shifted its attention to the ten F-14Ds of VF-31, which had returned to the NAG with *Abraham Lincoln* in early February 2003. Having ended its OSW commitment in December and then made it as far east as Fremantle, Western Australia, on its way home, the *Abraham Lincoln* battle group was instructed to turn around and head back to the NAG to help bolster the coalition forces massing in the region. Although VF-31 was the last unit to receive the D04 mission tape, this had some benefits for the squadron, as CVW-14 staff officer Lt Cdr Jim Muse (who regularly flew with VF-31 as a RIO) explained:

Although the squadron had less time to learn the new system, the NAVAIR engineers that were installing the tapes knew exactly what they were doing by the time they got to us. When they had modified the VF-213 aircraft they had found that there were some interface issues between the tape and the jets' mission computers, but a quick-fix modification was put in place and the conversion timetable got back on track. Guys from both VX-9 and the Weapons Test Center also helped out, bringing their upgrade kits with them. By the time they got to VF-2 they had the modification pretty much down pat.

When the team came to us on *Lincoln*, they brought two fully functioning mission computers with them, which allowed the conversion of our aircraft to be accomplished very quickly. This was just as well, for in order for the engineers to upload the D04 software into the jet, they had to modify the aircraft's databus – the latter was a no turning back mod, which changed the databus for good. If anything had gone wrong with the upgrade, VF-31 could have been sidelined from participation in OIF.

All three F-14D-equipped units committed to OIF were keen to prove the jet's new precision-bombing capabilities in combat prior to the campaign commencing, and on the night of February 28, 2003, VF-2 got the chance to do just that. Flying the first F-14D ("Bullet 111" BuNo 159613) to drop a JDAM in anger was prospective squadron executive officer Cdr Dave Burnham:

I was leading a section in the Box when my RIO, Lt Justin Hsu, and I were instructed by our AWACS controller to carry out an RO strike on a military intelligence facility in Basra. This gave us the chance to use JDAM, which we had only been carrying in live form for the previous 72 hours. There was very little in the way of AAA opposing us that night, and I don't remember seeing anything exploding at our altitude – that may have been because I was so focused on not screwing up with this new weapon! We simply drove into the target area and hit the bomb release when the symbology cue appeared in the HUD. Afterwards, my RIO and I spoke about how anti-climactic the whole event was. We were both in agreement that the creators of JDAM had really taken the skill out of dropping a bomb, as you did not even have to see the target in order to score a direct hit.

The weather was good that night, so we got our own BHA footage of the weapon impacting the target through our LTS. There were other assets in the area filming this drop too, and satellite imagery showing the building both before and after was also relayed to the ship post-mission. There was much rejoicing following this mission both aboard ship and in the CAOC, as Fifth Fleet and CENTCOM were anxious to ascertain just how well the F-14D interfaced with JDAM. They needed it to work "straight out of the box," as the Tomcat was to play a major role in Shock and Awe.

Although this mission had proven that the F-14D could indeed employ JDAM effectively, crews found that the GPS-guided "wonder weapon" occasionally created more headaches than it solved when working with the CAOC. A senior NFO from VF-2 recalled:

We used JDAM to execute several ROs where various targets had been identified – usually by UAVs due to their ability to stay on station for a very long time with a low profile. Most of our OSW targets were mobile in nature because they were re-locatable (radar and communication vans, wheeled missile launchers, etc). Once a target was located, it was positively identified by the CAOC and the attack co-ordinates accurately measured for use with GPS-guided weapons. The Collateral Damage Estimate [CDE] was also performed in the CAOC to ensure that only the target would be damaged and no civilian casualties sustained. The RO was then called away.

One problem that we encountered was that the CAOC took a long time to work through targeting and CDE, and often the target would move. If it did – even only a few hundred meters – the whole process would start all over again. Aircrew were not relied upon to identify targets themselves at this time.

The CAOC loved JDAM because it supplied the identification of the target and the co-ordinates to the jet in the Box. JDAM turned the aircraft into a dump truck with no identification or aircrew guidance required. It basically let the CAOC be the shooter, not the aircrew. Unfortunately, JDAM did not then work for moving targets, and required new co-ordinates – which, in 2002–03, could not be generated by most TACAIR aircraft – if the target moved and then stopped. We spent a lot of time with LGBs on the aircraft and sensors on targets that we were not allowed to hit, or just dump trucking ordnance around waiting for a target to sit still long enough for the CAOC process to work. It was very frustrating for aircrew.

Once OIF started, many people up the chain of command just didn't get the fact that JDAM was a great weapon for fixed targets – buildings don't get up and move – but had very limited use for the fluid battlefield when trying to address the mobile and re-locatable target set. For the weapon to be of any use, the target had to be stationary and someone in the air or on the ground needed to provide high quality co-ordinates. The field-based systems capable of producing these co-ordinates

The plane captain of "Bullet 110" (BuNo 164345) gives the jet's canopy a final polish prior to declaring it ready to fly one of the last OSW missions ever undertaken. The aircraft is armed with both a GBU-12 LGB and a GBU-31V(2) JDAM, the latter weapon only being cleared for use by F-14D-equipped units following the installation of D04 computer software into deployed jets in February–March 2003. A Tomcat from VF-2 became the first F-14D to drop a JDAM in anger on the night of February 28 when the aircraft carried out an RO strike on a military intelligence facility in Basra as part of OSW. (US Navy)

were not widely available, or used. JDAM was good, however, if support was required on bad weather days, and it did allow the war to continue even if it was less effective against most CAS and SCAR targets [which tended to be dynamic in nature] in comparison with other weapons such as LGBs, Maverick/Hellfire missiles and 20mm cannon.

Conversely, we were tasked with hitting buildings with LGBs and then had to abort the mission due to poor weather between the target and the aircraft when JDAM was ideally suited for such strikes!

Final OSW TARPS Missions

Although JDAM now allowed F-14B/D-equipped units to get back into the precision strike business, Tomcat squadrons had continued to earn their keep throughout their unofficial bombing ban by conducting all-important TARPS missions over the future battlefield. VF-2 NFO Lt Cdr Michael Peterson flew a number of these sorties in the final weeks of OSW:

We generated approximately two TARPS missions every three days during OSW, and ramped up to one mission per day in the month preceding OIF. We did a lot of reconnaissance using our LTS pods and FTI, the latter allowing us to send LTS-generated pictures to ground stations via an encrypted UHF link, in addition to using our traditional wet film and digital TARPS capability.

A typical mission had up to 25 target areas planned for coverage. We would send either two TARPS jets or one TARPS and one LTS aircraft, depending on the target set and the environmentals. There were some common wiring and control panel locations for TARPS and LTS, so they were mutually exclusive configurations for the F-14. An aircraft could carry both pods in a ferry configuration but one of the systems would be inoperable. The TARPS/LTS combination was helpful in that the latter aircraft could find the target with the FLIR and direct steering for the TARPS aircraft to capture the higher-resolution footage with the pod's stand-off KS-153 camera. The KS-153 had great resolution, but a very small field of view. If the tasked co-ordinates were inaccurate, you could miss the target entirely.

On one particular OSW photo-reconnaissance mission flown just prior to the start of OIF, we launched as a section of TARPS aircraft. Our main tasking was to ingress via the Al-Faw peninsula south-east of Basra and flow up the eastern lines of communication to image various targets. We then flew south-west across several major towns and out through Kuwait to the east. The final targets would be the ABOT (Al Baṣrah Oil Terminal) and KAAOT (Khawr Al Amaya Oil Terminal) sites in the NAG.

The initial flow to the north focused on troop concentrations, and the assessment of their reactions to the increasing number of American and British troops and armor in Kuwait and naval vessels in the NAG. We wanted to see if Iraqi mechanized troops were massing to provide a focused resistance to the kick-off of OIF, or whether they were following the instructions that were dictated to them on leaflets dropped in the area in the preceding weeks. There were instructions printed on

This TARPS photograph, taken by the crew of a VF-2 jet during one of the last BHA sorties undertaken by the unit prior to OIF I, is typical of the photo-reconnaissance material that was captured by Tomcats overflying southern Iraq during a decade of such missions in OSW. Here, a series of command, control and communication bunkers have been knocked out by three JDAM dropped from F/A-18Cs assigned to CVW-2. This kind of BHA material was crucial for the CAOC when it came to assessing the effectiveness of the remaining Iraqi air defenses in the lead up to OIF I. (VF-2)

these leaflets that allowed us to assess the troops' willingness to comply once OIF started. It was a clear day, and we were able to image all targets in the area successfully.

We then turned south-west to line up our targets in the vicinity of As-Samawah and Nasiriyah. The next photo targets were particularly enjoyable in that they were located within known SAM rings. Sometimes, we had targets within the rings, and on other occasions the targets were the SAMs themselves. We used the TARPS imagery to ascertain the number of launchers and supporting equipment located within the rings and to see what their current configuration and operational capability might have been. We proceeded to "take frames and click ass" as we headed through the SAM rings, paying particular attention to our RWR and maintaining a good visual lookout within the section. With no "spikes" on the RWR, we pressed on toward our remaining targets on the way to Kuwait.

We would usually have both film and gas remaining as we arrived at the Iraq–Kuwait border, so we typically shot some additional targets of interest that had been "informally requested" by 1st Marine Expeditionary Force [MEF].

The competition for reconnaissance assets was intense in-theater during the pre-war build-up of forces. With all the UAVs and manned reconnaissance in the area, you would have thought that everyone would get their requests addressed. The reality was that many requests were dropped during the target nomination process as assets were assigned to targets at higher levels.

During one of our liaison trips into Kuwait pre-OIF to visit V Corps and 1st MEF in preparation for OIF, we discussed our TARPS and FTI capabilities, in addition to our FAC(A) plan for ground troop support. Both V Corps and 1st MEF mentioned that they were having trouble getting reconnaissance assets tasked to support their requirements. Specifically, they wanted reconnaissance footage of selected border crossing sites to determine the potential resistance they might meet, as well as photographic coverage of the Iraqi defensive posture in these areas. We mentioned that we flew over these areas on almost every officially tasked reconnaissance mission in-country, as well as during other LTS-equipped sorties. They gave us their "hot list" and we set up a back-door reconnaissance shop to get them the footage that they requested. We always shot the ATO-assigned targets first, but used any "extra" film to provide support directly to the Army and Marines outside of the normal channels. We provided this footage as often as we could.

On the way out of Kuwait during the flight back to the carrier, we went by the massive ABOT and KAAOT terminals in the NAG to shoot detailed footage of the Iraqi forces pre-positioned on them, and to get good overall pictures of platforms for pre-assault planning use by SOF.

Thanks to the close working relationship that VF-2 fostered with forces ashore in the final months of OSW, the unit proved instrumental in helping to set up the Kuwaiti Rapid Precision Targeting System (RPTS) station. The squadron also played a part in the integration of RPTS within the Coalition Forces Airborne Analysis Cell and Coalition Forces Land Component Command operations. The RPTS, data-linked with several mobile sites, allowed the sending and receiving of near real time TARPS and FLIR imagery from FTI-equipped F-14Ds of VF-2 and VF-31 undertaking sorties over Iraq.

Shared Workload

With the return of *Abraham Lincoln* into the NAG in late January, the OSW mission load was now split between CVW-2 and CVW-14. The staff officers from both air wings also began preparing for the invasion of Iraq as part of OIF, working out how best they could support the coalition ground assets in their quest to defeat the Iraqi Army. CVW-14's Lt Cdr Jim Muse was intimately involved in this late phase of OSW:

By the time we arrived back in the NAG, *Constellation* was also on station, so the workload for OSW was now split between two air wings. However, the CAOC was trying to expand our flight times in the Box. Previously, the air wing would send a single package of say ten to 12 jets into southern Iraq for between an hour-and-a-half and three hours at most, and that would be it for the day. In preparation for war, the CAOC wanted us to expand that footprint so that we would maintain a presence in the Box for several hours. That meant launching three complete packages during the time that we were the duty carrier, which was a lot of flying.

Rear Admiral John Kelly on *Lincoln* was designated as commander of Carrier Task Force 50 [which planned and conducted all strike operations mounted by carrier battle groups in the NAG]. A dedicated staff was then brought out to *Lincoln* and assigned the job of deconflicting the flight schedules, alerts, underway replenishments and all the other details to make sure that one of the carriers was ready to launch jets at any time. CVW-2 on "Connie" apparently volunteered to fly at night, so it switched to nocturnal missions early on in order to get everyone on the correct circadian rhythm. We were the day boat, and when *Kitty Hawk* showed up, the vessel's CVW-5 generally shared the daylight with us.

Following CV-63's arrival in the NAG, CENTCOM staged a conference to discuss the problems associated with airspace in the Persian Gulf, and how things would be changed leading up to the big fight. There were jets coming from Qatar, UAE [United Arab Emirates], Diego Garcia and Bahrain, as well as the three carriers in the NAG – literally hundreds of aircraft all heading north into Iraq on the first night of the war, and very little space in which to keep them apart. We didn't know if we'd get to use Saudi Arabian airspace at all. The "driveway" system employed in OEF would come into play again in OIF. This saw all aircraft using a 15-mile-wide corridor on which they were deconflicted vertically, laterally and by speed as if on a highway, with aircraft going north staying on one side and those going south staying on the other.

It took a while to iron this system out, as we all wanted to use the real layout as soon as possible, but with all of the civilian airliner traffic flying into and out of Kuwait and between Saudi Arabia and Iran, we couldn't fully implement the driveways we wanted to use during OSW. CAOC ended up "phasing in" the new scheme, which ultimately meant you never knew for sure if you were using the right one because the driveways were always changing! It drove me nuts because I was working on the "smart packs" for the air wing, and recreating the same "airspace" pages over and over again. I really felt bad for the guys on "Connie" because they were flying at night, and the airspace changes would usually take effect right in the middle of their vul window. It was pretty much a miracle we didn't have a mid-air collision over the NAG during the final weeks of OSW.

While the air wing staff officers were busily planning how carrier aircraft would conduct the coming war in Iraq, all three Tomcat units in the NAG (VF-154 had arrived with CVW-5,

Kitty Hawk heads west across the Indian Ocean on February 22, 2003, bound for the NAG – it would arrive on station four days later. A solitary Tomcat and Hornet sit over the waist catapults on Alert 5 readiness, both jets being manned and fully armed. The remaining aircraft in CVW-5 have been spotted on the flightdeck in a typical transit configuration, indicating that no flying operations were planned until the carrier reached its intended destination. Nine of VF-154's 12 F-14As can be seen in this photograph. (US Navy)

embarked in *Kitty Hawk*, on February 26, 2003) were involved in a series of precision strikes against targets in the Box as part of the battlefield preparation for OIF. On March 1, just 24 hours after VF-2 became the first F-14D squadron to drop JDAM operationally, the unit attacked a Roland SAM site and a truck-mounted P-18 "Spoon Rest" radar with four LGBs.

Two strikes in two days was a rarity for Tomcat squadrons at this last stage of OSW, as with so many aircraft now undertaking missions over southern Iraq, getting clearance to attack AAA and SAM batteries engaging aircraft in the Box was often difficult. Cdr Doug Denneny experienced this on March 11, noting in his VF-2 Commander's War Diary:

What a waiting game. Came close to dropping on a target in the H-3 airfield complex out west by the Syrian border. A long, long trip. Went after the Pluto EW [early warning] radar facility. We knew that we had to wait for clearance to drop – very frustrating after taking our jets hundreds of miles across enemy territory in lousy weather, which made tanking very challenging, only to fly over the target and return. After we left, the CAOC cleared USAF F-15E Strike Eagles to drop some LGBs. The Strike Eagles missed the target. Then they sent in some B-1Bs, who nailed it with JDAM. Probably my highest level of personal frustration this cruise.

We really hung our tails out there to get those bombs off, and we were turned around so that the USAF could blow it up (in my mind). The Strike Eagles blew it, so they covered their tracks with a JDAM strike and then flashed the pictures on the news. We have repeatedly gone all the way to H-3 and not had the opportunity to drop. We are thinking of making up some t-shirts that say, "I went all the way to H-3 and all I got was this lousy t-shirt!" At least we are keeping our sense of humor. Can't wait for this to get going.

Tomcats from VF-2 finally succeeded in bombing H-3 on the night of March 19, when aircraft from CVW-2, CVW-5 and CVW-14, as well as USAF and RAF jets, supported a crucially important long-range mission to both H-2 and H-3 airfields that also involved SOF elements on the ground. A VF-2 RIO who wished to remain anonymous gave the following account of the operation to the author:

> I was in one of two F-14Ds that launched off the boat to support a pre-planned strike against two airfields in the extreme west of Iraq. By the time it was over, the mission had lasted 8.6 hours from launch to trap. The operation included a ground [SOF] element, various TACAIR assets, including sections of A-10s rotating throughout the evening, F-15Es, F-14s of multiple flavors, AWACS and dedicated tankers.

VF-2's "Bullet 111" takes on fuel from a KC-10 during an OSW DCA escort mission for a TARPS jet from VF-2 in early March 2003. As the unit's Lt Cdr Dave Grogan explained, "The biggest limiting factor to the generation of sorties for OSW/OIF from a Naval Aviation perspective was the availability of airborne gas in-theater. We faced an extra challenge flying from 'Connie,' as it was positioned furthest south of all three carriers in the NAG. We were always tight on fuel, and in order to avoid running out, you had to be very efficient getting to the tanker for mid-cycle gas when heading both to and from the target." (US Navy)

> Our task was to show up after the initial ingress into the area and take over TACAIR assets control in support of the ground element as it headed toward its objective. We started to gain situational awareness of which phase the operation was in as soon as we could get radio contact a few hundred miles out. Having Link-16 JTIDS was invaluable, for we were receiving datalink to all the other assets in our package without having to use our radio or radar. It sounded like things were going according to plan, and that the assets were getting into position.
>
> As we got closer, we heard a VF-154 F-14 divert for unknown reasons. We weren't certain if he had been hit by ground fire or had suffered some other type of failure, but he had an immediate divert problem and left his wingman on station to wait for our relief. We quickly arrived on station and relieved the lone Tomcat as the FAC(A)s for the operation. He passed a turnover brief as he departed and we started getting a check-in from the other assets in the area.
>
> Our plan was to have at least one Tomcat on station to quarterback the operation, while the other one was headed to, or from, the tanker – located over the border – to refuel. If required, we would pass control to one of the F-15Es if we had both Tomcats off station. JTIDS was once again invaluable when it came to deconflicting all the assets in our package that were flowing in and out of our AOR [Area of Responsibility] as required and holding in the same relative area with their lights out. It also allowed us to release the sections on station when we saw their relief headed inbound without having to constantly query the AWACS as to the status of our support assets. JTIDS provided deconfliction with attacking assets while prosecuting targets on the ground and, most importantly, it let us accurately find the tanker when gas became critical so that we could drive him to us, if required.
>
> Initially, things went pretty smoothly for what seemed to be the first 20 minutes, so we headed to the tanker to top off and left our wingman on station to run the show. The tanker proved to be its own unique type of pain this evening. It was WARP [Wing Aerial Refueling Pod] configured, with a boom in the center. That way it could refuel the USAF assets with the boom or stream the baskets from the wingtips for the Navy jets, which probe-refueled. As we got to the tanker, we joined up

on the left side after it streamed the basket. We had to stay pretty well topped off on gas because we would need a significant amount to get to a divert field should we be hit, suffer mechanical difficulties or have a problem getting gas.

Three things soon became readily apparent as we closed on the tanker: 1) we were stuck with manual [non-boosted] throttles, so it was like driving a Mack truck without power steering, and therefore very fatiguing on the pilot who was trying to make the minor throttle corrections needed to get in the basket; 2) our probe light, which illuminates the otherwise pitch-black basket and lets the pilot see where he's putting the refueling probe in the endgame, was not working; and 3) there was a lot of turbulence at the tanker's altitude that was causing the wingtips to bounce up and down like a big lumbering bird. This caused the basket to sporadically rise and drop about 4ft [1.2m] at a time.

I stayed on NVGs in the back seat and tried to help talk the pilot into the basket, which was difficult to see without a probe light. Tanking on NVGs was also hard, as it seriously distorted your depth perception. My pilot got in on the third try, which made us very happy, as the closest divert field was nowhere as close as we wanted it to be.

As we returned to the AOR and switched back to the working frequency, things had obviously moved along to the next phase of operations. Our wingman was controlling the engagement of an AAA piece that had started opening up, and we had A-10s and F-15Es attacking vehicles that were moving in the general direction of the SOF units on the ground. We quickly arrived on station and received control from our wingman as he headed for the tanker. We were able to neutralize one of the targets pointed out by the ground element ourselves before the dust settled a little bit.

Now that the initial exchanges had subsided, we had a ten-minute break. The next target that the ground element wanted addressed was located in a revetment about one kilometer away from their position, but between them and their objective. We located and verified the target with the LTS and were able to see the ground unit with NVGs, but we could not see the target visually with the goggles.

The release cues from the LTS showed an aim point after we were past the ground element position. The SOF team wanted us to illuminate the target with an IR [infrared] pointer to confirm its position before dropping our LGB on it since it was close to their position, but we said that we couldn't visually put the IR pointer on it because we could not see it through NVGs, but that the LTS indicated that it was clear of their position. After making two passes, and asking for clearance to release, which we did not get because the SOF unit did not acquire us until we were past our release solution, we brought in an F-15E, which was operating a Litening pod. We talked the crew on to the target and verified the ground position with them as well.

The Litening-equipped F-15E had an internal IR pointer (unlike our handheld version, which required us to first see the target with NVGs), and its crew illuminated the target that they had located with their FLIR. We saw that the F-15E had the target about 800m [2,624ft] away from the friendly position, and we passed terminal weapon release control back to the ground element, which in turn authorized the F-15E crew to drop its LGB on its first pass, scoring a direct hit as verified by our LTS. We now headed for the tanker again for round two, as our wingman was on his way back after refueling. The next duel with the tanker took additional tries to get in the basket, and left my pilot with a serious cramp in his throttle [left] arm.

As we arrived back at the fight to join our wingman, the ground unit had reached its objective and was ready to egress the area, which involved moving across a major "enemy line of communication" – a road. We were tasked with sanitizing the area where the crossing was to occur, searching out any possible resistance. There were a few vehicles moving along the road, but none were presenting a direct threat to the ground element. We then caught sight of a fast mover on NVGs doing about 80mph [129km/h] – easily three times the speed of any other traffic observed – and headed directly toward the location of the ground unit. We passed the information to the ground controller, who told us to take the vehicle out.

We rolled in in the same direction that the target was heading and tried to lead it with an LGB, but he was seriously moving. The LGB hit about two vehicle lengths short but the blast forced the driver off the road. We observed four individuals get out of the vehicle once it stopped, and they ran into a ditch next to the road. The final person leaving the vehicle had a pretty good heat signature on the FLIR, and he appeared to be severely injured based on the way he rolled into a ditch on the roadside. We watched as the group continued to head away from the damaged vehicle and toward the ground unit. Before we could come back around to see how close the group was getting to the crossing site, an F-15E "cleaned up" the scene with his own LGB and this group was no longer a factor. We covered the rest of the egress by the ground unit, directing other fires as required until we proceeded off station.

There was one more trip to the tanker left, and if we topped off, we could make it all the way back to the boat at high altitude on a conservative profile. By the time we reached the tanker my pilot was flagging. It was early in the morning hours and still totally dark. We had been pumped up during the action and the flight had been fairly fatiguing. The basket was really bouncing and we were low on gas. We made the decision to take the jet down to the lowest possible fuel state before heading off to the divert airfield (where we could be stuck for days). After many attempts, and a severe forearm cramp from running the manual throttles all night, my pilot steadied the jet about 10ft behind the basket for about a minute, and I asked, "Hey, you know we've only got about 400lb of gas to play with here before we gotta go?" He replied, "Yeah, I know, but I've got to give it a little break or I'm not gonna make it in."

With 200lb to go (enough for another ten seconds of trying to get in), my pilot made one more try and got into the basket. The fuel gauge touched the bingo number just as it reversed and started showing good flow from the tanker into our tanks!

Now we had the long ride home to end in a Navy-specific event – the night trap. En route, we consumed two mocha-flavored power bars and a bag of spicy teriyaki beef jerky – a meal that would not be repeated during the war! Once we were feet wet, my pilot levelled with me and said that he was pretty tired – something that a RIO never wants to hear, but I appreciated the honesty. I implicitly trusted him with my life, and had little worry about his ability to get aboard. It might

Armed with a pair of GBU-16s, "Knight 103" (BuNo 161293) briefly dips below flightdeck level after launching from *Kitty Hawk*'s waist catapult four in full afterburner at the start of one of the last OSW missions on March 18, 2003. That same day, VF-154 sent five aircraft to the "beach" at Al Udeid for a 21-day deployment following a CENTCOM request to Fifth Fleet for Tomcats to support SOF in the coming invasion of Iraq. (US Navy)

just take us a few tries. I passed him two Red Bulls to see if they would help. He had one right away and saved the other for closer to the boat, and they really did the trick. I don't remember if it was an OK 3 wire, but we were aboard safely on the first pass.

Elsewhere that same night, two more Tomcats from VF-154 and eight jets from CVW-5's three F/A-18C squadrons (VFA-27, VFA-192 and VFA-195) hit Republican Guard barracks, HQ buildings, mobile missile launchers and AAA and fixed missile sites in and around Basra with LGBs (two GBU-16s from a single Tomcat) and JDAM. The pilot of the F-14 that dropped ordnance on the Basra targets was Lt(jg) Tom Lunsford, who also happened to be VF-154's least experienced Naval Aviator. Shortly after returning from the mission, he told *Stars and Stripes* reporter Kendra Helmer, "I was telling myself 'nothing's different [on this mission]; I'm going to do this like every other time I've flown over Iraq.' All we hit was pretty much equipment. It would've felt different had we seen personnel. It was so antiseptic up there."

Lunsford's RIO was Lt Cdr Jerry Morick, who explained to Helmer that "Just before we walked, we got passed new targets, but this did not bother Lt(jg) Lunsford. I was very impressed by how cool he was. We obviously know we're in a transition period now. We're prepping the battlefield for operations coming up."

One of the very last OSW sorties flown by Tomcats took place in the early hours of March 21, and during the course of the mission VF-2's Cdr Doug Denneny was finally cleared to expend ordnance on a target in southern Iraq. His diary entry noted: "Dropped for the first time, [flying] A/C 104 with [Lt Cdr Kurt A] Webstur Frankenberger. Went out to H-3 and levelled a radio HF/MF [high frequency/medium frequency radio] site outside a small city called Ar-Rutbah. The war is going to start later today – we are currently still doing OSW, but it will all change soon. Ground troops going over the line at 0600."

Less than 24 hours after attacking H-3, Cdr Denneny and Lt Cdr Frankenberger led a division of Tomcats from VF-2 into the SAM engagement zone protecting Baghdad. These jets, which were part of a strike package from CVW-2 that also included F/A-18Cs from VFA-137, VFA-151 and VMFA-323, were the first non-stealth coalition aircraft to target Baghdad during the opening night of OIF's "Shock and Awe." After more than a decade, OSW had finally ended.

Lt Cdr Jerry Morick from VF-154 checks the guidance vanes of the two 1,000lb (454kg) GBU-16B/B LGBs secured to his jet's bomb pallets prior to launching on one of the last OSW missions to be flown by coalition forces on March 19, 2003. Morick and his pilot, Lt(jg) Tom Lunsford, targeted an "Iraqi intelligence facility and mobile SAM launchers" with their ordnance during this mission, which also involved another VF-154 jet and no fewer than eight F/A-18Cs from CVW-5. After expending the first bombs dropped by VF-154 on this deployment, Morick told embedded reporters onboard *Kitty Hawk*, "We obviously know we're in a transition period now. We're prepping the battlefield for operations coming up." (US Navy)

CHAPTER 8

BALKAN BOMBERS

The Tomcat had first been blooded as a bomber over the Balkans in September 1995, and almost four years later the jet would truly prove its worth in the precision strike role – both by delivering ordnance and providing target guidance for other aerial assets – in the same theater.

Codenamed Operation *Allied Force*, the NATO bombing campaign against the Federal Republic of Yugoslavia was undertaken in an effort to stop human rights abuses in Kosovo. The strikes, which lasted from March 24 to June 10, 1999, were not approved by the UN Security Council. *Allied Force* was the second major combat operation mounted by NATO in the Balkans, coming almost four years after the bombing campaign in Bosnia-Herzegovina.

CVW-8 had been at the forefront of Operation *Deliberate Force* in August–September 1995, and, as luck would have it, the air wing would play a key role in *Allied Force* (US military operations during this campaign were codenamed *Noble Anvil*) as well. In 1999, however, thanks to the arrival of the LANTIRN pod in the fleet, the Tomcats embarked in *Theodore Roosevelt* would be the "stars" of the campaign, rather than the "walk on extras" they had been in *Deliberate Force*.

Robert K Wilcox summarized the role played by VF-41 in *Allied Force* in the introduction to his outstanding history of this campaign, *Black Aces High*:

Kosovo presented a new kind of conflict for modern fighter squadrons – a precursor to what would happen later in Afghanistan and in other terrorist countries the US, in its subsequent War on Terrorism, might bomb. It wasn't a conventional conflict in which fighters battled other aircraft for control of the skies, and air operations supported ground invasion and conquest. In Kosovo, fighter squadrons, usually pristine in their role as fencers, got down and dirty and dropped bombs. Kosovo marked the first time in the history of modern warfare that victory was won without the introduction of troops on the ground. Aircraft reigned supreme – but not as easily as has been portrayed in the press or by the pundits.

VF-41 "Black Aces" probably played a greater part in that 1999 victory than did any other squadron in-theater, whether air force or navy. Because of its superior performance, sophisticated equipment and two-man crews, who took it upon themselves to do something extra, the Tomcat and its flyers distinguished themselves. Basically, they proved in Kosovo to be the one US asset

VF-41's color jets (BuNo 161295 in the foreground and BuNo 162608) fly over hilly terrain as they go in search of targets in Kosovo in April 1999. Both aircraft saw considerable action during *Allied Force*, with BuNo 162608 also being used by VF-41 in combat over Iraq and Afghanistan in 2001. (via Robert K Wilcox)

that could both find hiding enemies and destroy them with smart bomb accuracy. And doing that eventually became the main mission of the navy there, despite daunting problems.

Flying ageing Tomcats and faced with having to locate Serbian troops operating covertly in a mountainous land much like Afghanistan, and with almost no help from ground spotters, VF-41 aviators spearheaded for the navy the creation of new ways to pinpoint, identify and destroy enemy troops and weapons. These were tasks that fighter crews had seldom had to do before. The "Aces" had to break rules and frequently go in harm's way in order to be successful. In the beginning, there had been resistance to their taking license. But they eventually had done so well that for the first time in aviation history, a fighter squadron – theirs – was awarded the Wade McClusky Trophy, the navy's premier bombing honor. The award, named for a World War Two dive-bomber pilot and post-World War Two admiral, had been won previously only by bombing squadrons. The award was quite a coup, but it was hardly indicative of the hard work, mis-steps, pain, sacrifice and dedication needed to win it.

VF-41's sister-squadron, VF-14, proved itself to be just as effective in the precision bombing role over Serbia. Indeed, according to its command history for 1999, "VF-14 provided only 2.4 percent of the strike assets in *Allied Force*, but was responsible for the assessed destruction of 12 percent of all targets hit in Serbia, through own ordnance delivery or the control of other aircraft ordnance." Both units benefited from recent updates to their venerable F-14As, as VF-14's Lt Jim Stanley explained on the eve of the 1999 deployment:

Although VF-14 and VF-41 flew primarily air-to-ground missions during the campaign over Kosovo, both units completed plenty of CAP and DCA sorties too. These were flown in protection of EA-6Bs flying anti-SAM missions, as well as Hornets and Tomcats undertaking precision strike sorties – TARPS escort was also routinely performed. NATO was concerned about Serbian fighters attempting to engage its strike aircraft, but this threat quickly evaporated after USAF F-15s downed a handful of MiG-29s in the early stages of *Allied Force*. F-14A BuNo 158637 of VF-41 is being carefully taxied forward on to the launch shuttle for bow catapult two prior to flying a CAP on April 17, 1999. After spending 25 years serving with training, test and evaluation and reserve units, BuNo 158637 finally reached the fleet in late 1998. The aircraft ended its long career literally with a bang by participating in *Allied Force* with VF-41 and OEF with VF-211 (in 2001–02). (US Navy)

Our Tomcats have received numerous modifications in the last year. The latest change is new software for the jet's AWG-9 radar control system, which has brought added capabilities to a system designed in the 1960s. One of the most exciting improvements has been the incorporation of the new DFCS that has replaced the analog system in use since the aircraft's inception. All pilots agree that the Tomcat is now more maneuverable and has a crisp response to pilot control inputs. Aside from the DFCS modification, our jets also have a new, more advanced Radar Warning Receiver that gives the "Tophatter" aircrew earlier and more accurate indications of enemy radar trying to detect and lock on to their aircraft, and BOL chaff dispensers to defeat SAM guidance.

Another significant improvement is the upgraded software package for the LANTIRN pod, which allows the F-14 to more accurately employ weapons, as well as record more accurate target co-ordinates. "This is perhaps the greatest new bit of technology for us to put to use," said squadron CO, Cdr Ted Carter. Using the newly installed Fast Tactical Imagery system, F-14A aircrew can transmit digital images captured from the LANTIRN pod video and send them to another Tomcat or to the battle group commander. These images can be used for immediate attack by another aircraft, for bomb-damage assessment, for locating targets of opportunity or simply for determining precise co-ordinates for targeting by other weapons.

Post-deployment, Lt Stanley produced the following report detailing VF-14's exploits in *Allied Force* for the unit's command history:

> Months of pre-cruise preparations and training finally paid off for the "Tophatters" of VF-14, who led the first navy air strike of Operation *Allied Force*. During the first week of strikes over the former Yugoslav Republic, navy forces met with significant opposition, including numerous launched surface-to-air missiles and AAA fire, but successfully countered the threat and executed each strike. In the weeks that followed, the battlefield picture changed and the NATO campaign moved to another level, where "Tophatter" FAC(A) aircrews took on the assigned task of striking Serbian military ground forces deployed throughout Kosovo. The experienced guidance of strike leaders, solid planning and respect for the capabilities of the former Republic of Yugoslavia's ground forces resulted in zero damage to navy aircraft and the smooth control of navy and air force aircraft over the hostile skies of Kosovo.

Knowing that NATO had ordered bombing to begin on Serbian forces in Kosovo in late March, no one was surprised when, just a week after leaving Norfolk [Virginia], the "Tophatters," with CVW-8, were stationed off the coast of Italy with live ordnance loaded on to their F-14A Tomcats. The first navy strike, led by VF-14's CO, Cdr Ted Carter, took place on the night of April 6 against the Pristina POL [petroleum, oil and lubrication] facility. Despite the heavy concentration of AAA and numerous SAMs launched from all around, the "Tophatter" aircrew, along with Tomcats from VF-41 and F/A-18Cs from CVW-8's VFA-15 and VFA-87,

Cdr Ted "Slapshot" Carter was the much-revered CO of VF-14 during *Allied Force*, his command of the squadron earning him the CNO's Atlantic Fleet Vice Admiral James Stockdale Award for inspirational leadership. A veteran of 125 combat missions over Bosnia, Kosovo, Kuwait, Iraq and Afghanistan, Carter accumulated 6,150 flying hours in the F-4, F-14 and F/A-18 and 2,016 carrier landings (a record among all active and retired US Naval Aviators and NFOs). (US Navy)

fought their way into the target area and delivered their weapons with deadly accuracy. All the aircraft made it safely out of the target area and returned to the waiting carrier later that evening. During the mission, the location of several SAM and AAA sites were identified, and this information was passed on to the second strike package that was launched that same night.

"Tophatter" aircrew subsequently stayed busy from early morning until late in the evening, planning strikes against fixed targets in order to degrade the former Republic of Yugoslavia's command and control structure. The success of those strikes was a direct result of careful planning by all the navy Strike Leaders. They used up-to-date intelligence reports in order to plan an efficient strike on their assigned targets, while at the same time minimizing the danger to navy aircrew. The LANTIRN pod was the critical asset most called upon for those missions. Capable of identifying targets from considerable distances, the LANTIRN pod was used to find targets in daytime and at night and guide LGBs to impact with incredible accuracy.

As navy strikes continued, the need to locate and target Serbian military forces became apparent. FAC(A) aircrews from VF-14 and VF-41 were sent over Kosovo to do the job, under the protective watch of the EA-6B Prowlers of VAQ-141. "Tophatter" F-14s routinely managed the battlespace while they located hostile targets. Once they found one, they called up sections of F-14s or F/A-18s from CVW-8, or air force aircraft such as A-10 Thunderbolts and F-16 Fighting Falcons, to deliver their weapons. The ability of the Tomcat FAC(A) aircrews to quickly pass targeting information to other aircraft, and to efficiently flow those aircraft into a target area, was the result of many hours of pre-flight planning and airborne flexibility. The value of a Tomcat FAC(A) quickly became evident against rapidly advancing Serbian forces in Kosovo, as Cdr Ted Carter stated:

> The Tomcat has truly come of age. After 27 years of service to the fleet, this has been the aircraft's and the aircrew's greatest contribution. Our success in *Allied Force* was a total team effort that has proven the value of a two-seat cockpit. We flew in sections, one aircraft serving as an escort for the

CVW-8 flew a fair percentage of its strike missions during *Allied Force* at night, with Tomcat and Hornet pilots relying on NVGs to aid them in the darkness as author Robert K Wilcox explained in *Black Aces High*. "Night was their ally. They had the new light-enhancing 'cat's eye' NVGs, small, cylindrical, opera glass-type protuberances that dropped down from the top of their helmets like dark glasses on a baseball player's cap, and gave them a bug-like, robotic appearance. But they intensified ambient light. Night became day, impressively so. Their target-finding display screens, like those in the new LANTIRN systems in the Tomcats, used more-sophisticated infrared technology to see up to 20 miles (32km) through the darkness and lock on their PGMs with accuracy described as 'pin-point'." This dramatic photograph was taken on April 12, 1999, at the start of *Theodore Roosevelt*'s second week on station in the Adriatic Sea. (US Navy)

other. Each F-14 usually carried four bombs, which we used for both striking a target ourselves and for marking a target for other strike aircraft. The FAC(A) was like a quarterback on a football team, seeking out and identifying targets, ushering strike aircraft to the scene, recommending the type of ordnance for a particular target, ensuring the pilots under their control recognized potential terrain hazards, and providing them with run-in and recovery headings.

"Tophatter" Tomcats expended more than 395,000lb [179,168kg] of ordnance on various targets in support of *Allied Force*. Cdr Carter added, "Our tireless ordnancemen loaded more than 800,000lb [362,873kg] of ordnance in two months in VF-14 alone. Combined with an outstanding maintenance effort, the work of our troops was most impressive. 'Tophatter' aircrews led CVW-8 in strike missions as FAC(A)s. By using their state-of-the-art LANTIRN pod to the fullest in this conflict, the 'Tophatters' of VF-14 helped to prove that the F-14 Tomcat is one of the most accurate and lethal aircraft in the air-to-ground arena, and truly the Commander-in-Chief's airborne platform of choice."

The command history also highlighted a number of VF-14's "firsts" and successes from its 1999 deployment as follows:

VF-14 FAC(A) aircrews inflicted significant damage to Serbia's deployed forces in Kosovo. Unrelenting and effective targeting while operating in known SAM and AAA envelopes was key to bringing about an end to the Kosovo campaign. VF-14 developed stand-off weapons delivery tactics to provide an additional safety margin for shooter aircraft. VF-14 supported 190 external weapons, including LGBs and LMAVs, guiding them to direct hits.

The "Tophatters" developed innovative and exceptionally effective FAC(A) tactics over Kosovo, which were subsequently employed in southern Iraq during OSW. These were duly exported to other battle groups. It perfected mixed section employment tactics using the F-14 LANTIRN targeting system and the F/A-18's LMAV. VF-14 recorded a perfect 100 percent success rate while guiding 35 LMAVs. These tactics were passed down and employed extensively by the *Constellation* and *Kennedy* battle groups, which have recorded similar results in OSW.

VF-14 also provided the first operational combat testing of the Infrared Zoom Laser Illuminator Designator pointing device by FAC(A) crews during *Allied Force*. It proved the usefulness of this device by demonstrating instantaneous designation of tactical targets to NVG-capable platforms, resulting in immediate target identification and weapons delivery at night. The "Tophatters" were also the first Tomcat squadron to use gyro-stabilized binoculars for enhanced visual acquisition and target area description. FAC(A) aircrew validated their effectiveness by locating tactical targets quicker, achieving rapid targeting and providing more available time for armed reconnaissance and weapons employment.

The squadron was responsible for the most sorties and hours logged with LANTIRN-equipped jets during a deployment, recording the best deployed pod availability.

Theodore Roosevelt underway at speed during the vessel's 1999 Mediterranean/NAG deployment that saw CVW-8 attack targets in the Balkans and southern Iraq. The vessel has experienced more combat than any other carrier serving in the modern US Navy since being commissioned in October 1986. (US Navy)

The "gunner" from VF-14 (squatting at right) keeps a close eye on his "ordies" as they check the guidance vanes of a GBU-24A/B Paveway III LGB recently shackled to the port BRU-32 bomb rack of a "Tophatters" F-14A on April 14, 1999. First dropped from a Tomcat in combat by VF-32 and VF-213 during *Desert Fox* some five months earlier, the "bunker-busting" GBU-24 was developed for use against well-defended, high-value targets. Fitted with a more expensive guidance kit than that seen on other LGBs cleared for use by the F-14, only ten GBU-24s were dropped by VF-14 in *Allied Force*. By comparison, the unit expended 181 GBU-12s, 114 GBU-16s and ten GBU-10s, as well as 116 Mk 82 and 18 Mk 83 "iron" bombs. (US Navy)

VF-14 was the first squadron to operationally test and utilize the Tomcat Tactical Targeting LANTIRN software. This led to the recent clearance to use designated target co-ordinates via LANTIRN for GPS-guided munitions, allowing for real-time targeting of high threat mobile tactical targets with GPS-guided stand-off munitions previously reserved for fixed targets.

The "Tophatters" established the F-14 as the FAC(A) platform of choice in two separate theaters of operation during CVW-8's 1999 deployment. Effective use of LANTIRN to locate and identify enemy armor and emplacements and superb co-ordination of all airborne strike assets combined to showcase the success of the F-14 Tomcat over Serbia and Iraq. Given the task-management capability of a two-seat platform, airborne results greatly exceeded those achieved by sister service USAF A-10 and F-16 FAC(A)s. In two separate warfighting areas of responsibility, the superb capability of the F-14 was demonstrated.

By creatively managing the daily combat flight schedule, VF-14 increased the availability of critical FAC(A)-qualified aircrew. Thoughtful scheduling enabled VF-14 to provide the maximum number of FAC(A)-qualified sections, resulting in significantly enhanced effectiveness. In addition

An armorer from VF-14 fits wire lanyards to a GBU-12 attached to a jet's starboard BRU-32 rack on May 4, 1999. One lanyard would go forward to the front of the bomb to activate the laser guidance package when the weapon was released from the jet, and a second wire was the strake delay deployment line. The latter prevented the GBU-12's pop out-fins from extending while the bomb was still attached to the aircraft or within its laminar flow immediately following release. When the LGB fell away, the lanyards (wired to the pylon) would be pulled out, activating the mechanical links within the weapon governing target guidance and fin deployment. (US Navy)

VF-14 flew 1,170 sorties with pod-equipped F-14As, maintaining an impressive 94.8 percent pod reliability rate.

VF-14 developed enhanced LGB tactics to counter the inclement weather often encountered in the target area during *Allied Force*. It modified the "Trombone" [target circling] tactic to maximize stand-off and minimize podium effect by executing a left-hand Trombone. This modification means that the LANTIRN pod has complete and uninterrupted coverage during the entire weapons delivery phase and eliminates masking concerns. The tactic was debriefed and submitted as a change to the F-14 Tactical Manual.

FAC(A)s perfected five variations of co-ordinated "buddy bombing" and successfully executed all of them in combat. This allowed VF-14 to act as a force multiplier by using LANTIRN to guide LGBs into tactical targets that other aircraft were unable to locate with their own onboard sensors. With the mark on the deck, the FAC(A) provided talk-ons for other aircraft, enabling them to follow up with visual deliveries on additional targets.

to aircrew management, aircraft were also configured to maximize efficiency. Two aircraft were designated FAC(A) platforms and configured as "quad-bombers." These aircraft maximized the number of LGBs a FAC(A) could deliver (four) as a "mark" and maximized bring-back capability. The remaining squadron aircraft were configured as "dual bombers." The combination of the FAC(A) with four GBU-12s and the escort with two GBU-10s resulted in unparalleled airborne flexibility.

The unit managed ordnance loading based on a real concern for ordnance bring-back and preservation of ordnance assets. Jets routinely carried two 2,000lb [907kg] GBU-10s, and aircrews were never forced to jettison ordnance for weight considerations. VF-14 carried and delivered more GBU-10s than any other navy squadron, and 50 percent of those carried were recovered. VF-14 recorded an 88.1 percent combat success rate with the GBU-10. Due to a detailed LANTIRN and delivery techniques training course conducted within VF-14, the squadron enjoyed an increasing ordnance delivery success rate throughout its deployment. The initial overall rate was 65 percent during the first week of *Allied Force*, and this eventually swelled to more than 82 percent.

FAC(A) Focus

The FAC(A)s running SCAR missions had assumed prominence in *Allied Force* when it was realized by NATO that its original strategy of bombing non-vital fixed or stationary targets in and around the Serbian capital of Belgrade was not stopping the massacring of ethnic Albanians in Kosovo. Robert K Wilcox recounts in *Black Aces High*:

> In view of the failure to stop the slaughter, there was increasing talk among the war-runners about going after the Serbs' "fielded forces" – their troops, tanks and mobile artillery in Kosovo proper. The air force, with generally more aerial assets than the navy, was concentrating on the larger targets in the north. Naval air [CVW-8], now that it was there, was being sent to Kosovo.
>
> VF-14 and VF-41 had three FAC(A)-trained aircrews each, and like their CAG, Capt Dale Lyle, and the strike leaders in CVW-8, they all wanted to continue flying the large Alpha-type strikes against fixed targets they had initially mounted when they first arrived in the Balkans. They had trained hard to perfect these strikes at places such as Fallon as an air wing, and they felt most comfortable flying them. They did not see how, given the easy-to-hide-in Kosovo terrain and the NATO-imposed altitude floor of 20,000ft, below which they were not allowed to fly, they would be able to even find the smaller, mostly mobile targets, let alone bomb them. At 20,000ft (imposed to minimize NATO air losses to SAMs and AAA), a tank, even in the LANTIRN scope with the FLIR's considerable magnification abilities, is little more than a spot. A soldier would be imperceptible. And even if seen, such targets could run and hide.
>
> But CAG had been directed to explore the mission. In other words, it was his problem to solve. He therefore tasked Lt Cdr Brian Brurud, VF-41's bombing expert, to fly with his skipper, Cdr Joseph Aucoin, that night in what was technically called a SCAR mission but was listed on the squadron's April 7 daily flight log as a "QB", probably because of its preliminary nature. The two letters stood for quarterback – go out and find the Serbs' fielded forces and then "quarterback" their destruction with bombers that would be put at the QB's disposal.
>
> The air force was doing some SCAR already, but obviously with little success. The A-10s were hampered by the altitude restrictions and lacked sophisticated detection and bomb-targeting and guidance gear like the F-14's LANTIRN. The Serbs had deployed SA-6 and SA-10 SAMs and AAA into Kosovo, [a location that] was stormy, mountainous, covered with trees and pocked with caves, villages and farms, any one of which could conceal tanks and SAM launchers. Finding and destroying the Serb fielded forces, if it could be done at all, was going to be both hard and dangerous.

Results for the Tomcat FAC(A) crews were not encouraging to start with due to a combination of poor weather, restrictive operating heights and a general unfamiliarity with the mission in CVW-8. However, according to Robert K Wilcox:

> Brurud and Aucoin resolved anew to make SCAR work – this despite continued resistance from those in the battle group who either still believed it was impossible under current conditions, or resented the departure it represented from established navy ways of doing business.

Both VF-14 and VF-41 adopted the "quad bomber" configuration once they started flying SCAR and CAS missions in *Allied Force*, with aircraft usually carrying two GBU-16s (seen here to left) and a pair of GBU-12s (one visible to right). Unguided Mk 80 series bombs were also routinely used as well. (US Navy)

None of this bothered the "Black Aces." Their main problem at this early stage in *Allied Force* was that little in the way of SCAR–FAC(A) procedures or tactics had been standardized or even worked out. This was forging on the fly, trial by error, test and discard. They had bad intelligence. Their equipment, like the LANTIRN pod, while good, was not what was specifically needed for the hunting. The normal picture in the FLIR didn't have enough magnification to see troops and distinguish vehicles at the heights they were restricted to. And the narrow, magnified setting was like looking through a straw. If you lost the target, it was terribly hard to find again.

They only had two pairs of gyro-stabilized binoculars, which meant that only one or two aircraft (if they wanted to split the pairs) could hunt with them at one time. SAMs remained a very real danger, and inherent in SCAR was the loiter. You had to circle, stay in the area, check things out. Speed and maneuverability – the basis of SAM avoidance – were not an option when looking for targets. You had to get slow, dip down, hang around. "My biggest concern was that we were going to get bagged out there as FAC(A) escorts," recalled VF-41 XO, Cdr Jim Bauser, who, not being a FAC(A), would do a lot of the escorting. "It was a risky thing. This wasn't swooping in, dropping bombs and getting out. We're in the same piece of sky – very predictable. The FAC(A)s and FAC(A) escorts hung it out – no ifs, ands or buts about it."

Fellow "Black Ace" Lt Brian Fleisher (also not a FAC(A)) wrote in a letter home, "Going after the troops and the tanks is a totally different ball game. Those things move around a lot and it is hard to know where they are, so we have to do what is called close air support with forward air controllers. Basically, now we have to send someone in to look for the tanks and APCs and then they call in the strikers to take them out – very difficult to do from high altitude. Initially they thought the Tomcats could do it without LANTIRN pods, but we found it to be harder than we thought. So, what we have to do now is go lower into the SAM envelope. Sounds like fun."

The plan eventually devised by Brurud and Aucoin for performing SCAR missions over Kosovo involved giving two FAC(A)-manned Tomcat sections (totaling four jets) strike packages for use as bombers. The strike packages, medium-sized with four to six jets each, would be made up of the air wing's Hornets. They would arrive at different times and be nearby in order for the FAC(A)s to have at least one package available to them at all times. They'd also always have secondary targets assigned in case the FAC(A)s couldn't find any for them to bomb at the FAC(A) location. The FAC(A)s would be protected by a Prowler, as were the strike packages. If they found a target, they'd quarterback the strikers in. If they didn't, the strikers would go on to their secondary, fixed targets.

Typically, the FAC(A) sections would launch for separate two- to three-hour hunting missions, with the first section being relieved by the second so that there would be continuity. The section returning would update the section arriving. They'd start within southern Kosovo, where the killing was concentrated. FAC(A) crews would use "whatever means available" to identify the target, which meant UAVs, binoculars, LANTIRN, other aircraft in the area and secret sources. They'd need stacked strikers for when they found targets and Prowlers for defense.

VF-14's "Camelot 201" (BuNo 158624) launches on a DCA/CAP mission for an air wing strike on April 8, 1999, CVW-8 hitting fixed strategic targets in and around the Kosovan capital Pristina during the first week of combat operations. As with a number of the F-14As flown by VF-14 and VF-41 at this time, BuNo 158624 was an early-build airframe with few flying hours following years of service with test and evaluation units followed by assignment to Reserve-manned VF-201. When the latter unit transitioned to F/A-18As in early 1999, six of its jets were sent to CVW-8, including "Camelot 201". Aside from seeing combat in *Allied Force*, this aircraft dropped bombs with the unit in OSW and OEF in 2001 and then participated in OIF I with VF-154. (US Navy)

Airfield Strike

Before the SCAR mission really got into full swing, however, CVW-8 was presented with the opportunity to carry out another traditional Alpha strike. The US Army had sent 24 AH-64 Apache attack helicopters to Albania in early April in the hope that they would act as a deterrent to Serbian forces in Kosovo. This move provoked a response from the Serbs, as Vice Admiral Daniel J Murphy Jr, commanding Sixth Fleet, explained to the Senate Armed Services Committee on October 13–14, 1999:

"The Serbs moved a significant number of air-to-ground aircraft into Podgorica's [Golubovci] airfield [on April 14]. This was only 30 miles from the Albanian border. When we detected that move, Gen [Wesley] Clark [NATO's Supreme Allied Commander Europe] said, 'I have to have that airfield taken out now. We cannot afford a strike, even an ineffective strike, against Task Force Hawk [the codename for the Apaches]'."

When the USAF stated that it could not hit the airfield in less than 72 hours after it had been designated a target for attack, Vice Admiral Murphy confirmed that CVW-8 could undertake the mission before dawn the following day (April 16). With US Navy pride on the line, Capt Lyle set about planning for a strike on a major airfield. "There was clearly a sense of urgency," he recalled. "My impression was that the mission was politically driven to some degree. Maybe there was some concern [within Sixth Fleet] that they were going to lose approval at some point when someone [at NATO] lost the stomach for it for whatever reason."

VF-41's CO, Cdr Aucoin, would lead this mission, and because it had been so hastily ordered it was not on the daily ATO (which had to be approved by all NATO member countries) that was written days in advance. According to VF-41's XO, Cdr Jim Bauser, many in CVW-8 believed that the Serbs "had our basic air plan of when we were flying and when we weren't, because when the air wing wasn't flying, they'd take off in their [G-4] Super Galebs, their air-to-ground attack planes, and go whoop it up [in Kosovo]. And then as soon as we got airborne, they'd run like hell and hide."

Robert K Wilcox described Golubovci airfield as:

... a fortress. It sat at the base of a mountain. Its runways and aprons were home to MiG-21s, the little Super Galebs – two-seat, single-engined jet bombers that resembled navy trainers – and possibly other aircraft. There was an SA-6 school at one end, four large hangar buildings near its center and fuel and ammunition storage facilities. It was guarded by SAM batteries, some with target-tracking radars. But the crown jewel of the base was a large underground hangar encased in reinforced cement and dug into a mountain. It was approximately half-a-mile long and curved in the middle, with giant cast-iron and steel doors at both ends and small shafts for ventilation visible on the hilly mountainside. The planes that had recently been moved to the field, mostly Super Galebs, had been parked nose to tail in the underground hangar. It was believed that there could be as many as 35 in the mountain vault, along with stores of fuel and ammunition. An army of technicians and workers attended the planes and stores inside.

Having experienced the adverse climatic conditions prevalent in the Balkans since commencing *Allied Force* missions on April 6, Cdr Aucoin decided against a night strike. "Traditionally, what

Podgorica (Golubovci) airfield was successfully bombed by VF-41 on April 15 in an attack that reportedly destroyed 24 G-4 Super Galebs packed into an underground hangar. This pre-war photograph was used to brief crews tasked with striking the base at dusk. (US DoD)

happens there, when the sun goes down, with the mountains and all, you have a temperature drop. Not only do you have clouds, but there's a fog that rolls in." CVW-8 operations had been badly affected by such fog since the start of the campaign, and more was forecast for the night of April 15–16. Having been cleared to plan the mission as best he saw fit, Aucoin decided to hit the airfield at sundown. Robert K Wilcox noted:

Only the A-Team would drop, or fly other important positions. They'd throw every available plane into the package. They'd over-ordnance the jets on the theory that more blasting power would be better than less. The core strikers [in F-14s] would use GBU-24 "bunker busters" to attack the mountain tunnel [leading to the underground hangar], the primary target. They'd hit it at both entrances and also try to put penetrators through the ventilation shafts. Because the various targets at the base were relatively close together, and lots of airplanes over the field at the same time could cause midair collisions, they decided to send four waves from two different axes, one after the other. Each wave would have a different objective, swooping in at the precise time the prior wave exited. The result would be sustained pounding for over 15 minutes.

To get to this point, the various units assigned to the strike would have to arm their aircraft in a fraction of the time usually available for such a mission. "I can remember briefing my gunner [CVW-8 ordnance officer] on the flightdeck, and I thought he was going to throw up on my shoes when he saw [what I wanted – and how fast]," explained Capt Lyle. "It was the most awful look I'd seen in my life."

Cdr Aucoin detailed the task now facing CVW-8's armorers, who had initially been preparing jets for a scheduled ATO mission to Pristina on April 16:

The Pristina operation was a large strike but not on the magnitude of this. Podgorica [airfield] required a different ammunition load-out. These things normally take three days to get ready. You gotta put in your order, you gotta put in your weapons load, and the weapons guys have got to dig this stuff out of the magazines seven floors down [in the carrier] and then arm and build up these bombs. The GBU-24 takes quite a long time to build, after which the seeker head and fins are put on. All the safety checks have to then be done to make sure the thing is ready to go.

While Aucoin oversaw the whole strike from an air wing perspective, his XO, Cdr Bauser, had to deal with getting VF-41's aircraft ready for launch:

We have airplanes all over the flightdeck, and we've got to bring some up from below. And we're moving them around to get them situated for launch. That in and of itself is a big job, not to mention that we have bombs to load. And a lot of the airplanes are parked tail over the water to make room. So, their ass is hanging overboard, but you gotta move them forward so you can load the bombs and then move them back again. Then you gotta develop a launch-sequence plan – who's going first, and co-ordinating that through the tankers, and then a tanker plan. And then, of course, there's the tactics of actually how we're gonna do this hop.

Despite the limited time available to get the mission aircraft armed and ready and the crews briefed on the targets to be attacked, the strike package made its prescribed launch cycle. "It was pretty unbelievable," Aucoin recalled, "because I saw the CAG and the admiral [Rear Admiral Winston Copeland, Commander, Carrier Group 8] walking around [a rarity for a launch] just watching guys build up bombs and putting them on the flightdeck, and they gave me a thumbs-up. I couldn't believe this was happening. It really was like going to war with the Russians."

VF-41's XO, Cdr Jim "Dog" Bauser, enjoys a cold beer during *Theodore Roosevelt*'s R&R port call to Palma de Mallorca shortly after completing its *Allied Force* mission. Bauser, who had switched from the US Marine Corps to the US Navy in the 1980s when the latter was looking for recruits to be trained as pilots, would assume command of VF-41 from Cdr Aucoin shortly after the unit returned to Oceana at the completion of the cruise. (via Robert K Wilcox)

The launch went smoothly, with 40 of the 55 aircraft on *Theodore Roosevelt*'s flightdeck at that time getting airborne. The aircrew not assigned to the mission noted that they had never previously seen the carrier so empty when the air wing was embarked on deployment. After the aircraft had departed, Capt Lyle turned to Rear Admiral Copeland and said, "Admiral, you know it's a good thing nobody told them they can't do the impossible."

It had started raining prior to the launch, and as the aircraft flew almost due north over the Adriatic Sea in the direction of Podgorica (which was about 30 miles inland from the coast of what was then the Federal Republic of Yugoslavia) conditions grew progressively worse. "We were going in and out of the clouds," said Aucoin, "and that's one of the hardest things – going up to the tanker, getting gas, and then going to the rendezvous points. We were looking up north and it was just solid clouds. We've done everything we can, and now the damn weather had closed in – it was looking doubtful."

The formation was led by a section of four VF-41 jets, each one armed with a single GBU-24 penetrator LGB. They were all tasked with hitting the mountain tunnel that led to the hangar. A fifth Tomcat from the unit carried two smaller GBU-10s that were to be used to destroy the airfield's control tower, and a sixth jet was armed exclusively with air-to-air missiles and assigned a distant CAP. The second wave of Tomcats to target Golubovci airfield was led by the CO of VF-14, Cdr Ted Carter.

With the lead section nearing the target, Aucoin was told by a USAF E-3 AWACS controller overseeing the mission that there were MiGs nearby:

"The E-3 came up and said there were two MiGs orbiting south of the field. I vectored two fighters on to them. They [AWACS controller] asked me if I was gonna delay, and I said 'No'. To tell you the truth, we were closer to them than the vectored fighters. I said, 'It doesn't matter, we're gonna run over them. If they're in the way, we'll shoot then down.'"

"Camelot 200" (BuNo 162698), bearing 80th anniversary titling on its vertical fins and toting a single GBU-16 on one of its centerline racks, returns to *Theodore Roosevelt* at the end of a mission over Kosovo on May 5, 1999. Issued new to VF-33 in September 1986, this aircraft saw combat in *Desert Storm* as the unit's CAG jet in early 1991. With the "Starfighters" disestablishment on October 1, 1993, the jet was transferred to VF-32. When that unit transitioned to the F-14B in 1996, BuNo 162698 was passed on to VF-14. A veteran of combat in *Allied Force* in 1999 and in OSW and OEF in 2001, the Tomcat served with the unit as its CAG jet for five years until sent to AMARC on February 1, 2002. (US Navy)

Due to a distance miscalculation, the lead aircraft were 12 miles ahead of where they should have been at that point in the mission, which meant they would arrive over the airfield too early. The HARM shooters dueling with SAM launchers and the successive waves of strikers would also reach their ToT too early as a result. In order to ameliorate this issue, the VF-41 jets flew a series of "S-turns" over the Adriatic to burn off 72 seconds prior to heading inland to the airfield. It was at this point the strike packages split up in preparation for their attack runs from the two pre-briefed axes. Aucoin's "heavy" division (because the four jets were armed with GBU-24s) maneuvered so that they could ingress to the target from the west, placing the setting sun behind them.

As the aircraft raced towards Golubovci airfield, AAA opened up in their direction. "Now they know we're coming," Aucoin told his pilot, Lt Cdr Brian Brurud. The first of at least 12 SAMs was launched at this point, and EA-6Bs supporting the strike fired HARM rounds in return. Aucoin was oblivious to all of this, however, as he was busy using the FLIR capability of his LANTIRN pod in an attempt to find cues for his target. Cloud cover was making his job difficult.

"There's so much shit going on, but I'm relying on a lot of experience, and I know I need to act calm and all that, so nothing's fazing me. We're just going on in there. I'm feeling kind of down because the weather's so bad. You can't see anything, and the sun's getting ready to set. The glare off the clouds is pretty bad and the fog is starting to come in."

Fortunately, as Aucoin closed to within 20 miles of Golubovci the clouds moved away and he could see the target clearly through the FLIR. "It was unbelievable, because up to that point it had been solid cloud cover. We go through this last set of clouds, and 'Oh my gosh, I can see the mountain and the airfield.' I thought to myself, 'God, we may have a chance.'" He then broke radio silence and told the rest of the strikers, "Target area clear. Target area clear."

Having achieved poor results in previous attacks with GBU-24s during *Allied Force*, the four VF-41 crews were determined to make amends. Following a disappointing mission to Pristina on April 8, Lt Cdr Dave Lobdell, who had flown F-117s on an exchange tour with the USAF, made a telephone call from the carrier to crews he had served with who had successfully dropped GBU-24s from the stealth fighter and the F-15E. According to Robert K Wilcox:

He [Lobdell] had learned that a certain dialed-in setting for the bomb the navy flyers had used was probably wrong. The air force had had trouble with it too. When the air force had switched to a new setting, they had gotten much better results. Now Joey Aucoin's A-Team was using the new setting. They were also very mindful of the aircraft-steering errors that had been made at Pristina, and the fact that they had to manually correct the range from the target the LANTIRN was showing. And to make sure the bomb would stay within its very narrow working envelope, they'd decided to release it at the exact distance from the target, given their speed and altitude, that allowed it the greatest margin of error for success. Each would go for the absolute perfect shot – not just the approximate envelope as had been used at Pristina.

More SAMs were fired at the lead division of Tomcats, which was closing on the target at 30,000ft. Some were shoulder-mounted weapons and other missiles were shot ballistically, as Serbian missile crews feared being hit by a HARM in return. Nearing the airfield, the F-14s descended beneath 20,000ft as they passed the 15-mile (24km) point to the target. All four RIOs picked up their tunnel target at a distance of 13 miles (21km), and they started calling off every mile until they reached the precise point at which they needed to release

Lt Cdr Brian "Bru" Brurud and Cdr Joseph "Joey" Aucoin perfected the FAC(A) mission during *Allied Force*, despite initial resistance from senior staff officers in CVW-8 and Naval Aviators and NFOs within the air wing. Brurud, in particular, was the driving force behind VF-41's embracing of the air-to-ground mission, having initially received orders to fly the A-6. Completing the MAWTS-1 course with the US Marine Corps in the early 1990s and a posting as an NSAWC instructor specializing in the FAC(A) mission, Brurud was one of the most qualified "Bombcat" exponents in the fleet by the time of *Allied Force*. He and his CO, Cdr Aucoin, flew myriad missions together as pilot and RIO during the campaign, proving the Tomcat's worth in the FAC(A) and SCAR roles. (via Robert K Wilcox)

their bombs in order to give the GBU-24s the best chance to guide. Brurud and Aucoin were the first to drop, with the latter noting: "We drop this thing, and it's like 'Ka-dunk!' I mean this is a big cow bomb coming out. And then my wingmen, Dog [Cdr Jim Bauser] and Wog [Lt Cdr Steve Carroll], drop theirs. You see this big thing come off, and it sort of goes with you for a while. It's got big fins, and it hangs up there, sort of like having a little wingman right beside you."

While the pilots maneuvered away from the target, the RIOs began lasing as the GBU-24s built up speed diving earthward. Aucoin chose to keep his FLIR in a wide field-of-view as he lased because "it was still kind of hazy. Wait until about the halfway time of flight [for the bomb] and then go to narrow field-of-view, and I can see the access road going right up to the bunker door." As he put his cursor on it, "There's talk on the radio about stuff going on, but I don't listen. The only thing that's in my mind is to put that bomb through the door. The time of impact goes down to zero, and nothing happens. I thought, 'Oh shit, did I screw it up?' One or two seconds later you see this massive 'blimp' [on the LANTIRN display]. A smoke puff comes out – 'Hey, shit-hot!' It went all the way through the door, and this stuff starts spewing [out]."

Seconds later, the GBU-24 dropped by Bauser and Carroll also hit the mountain roughly halfway up the hardened tunnel. Carroll remembered: "The skipper's [bomb] went into the door and just blew it to pieces – lifted the steel plate that's in the side of the mountain that they [the Serbs] had covered with dirt. As I see that happening, I'm going, 'Okay, ours should be hitting about…' and then this little puff underneath my crosshairs is my GBU-24 going in. It didn't look like it penetrated. It just looked like it blew up. So, we're going, 'Oh, I hope our bomb went in'."

The remaining two crews missed their target – a protruding shaft at the far end of the tunnel, which proved very hard to find. The single jet attacking the tower also missed when cloud cover broke the laser lock after the bombs had been dropped. Nevertheless, by the time the VF-41 F-14s had cleared the airfield, the tunnel was burning fiercely, with secondary explosions "hurling chunks of concrete in the air the size of my desk," Cdr Bauser recalled. Billowing clouds of smoke and flame, the latter caused by detonating ammunition, could be seen from a distance of 70 miles (112km) – according to Lt Cdr Brurud, it looked like "a nuclear explosion." The smoke was so thick that RIOs in the division of VF-14 Tomcats that followed the VF-41 jets could not see their target (the tunnel's rear entrance) through their FLIRs. They dropped their LGBs regardless. All 40 CVW-8 aircraft involved in the mission subsequently returned safely to *Theodore Roosevelt*.

According to the Joint Analysis Center report on the April 15 raid, a single GBU-24 penetrated the underground hangar and exploded. The detonation of the weapon started a fire that destroyed 24 G-4 Super Galebs tightly packed into the hangar. The report noted that "this single bomb hit resulted in the greatest loss [suffered] by the Serbs during the entire conflict."

"Podgorica [airfield] was a huge success," noted Robert K Wilcox in *Black Aces High*. "The admiral was thankful. The battle group had proven its worth in a single afternoon. No other US force could have mounted a successful raid so fast. And Joey [Aucoin] and the 'Black Aces' had led the charge. The next day, April 16, US Defense Department spokesman Kenneth Bacon told members of the press at the Pentagon briefing, 'We hit the airfield in Podgorica last night with naval air and did a great job. I'll have some video of that tomorrow'."

Back to SCAR

Having demonstrated the F-14's effectiveness as a precision bomber during the Golubovci airfield attack, VF-41 at last got the chance to prove the squadron's worth in the SCAR role on April 19 when unusually timely intelligence revealed the location of a 9K52 Luna-M short-range artillery rocket system to CVW-8. Codenamed "Frog-7" by NATO, the Scud-like, truck-borne, surface-to-surface missile was being used by the Serbs to decimate Kosovar

An aviation ordnanceman prepares to "pull the pin" to arm an AIM-9M fitted to "Camelot 200's" starboard LAU-138 Sidewinder launcher rail on June 1, 1999. Immediately below the weapon is the jet's all-important LANTIRN pod, which allowed both VF-14 and VF-41 to conduct precision strikes on Serbian targets in Kosovo. (US Navy)

Albanian villages. Robert K Wilcox detailed how the squadron dealt with this target in the following extract from *Black Aces High*:

The "Frog" happened to be in the general area that Brurud and Aucoin were planning to reconnoitre, and they were given a photograph of it shortly before they were due to depart "TR". They launched into an anomaly – reasonably good weather. It appears they had some trouble finding the "Frog" site, possibly because of problems with the GPS co-ordinates. But they kept hunting, using both the gyro-stabilized binoculars and the FLIR's wide view. "We were just flying around the countryside", said Aucoin, "when I said to 'Bru', "That area looks real similar to the photo." The picture had some farm-like buildings in it, and an adjacent field. The area below them had similar features in roughly the same pattern. They were at 15,000ft – too high to see details like the "Frog." They decided to take a risk and go lower.

They had already been dipping down and popping back up as a cautious means of entering and exiting SAM envelopes. Now they swooped down lower than ever before. It paid off. "The buildings matched up," said Aucoin. It was a barn-like structure with other dwellings nearby. Near the barn was a "parking lot" with vehicles. And in the field was the missile launcher! "It was on a truck," said Aucoin. "We suspected they stayed within the barn-type building and just came out to fire. Then they would run back in the barn."

They zoomed back up to safety. Excited about what they had found, they got confirmation to drop. It's not clear whether they hit the "Frog" themselves or brought in Hornets to do it. Aucoin's recollection is that they used their own GBU-12s on the launcher and then brought in the Hornets for the other targets. Brurud says they decided to wait for Hornets to do all the bombing because of the impact it would have on the SCAR debate. "Anybody probably could have gone out with what we had and found the target and bombed it," Brurud said. "The difficult part comes when you've got different areas or targets and an allocation of airplanes with say a 20-minute window. I've got 20 minutes to go up there, circle around, confirm the target here, confirm the target there, get authorization, prioritize, deconflict and start directing the bombing."

Regardless of whether they hit the "Frog" themselves or not, they did have plenty of other targets to parcel out on the site. Brurud wanted to demonstrate how good the F-14 was in the SCAR role, and Aucoin agreed. They had Hornets standing by, and they were probably about eight to ten miles away. One of the advantages of using the Tomcats was that their LANTIRN system was stronger than the F/A-18's Nite Hawk targeting pod. The LANTIRN could see farther and provided a larger scope picture to its crew than the Nite Hawk did. With the Tomcat's sighting of the target, the Hornet's didn't have to get close to it as they otherwise would.

The Hornets were carrying laser-guided Maverick air-to-ground missiles. These honed in on laser reflection and were unique in that they could be fired and then the Hornet could fly away. Nothing else was required of the pilot. As long as the target was illuminated, the Maverick, which had wings, fins and a motor, would fly to the laser spot on its own. It was an easy-to-use weapon for what would come to be termed "buddy lasing" – using the laser of one aircraft (in this case the Tomcat) for the weapon of another (the Hornet). The Maverick was also specially made to destroy hardened targets such as tanks and armored vehicles. It could easily penetrate farm buildings.

One by one, Brurud and Aucoin called in the Hornets, which stayed out of visual range. They gave the Hornet pilots the target co-ordinates to aim the Maverick and the LANTIRN laser code so the missile could recognize and be guided by it. Once Aucoin had the laser aimed at the spot, the Hornet pilot would make his run, still staying a safe distance away. When the Maverick located the laser, probably about five miles from the target, the Hornet pilot would fire the missile. It would then fly to whatever Aucoin was illuminating.

They decimated the area. This mission provided a definite sense of power seldom experienced in normal bomb runs. It was almost point and shoot. The laser operator wielded a beam of concentrated energy that doomed everything it touched, with the rain of devastation relentless and seemingly unending depending on the number of bomb-carrying strikers being directed. The concentration of one of the Hornet pilots sticks in Aucoin's memory. "A young lieutenant, I can't remember his name. He couldn't pickle [fire] on his first run so he really bore down on the second." He didn't realize it, but he was pointed almost straight down, scorching the air at about 18,000ft, approaching 600mph [965km/h]. "All that is out of his scan. He just wants to get the missile off. When he launches, he's almost at Mach 1. It comes off and blows up the 'Frog's' support vehicle – just blew it to bits." Not until the lieutenant got back and viewed his video did he see the numbers recorded by his sensors showing such a thrill ride. He'd been oblivious. He had just not wanted to screw up.

The April 19 mission was VF-41's first bona fide SCAR success – roughly two weeks after the unit had started trying to perfect it. And although it by no means ended the "Black Aces'" SCAR–FAC(A) problems, the mission had an immediate impact. "Once they did it," said Brurud, referring to the Hornet pilots, who were among the most skeptical of the mission in CVW-8, "they suddenly came aboard." Not every Hornet pilot was out there on that particular mission. Far from it. Nor were the skeptical F-14 crews or their doubting higher-ups. But the seed had been planted. More importantly, CAG had got a much-needed demonstration. According to Aucoin, "Once Capt Lyle saw the video – and if we hadn't had the videos he wouldn't have believed it because we needed them for confirmation and all that – he felt really good about it. Hey, we can actually do this."

Inspired by their success, Brurud and Aucoin increased their SCAR efforts. They began an almost round-the-clock schedule of SCAR activity usually involving two missions a day, which was about the limit they could fly. They'd jump into a jet in the early morning for two to three hours of hunting and return near noon. Still in their sweaty flight gear, they'd get a brief while munching a candy bar for lunch, pick up the best intelligence-supplied pictures or co-ordinates available, and launch back out for an afternoon-through-night hunt.

Little by little they began to refine and create their SCAR techniques. For instance, where before they were dipping down and then quickly returning to safe altitude in order to check on possible targets, they

Above: "Fast Eagle 100" (BuNo 161295) rapidly approaches the runway at Aviano air base, in Italy, on May 2, 1999, the aircraft leading in "Camelot 205" (BuNo 158633) directly from *Theodore Roosevelt*. VF-41's CAG jet would subsequently be adorned with an impressive bomb tally denoting the APCs, artillery pieces, tanks and mobile missile launchers knocked out by the squadron. Passed on to VF-211 in December 1999, the aircraft was marked with six bomb symbols following its participation in the OEF campaign in 2001–02. (Sergio Gava)

Right: With its starboard gull panel hinged open, "Fast Eagle 100" has the avionics boxes contained within the exposed bay fettled by a maintainer on the VF-41 ramp at Oceana in the fall of 1999. The kill tally worn beneath the cockpit was applied to the aircraft after *Theodore Roosevelt* had completed its OSW commitments and was headed home. (Mike Kopack Snr)

began relying more on their wingman, Prowlers and the E-2 Hawkeye in the area to signal danger and fight it. The wingman would be their defensive eyes, checking the ground and the immediate airspace for threats. The Prowlers would jam and missile nearby enemy batteries that activated. The E-2 would alert them to distant hostile threats. All this meant that they could stay lower longer, concentrating on the search. The rules about how low they could go were not yet defined. Strikers were restricted to a 20,000ft floor, but the SCAR floor hadn't yet been posted. "It was one of those deals where we're not exactly sure we want to ask the question," said Brurud. "So, the strikers kept up there. But as FAC(A)s, I know I went lower. It was more just doing it than asking about it."

Eventually, they would fly into the heart of the enemy's defensive envelopes – 10,000ft and lower. They'd do so with wingmen who usually remained on guard at a higher altitude and were also FAC(A)s. The idea was to always have a trained FAC(A) crew in the area in case Brurud and Aucoin went down. In the meantime, the wingmen, usually JOs, hopefully would learn from them. One of the pilots protecting the FAC(A)s was Cdr Jim Bauser. "While they were totally in the cockpit looking at sensors or looking at the ground, our job was to do top cover on them, call out SAMs and defeat the SAMs that might elude the Prowlers – protect them. I guarantee you, if they were heads down looking for things and there was a junior crew in the other jet, they wouldn't feel so good."

VF-41 expended its last ordnance of the campaign on June 9, and three days later NATO's Kosovo Force of peacekeepers entered the country with the backing of the UN following the enforcement of a ceasefire and the withdrawal of Serbian troops. During *Allied Force*, VF-41 had undertaken 384 combat sorties totaling more than 1,100 flying hours. Its aircrew had expended 160 tons of LGBs, achieving an 85 percent success rate.

The "Black Aces" had relied heavily on its FAC(A)-qualified crews to take the fight to Serbian fielded forces. Amongst this elite band of Naval Aviators and NFOs was RIO Lt Clay Williams, who summed up *Allied Force* from VF-41's perspective in his final diary entry for the Kosovo campaign:

This mission was very difficult. Being a FAC(A) for troops on the ground would have been so much easier. The Marines on the ground would have done most of the target recce for us. Our primary challenge would have been not to endanger the Marines we were supporting. But without the ground troops, the entire problem of reconnaissance and target identification fell to us. It was an almost overwhelming task. So why did we succeed?

It became apparent fairly quickly that the success of the navy in Kosovo and the success of the F-14 FAC(A) program relied on us being successful at targeting the fielded forces. We worked very hard at cracking this nut. If we found revetments and got ordnance off the strikers, it was tremendously relieving. Conversely, if we did not find any targets we felt like failures.

I am convinced that it was the air campaign that brought President Slobodan Milošević to the table. I don't believe that it was the threat of a ground offensive because I do not believe that threat was credible. We were not going to commit ground troops, plain and simple. We may not have destroyed very many pieces of his war-fighting machine. This is largely due to the fact that almost all of his army was cowering in holes and hiding spots, which can be seen as a success, if you like.

As noted in the previous chapter, following the completion of combat operations in Kosovo, *Theodore Roosevelt* sailed to the NAG to support OSW. In addition to its regular missions of enforcing the southern no-fly zone over Iraq, CVW-8 executed a handful of strike missions on Iraqi targets in response to being fired upon by surface-to-air threats. With the experience gained from their combat time in Kosovo, VF-14 and VF-41 completed their missions successfully, thus helping to make CVW-8 the first air wing in many years to see combat in two different theaters of operation during the same deployment period – it would repeat this feat in 2001.

"Fast Eagle 101" (BuNo 162608) carried similar, but not identical, mission markings to "Fast Eagle 100," although its tally had been started while VF-41 was still flying *Allied Force* missions. Cdr Bauser had assumed command of the unit by the time this photograph was taken in the fall of 1999, his place as XO being filled by Cdr Brian Gawne. The *TOMCAT FASTFAC* titling and insignia on the nose had been applied prior to VF-41 embarking onboard *Theodore Roosevelt* on March 26, 1999. (Steve Czerviski)

Still boasting *TOMCAT FASTFAC* titling on its nose, but with the *Allied Force* scoreboard having been painted over, "Fast Eagle 101" rolls back after landing onboard *Enterprise* in order to free the arrestor wire from its tailhook. The aircraft was flying from the carrier off the Virginia coast during an Independent Steaming Exercise in April 2000 – one of five times CVW-8 embarked in *Enterprise* during the course of that year. Note the *McCLUSKY* titling on the fin tip, which denoted that VF-41 had been awarded the Rear Admiral Clarence "Wade" McClusky trophy, presented annually to the "finest attack squadron in the Navy," in 1999. This award had previously been won by strike-fighter and bomber squadrons only. (US Navy)

FAC(A) Training

A handful of Tomcat aircrew had taken it upon themselves to seek out FAC(A) training when their F-14s started to routinely carry bombs in the early 1990s. The conflict in the Balkans at the same time underlined the need for this when American peacekeeping forces in Bosnia-Herzegovina complained that US Navy jets committed to the operation did not have airborne air controllers to spot targets and aid in the hitting of them – the aircraft also lacked the LGBs needed to support troops on the ground with precision strikes.

Pilots such as Lt Cdr Brian Brurud of VF-41 succeeded in getting a place on a six-week FAC course run by Marine Air Weapons and Tactics Squadron (MAWTS) One at Yuma. US Marine Corps aviators have a fierce reputation when it comes to providing CAS, as their primary mission is to support Marine ground troops. FAC from a Marine aviator's standpoint involves finding enemy troops and weapons while in the air, usually at the direction of an aviator on the ground, and then destroying them without hitting his own troops – no easy task given the altitude and speeds involved. The MAWTS course included FAC(A) training, which taught aircrew how to do everything when it came to providing CAS. This meant that there was no need for a ground controller.

After completing his course and finishing his fleet tour with VF-41, Brurud joined NSAWC at Fallon just as it was establishing a Joint Tactical Air Control Course (JTACC) under the direction of one of the US Navy's most experienced ground FACs, SEAL Sniper Senior Chief Andy "Senior" Nelson. Instructors running JTACC provided expertise that allowed the FAC(A) mission to be expeditiously ushered into the Tomcat community at the same time as LANTIRN pods started to reach the fleet. This meant that when the conflict in Kosovo commenced in March 1999 there were at least four trained FAC(A)s within each F-14 squadron.

An early graduate of the NSAWC JTACC was Lt John Saccomando of VF-213, who would subsequently use his training to the full over Afghanistan during OEF:

Lt John Saccomando was an early graduate of NSAWC's JTACC for F-14 crews, and he put his training to good use during VF-213's OEF deployment in 2001 when he undertook 52 combat missions over Afghanistan. As the squadron's air-to-ground training officer, Saccomando was responsible for ensuring that 28 Naval Aviators and NFOs in VF-213 were mission-capable prior to OEF commencing. Thanks to the effectiveness of his training, the "Black Lions" ultimately achieved a CVW-11-leading 95 percent bombs-on-target accuracy during OEF. (via Rear Admiral John Saccomando)

The FAC(A) training course lasted two months and was only available to three or four crews per squadron. It was pretty competitive to get selected for the course because it required an elevated level of situational awareness [SA] while flying low to the ground. The first phase was to get qualified as a ground FAC, Tactical Air Control Party [TACP]. That part of the course ran for several weeks, and it was taught at either Dam Neck, in Virginia, or at Coronado, in California.

The last week of the TACP training was spent in the field controlling live ordnance dropped from jets while you were in close proximity [within visual range] to the target. We had to set up ToTs in sequence for the strikers, have all of the jets deconflicted and holding at different altitudes, and then give them run in points that kept them deconflicted. We also had to check them in and figure out what ordnance they had and how much of it they were carrying. Once we had their time on station, our final task was to prioritize the targets according to what ordnance was available.

Upon becoming a qualified FAC, you would be allowed to commence the FAC(A) syllabus taught by the East Coast Weapons School [SWATSLANT]. It consisted of a week of ground school followed by several weeks of flying and controlling ordnance. The instructors were really good about getting Marine Harrier IIs and Cobras and USAF F-16s and A-10s to train with us.

Our responsibilities were to deconflict the stack [the holding stack that saw waiting strikers circling near the target at different altitudes], then send the jets in in sequence for different ToTs. Simultaneously, we would be maneuvering our jet so as to get into position to "look through their tails" as the strikers rolled in. It was up to us to clear them hot once we were sure they were rolling in on the correct target. Once cleared, I would immediately maneuver to get in trail of the next jets as they came into the target area – strikers usually only had about a minute's separation.

Performing FAC(A) missions both in training and for real were some of the most challenging, dynamic and rewarding flights I've ever done. Lt Cdr Michael Peterson – who was one of my weapons school instructors – and I were the first FAC(A)s in Afghanistan, controlling ordnance the first night American troops were on the ground. It was both exhilarating and terrifying at the same time. Indeed, it was one of the most rewarding flights I've ever participated in.

CHAPTER 9

OPERATION *ENDURING FREEDOM*

By fall 2001, the venerable F-14 had been developed into a truly multi-role fighter-bomber, with more mission taskings than any other aircraft then embarked in a US carrier. It was now set to play a leading role in the conflict over Afghanistan, rather than being the bit-part player it had been in *Desert Storm*.

The first operations conducted by the F-14 during the final phase of its service life with the US Navy occurred just hours after the "twin towers" and the Pentagon had been attacked by al-Qaeda terrorists in hijacked airliners. That morning, VF-11 and VF-143 were preparing to embark in *John F. Kennedy* as part of CVW-7's pre-cruise work-ups off the Virginia coast. North American Aerospace Defense Command (NORAD) contacted the US Navy soon after the south tower was hit and asked for its help in securing the airspace over the eastern seaboard. Both *John F. Kennedy* and *George Washington* were at sea in local waters under Second Fleet control, and the vessels embarked a handful of Tomcat and Hornet squadrons from NAS Oceana.

VF-11 and VF-143 were sent to *John F. Kennedy*, and Naval Aviator Lt(jg) Joseph Greentree from the latter unit subsequently flew several missions in support of the NORAD-controlled "air shield" that had been hastily established off the coast of New York: "For the first 72 hours that VF-11 and VF-143 were embarked in *Kennedy*, we flew round-the-clock CAPs up and down the eastern seaboard. The skies remained eerily empty during this time, with all civilian air traffic having been grounded. After three days, Second Fleet told us to abandon the CAP missions and commence our work-ups."

VF-41's "Fast Eagle 105" (BuNo 161615) keeps station off the right wing of a KC-10A while an RC-135V of the 55th Wing tops off its tanks midway through an ELINT mission along the Afghanistan–Pakistan border during early October 2001. VF-41 was charged with providing the fighter escort for a number of ELINT sorties leading up to the aerial assault on Afghanistan. (VF-41)

With al-Qaeda directly linked to the September 11 attacks, the US government turned its attention to the terrorist group's home in Afghanistan. Less than three weeks after the atrocities in New York City and Washington, D.C., carrier-based aircraft would be in the vanguard of a joint operation to remove the Taliban from power and destroy the organizational infrastructure that al-Qaeda had established in Afghanistan.

The carrier closest to this land-locked country was *Enterprise*, with the F-14As of VF-14 and VF-41 embarked. These units were nearing the end of their last cruise with the Tomcat, and they had seen action in Iraq during five weeks of OSW patrols. Also steaming towards the Northern Arabian Sea from the Indian Ocean was *Enterprise*'s OSW relief, *Carl Vinson*, with the F-14Ds of VF-213 embarked. These three Tomcat units would be in the vanguard of what was codenamed Operation *Enduring Freedom* (OEF) by Pentagon planners.

Enterprise was the first carrier to arrive on station in the Northern Arabian Sea off the coast of Pakistan, from where all such vessels would conduct combat operations during OEF. It then spent more than three weeks sailing in large circles as CENTCOM planned for the campaign. With so little known about potential targets in Afghanistan, both VF-14 and VF-41 had their work cut out for them when it came to strike planning. VF-41 CO Cdr Brian Gawne explained the task facing the "Black Aces" in the lead up to OEF:

Afghanistan was an immature theater, so when it came to strike planning, we were starting from scratch. Just weeks earlier we had been performing strikes into southern Iraq, and when preparing for such missions we could reference charts, planning tools and target imagery, as well as rely on the experience of senior Naval Aviators and NFOs who had been flying OSW sorties for almost a decade. The value of crews studying digitized imagery of the target area had been reinforced during operations over Kosovo in 1999, and that is exactly what we had done during OSW. We did not have sufficient imagery for such study prior to OEF, however.

VF-213's crews were similarly in the dark in respect to the threat posed to them by the Taliban. The unit's Intelligence Officer, Lt(jg) Nate Bailey, quickly set about trying to rectify this situation:

Combat operations in Afghanistan were something we had not even considered. Our primary focus, both during our work-ups and while sailing from San Diego across the Pacific and Indian Oceans, was our role in the NAG, not with attacking targets in a land-locked country in central Asia.

As part of our jobs, everyone in CVW-11's Intel department was tasked with being a Subject Matter Expert [SME] on different countries/issues in the Western Pacific and Middle East areas. Reflecting the threat level associated with Afghanistan pre-9/11, its SME was a very junior sailor. However, by the end of September, we had all became SMEs on Afghanistan. While there wasn't very much up-to-date information out there on the state of the country's air defenses, we would analyze anything we could get our hands on – old Soviet accounts of their campaigns and lessons learned, imagery of fixed SAM sites, history of the Taliban and al-Qaeda's presence in Afghanistan, etc.

The fact of the matter was that while the air-to-air threat – fewer than 50 MiG-21s and Su-22s at most – was not great compared to other potential adversaries, the surface-to-air threat was still a bit of an unknown. We knew that the Taliban had fixed SAM sites around their main strategic centers, but the proficiency of the operators was certainly a question. In my opinion, the more immediate threat was the AAA that would be located at higher elevations (mountain passes, etc). This tactic was observed during the Soviet occupation, and we briefed that we could expect to see the same thing.

The other key part of intelligence, besides briefing the threat, is supporting the targeting process. For the first phase of the air war, the CAOC produced a prioritized target list. The Intel folks on the carrier then turned that around into target folders – working closely with members of the strike teams – in order to provide the best situational awareness of the target area and desired impact specifics.

Targeting, in my opinion, is certainly the most "satisfying," and yet the most sober, part of being an intelligence officer in this environment. We had guidance to be aggressive and "take down the Taliban," and yet the specifics of that often fell to junior officers like myself. All the target

With their pre-OEF CAP mission over, Lt Cdr Van Kizer and his RIO, Lt Dave Bailey, keep their hands in the air, and away from the weapons activation panel, while armorers pin the pylon firing mechanisms for the ordnance attached to VF-14's "Camelot 200" (BuNo 162698). A 40-mission veteran of Operation *Allied Force*, Lt Cdr Kizer would fly another 20 sorties in OEF. (VF-14)

nominations I submitted – from TARPS imagery, for example – to the CAOC were approved. While we were aggressive, we also realized that we were dealing with real people on the ground. A surreal balance to be sure.

Like both air wings in-theater, the CAOC was desperate for up-to-date photographs of potential targets such as airfields, SAM and AAA sites, army barracks and terrorist training camps, and it authorized both VF-41 and VF-213 to fly a series of TARPS missions over Afghanistan starting in late September – VF-14 had no TARPS capability. One of the pilots to provide the fighter escort for several of these missions was VF-41's Lt Marcus Lopez:

> My first missions over the beach were flown about a week prior to OEF kicking off, when we provided cover for TARPS jets undertaking reconnaissance runs over key targets in south-eastern Afghanistan. We were not opposed on these flights, entering enemy airspace "high and fast." The TARPS sorties also allowed us to plan where we needed the tanker tracks to be established once hostilities commenced. We also used the flights to gather Electronic Intelligence [ELINT] data on the enemy's SAMs and radar-guided AAA – a Prowler flew with us throughout the missions, monitoring the Afghan response to our incursions into their airspace.

Due to the extreme distances involved in these flights, VF-41 could not venture beyond Kandahar, in south-eastern Afghanistan, at this time due to the limited nature of the tanker support then in-theater. Indeed, CVW-8's VS-24 was the only unit capable of providing aerial-refueling assets for the air wing during the early reconnaissance runs. Vikings dragged the Tomcats and Prowlers as close to Afghanistan as they could manage, then briefly held on the Pakistani side of the border while the TACAIR jets topped off their tanks, before returning to *Enterprise*.

By the time the reconnaissance package had headed out of Afghan airspace at the end of its TARPS run, replacement tankers were on station to provide the jets with sufficient fuel to make it back to the ship. VS-24's support was critical, as the F-14 used a lot of fuel conducting a TARPS mission – VS-29 performed identical missions for VF-213's TARPS flights pre-OEF.

VF-41 RIO Lt John Kelly was also heavily involved in CVW-8's intelligence-gathering push pre-OEF:

> We sometime flew two or three TARPS missions a day during this period, and in the flights in which I participated, I got the feeling that virtually no one in Afghanistan knew that I was overflying the country at 20,000ft. There was no threat, and our RHAW [Radar Homing And Warning] gear only detected the odd random search radar hit here and there. We had no idea what we were taking photos of, simply using the big KS-153B long-range camera to get stills of various "lat/longs" that we had been instructed to overfly.

These were classic Tomcat TARPS missions in respect to the fact that the admiral and the CAG staffers needed to get hot prints on deck for use by our own intel folks, as well as target analysts in the CAOC. In order to satisfy this demand, VF-41 would have photographer's mates pulling the film canisters out of the TARPS while the jet was still winding down on the flightdeck. The negatives would then be hastily developed and prints run off and dispersed. Those going beyond the ship to the CAOC had to be digitized for transmitting.

We also manned CAP stations for USAF RC-135V/W "Rivet Joint" and US Navy EP-3E ELINT/SIGINT [Signals Intelligence] platforms immediately prior to OEF commencing. After flying these missions for four or five days in a row, CAP sorties became very tedious very quickly. It was plainly obvious to us that the Afghan air force was going to offer no aerial resistance at all in the coming conflict.

Aside from the TARPS sorties that actually overflew southern Afghanistan, both air wings also performed several mirror-image strike missions from the Northern Arabian Sea up to the Afghanistan–Pakistan border during the first week of October. These sorties, involving divisions of Tomcats and Hornets carrying representative bomb loads, as well as Prowlers, Vikings and Hawkeyes in support, were flown along the newly established "driveways" through Pakistan that both CVW-8 and CVW-11 would strictly adhere to once the war started. Similar missions of an identical duration were also flown over open water to the south of both carriers.

All three Tomcat units started performing round-the-clock CAP missions along the Afghanistan–Pakistan border from October 3, for by that time the first big wing tankers had started venturing into the area from bases in the Middle East.

By then CVW-8 had been operating as the night carrier for about a week, the *Enterprise* battle group commander, Rear Admiral Henry Ulrich, having told Central Command Air Forces (CENTAF) that his air wing was more than happy to perform the midnight to noon slot. His opposite number in the *Carl Vinson* battle group, Rear Admiral Thomas Zelibor, who also led Fifth Fleet's Commander Task Force (CTF) 50 in the Northern Arabian Sea, readily accepted. CVW-8 made the gradual switch to night operations two weeks prior to OEF commencing. When performing the midnight to noon shift, the air wing would launch its first wave of jets between 2300–2330hrs. Typically remaining aloft for six to eight hours, they would land back aboard the ship in total darkness. The second wave launched at 0200hrs and landed just after dawn, while the third and final wave left the carrier at 0400hrs and recovered shortly before noon.

With the US military being politically prevented from using nearby land bases in the NAG and India, and unwilling to overuse frontline airfields in Pakistan, Uzbekistan and Tajikistan, aircraft carriers were initially the only way to bring tactical air power to bear in Afghanistan. The strike fighters of CVW-8 and CVW-11 would duly hit terrorist training camps, Taliban barracks, airfields and SAM/AAA sites in the longest carrier-launched attacks in history. Tomcat, Hornet and Prowler units prepared themselves for missions that would see them routinely operating more than 700 miles from their carriers during sorties lasting between six and ten hours.

On October 4, with OEF mission planning now all but complete, the CAOC instructed the air wing commanders to brief their squadron COs on the opening phase of the campaign. VF-14 boss Cdr Bruce Fecht was present at the meeting aboard *Enterprise*:

CAG [Capt Dave] Mercer called the COs together in his office to provide direction on the upcoming campaign. It was only then that we knew OEF was likely to go ahead. It was time to get specific strike package assignments, targets and plans together. Initially, it was decided that VF-213 would cover the northern high-altitude targets with its Tomcats due to the fact that it had the increased capability 40,000ft LANTIRN pods. As it turned out, however, CVW-8 was also assigned three high-altitude targets, which were to be hit in the second wave of strikes – an effort towards Kandahar, a second effort to the south-west against known terrorist camps, and a third effort, which I was part of, towards Kabul.

Conflict

OEF commenced on the evening of Sunday, October 7, 2001, when, according to USAF Gen Richard Myers, Chairman of the Joint Chiefs of Staff, "about 15 land-based bombers, some 25 strike aircraft from carriers, and US and British ships and submarines, launching approximately 50 Tomahawk missiles, have struck terrorist targets in Afghanistan."

Although the F-14s, F/A-18s and EA-6Bs of CVW-11 and, several hours later, CVW-8 would be in the vanguard of the action, the first aircraft sortied were actually USAF types dispatched from bases thousands of miles away in the American Midwest. B-2As, along with five B-1Bs and ten B-52Hs flying from the US Naval Support Facility on Diego Garcia (an atoll in the Indian Ocean), hit Taliban early warning radar sites and other air defense-related targets, as dictated by the CAOC's Master Air Attack Plan.

Charged with both attacking targets and providing protection for USAF heavy bombers while over Afghanistan, VF-213's F-14Ds would go into combat on the first night of the campaign in true strike fighter configuration. The unit's "gunner" was veteran ordnanceman CWO3 Michael Lavoie, and it was his job to oversee the arming of the four Tomcats that led CVW-11's opening OEF strike. Despite having worked carrier flightdecks for more than 20 years, Lavoie had never seen such heavily laden F-14s launched from a carrier prior to October 7:

On the first night of OEF, our aircraft were hitting pre-planned targets, so we knew what ordnance was required. We had four jets heading out, and we uploaded bombs in quantities that we had never previously hung on a VF-213 aircraft – two of the Tomcats carried pairs of 1,000lb [454kg]

The crew of "Black Lion 103" (BuNo 163899) prepare to launch from *Carl Vinson*'s waist catapult three on the opening night of OEF. VF-213 would spearhead CTF-50's nocturnal strikes on Kabul and Herat on October 7–8, 2001, with a section of F-14Ds being sent to attack targets in both cities. VF-213 had six "40K" LANTIRN pods for its aircraft, and these proved vital right from the word go in OEF. This particular jet was paired up with pod "22", which was responsible for lasing 55 LGBs during the unit's time in-theater. Some 44 of these bombs achieved near or direct hits on their designated targets. (Lt Andrew Mickley)

GBU-16s, and the remaining jets were armed with 500lb [226kg] GBU-12s. We also armed each of them with single AIM-54C Phoenix and AIM-7M Sparrow missiles, as well as two AIM-9L Sidewinders and 678 rounds for their 20mm cannon. The jets were prepared for anything, as we had little idea about what kind of air threat would be opposing us.

I truly wondered whether these aircraft were going to get off the deck laden down with all that weaponry. We had spent the previous weeks before going to war precisely weighing all the ordnance that might be hung from a Tomcat in OEF, and then calculating what the aircraft could carry. We had a program on our computers that would allow us to accurately calculate the weight of the various weapons configurations. We ran through literally hundreds of potential mission loads, and then gave the weights to the CO and CAG so that they could devise their strike plans. Weight is a critical thing for carrier-based jets, as the catapult has to be correctly adjusted to provide adequate thrust to ensure that the aircraft will fly after being launched.

OPERATION *ENDURING FREEDOM*

Despite us operating the bigger-engined F-14Ds, I watched with anxious fascination as we launched jets armed with two GBU-16s, four missiles – one of which was a 985lb [446kg] AIM-54C – and all the gas that could be squeezed into them to the point where the aircraft weighed in at 72,000lb [32,658kg] apiece. It was an amazing sight to see the Tomcats stroking in their afterburners as they accelerated down the catapult and then powered away from the carrier, despite these awesome loads.

We had flown with live ordnance during previous OSW cruises, but typically the jets would only be configured with a single CBU-99 Rockeye, a GBU-12 and possibly two or three air-to-air missiles at most – barely 2,000lb [907kg] of ordnance. These jets were laden down with more than twice this amount.

Strapped into VF-213's "Black Lion 101" on October 7 was squadron CO Cdr Chip King, who had been selected by his CAG, Capt Thomas C Bennett, to lead the CVW-11 strike:

CAG chose me to head up this mission because I was the senior strike lead in CVW-11 at the time, and the F-14D was the premier precision-bombing platform in the air wing. Unlike other Tomcats in-theater, our jets had all the "toys," with Link-16 JTIDS being the most important. This system allowed us to use the datalink to deconflict with all other assets in our package without having to use our radio or radar, as well as letting us find the tanker expeditiously when gas became critical. And with the LTS pod attached to my F-14D featuring the new "40K" laser, we could drop our LGBs from heights up to 40,000ft above ground level [AGL] – well outside of the Taliban AAA/SAM threat envelope.

The plan for the opening strikes of OEF called for a multi-prong, nearly simultaneous attack by Tomcat-led packages on western and eastern military targets within Afghanistan, followed by the final arm – all Hornets – striking the airfield at Kandahar. The lead element, consisting of two F-14Ds in the swing strike fighter roles, two F/A-18Cs, an EA-6B and two B-1Bs, would attack the target in the east – an SA-3 "Goa" SAM site and support facility near Kabul's international airport.

Right: CVW-11 OEF (2001)

Below: "Black Lion 101" (BuNo 164603) led the first manned strike of OEF, when Tomcats and Hornets from CVW-11 hit an SA-3 SAM battery, and its attendant target acquisition and guidance radars, near Kabul's international airport on the night of October 7–8, 2001. The second-to-last Tomcat ever built, this aircraft was originally delivered to VF-124 on May 29, 1992, and subsequently became one of the first D-models assigned to VF-2 in June of the following year. Transferred to VF-213 in late 1997, the jet remained with the "Black Lions" until passed on to VF-101 in early 2002. BuNo 164603 returned to the fleet in the summer of 2003 when the aircraft was sent to VF-31. Completing a further two cruises with the unit, the Tomcat had the distinction of making the last flight by a US Navy F-14 on October 4, 2006, when it flew from Oceana to Republic Airport, in Farmingdale, New York. The aircraft was duly put on display as a memorial to all Northrop Grumman workers at nearby Bethpage. *(Lt Tony Toma)*

We briefed the overall strike co-ordination aboard *Carl Vinson* in the early afternoon of October 7, with all the senior leadership in attendance. My two-F-14D element had two missions to perform. The first was to "sweep" the airspace in front of an element of B-1Bs to clear it of any enemy fighters that were foolish enough to launch against our strike package. Afghan air force MiG-21s reportedly almost never flew at night, but we nevertheless hoped that they would launch and give us a chance for an air-to-air kill.

Our second mission was to deliver additional ordnance on the SA-3 site, which was due to be hit by TLAMs minutes prior to our arrival. The LGBs that we dropped would ensure that the SAM battery, and its "Spoon Rest" and [SNR-125] "Low Blow" target acquisition and guidance radars, remained out of commission for future strikes on Kabul. The Hornets were to launch and guide a single AGM-84 Stand-off Land Attack Missile – Expanded Response [SLAM-ER] from a significant distance away from the target to impact in the same area, while the Prowler provided radar-jamming support to mask our approach and departure from the area.

Following the overall brief, we broke down into individual element briefs, which concluded approximately two hours prior to the planned launch. Rear Admiral Zelibor reminded us that we still did not have an executive order, so as we began to ascend to the flightdeck, we still did not know if the mission was a go or not.

The scene on the flightdeck was almost surreal – many of the crews had never been involved in a strike operation of this magnitude. There was a tremendous amount of activity around all the aircraft, and lots of excited conversations. Each of us approached our plane captains and began the walkaround. It was now roughly 1740hrs, and the sun had just set. It was at this point that I remember seeing numerous flashes on the horizon all around us, and within moments there was a hush as everyone was transfixed by this activity. We all seemed to have realized what those flashes were at the same time – the "small boys" [cruisers and destroyers] and submarines were launching their TLAMs. The Tomahawks provided the opening salvos in the Global War on Terror.

As the missiles climbed into the dusk sky, a loud cheer resonated across the flightdeck, which was followed almost immediately by an eerie calm as the Air Boss called away to start our engines and prepare to launch. The launch progressed flawlessly, and our strike package rendezvous and transit to the tanker was also uneventful.

An E-2C crew from CVW-11's VAW-117 performed excellent co-ordination while we transited north through Pakistani airspace, and we conducted a package roll call on the way to our front-side tanker. On the first night of the war, there were no AWACS or tanker assets located over Afghanistan. The corridor of airspace negotiated for coalition use with Pakistan had an east and west boundary that required us to meet a drogue-configured KC-135, receive our allotted gas and then fly a 500-nautical-mile round trip to Kabul and back to the tanker assigned to our package. We would then get our backside gas, before returning to the carrier.

As we joined up on the KC-135 on NVGs, we could see that something was very wrong. There was a solid stream of gas trailing from the "knuckle" where the drogue hose connected to the hard refueling boom. While the fighters, given their probe placement to the right side of their windscreens, were still able to take gas, the EA-6B pilot could not tank without obstructing his vision because his probe was located almost directly in front of his windscreen. I decided that given the minimal threat [opposing the strike] – a lone SA-3 "Goa" SAM, AAA and shoulder-launched MANPADS [Man-Portable Air Defence System] – the EA-6B should return directly to the ship.

We were supposed to get about 8,000lb [3,628kg] of gas each, but the boom operator told us "offload complete" when we had barely received 3,000lb [1,360kg] due to the amount of gas streaming out of the leak. To make matters worse, this KC-135 was also supposed to be our backside tanker when we came out of country. However, because of the leak, it would not be available upon our return. I requested that the USAF AWACS controller who was now handling us find my strike package a spare or alternate tanker – he would have approximately one hour to make these arrangements. We pressed on to Kabul, not knowing what our refueling options would be off target.

VF-213 OEF (2001)

Manning the second jet in Cdr King's section were Lt John Saccomando and Lt Cdr Michael Peterson, and the latter recalled:

We were able to maintain regular [UHF] communications with the AWACS in Pakistan for a while, but soon we only had contact using our JTIDS voice channels. Eventually, we could only talk with the members of our package, having received a check-in from the B-1Bs confirming their position. At least the RoE was easy for us to understand. If anyone launched in front of us, they weren't friendly and could be engaged after meeting a few other simple criteria.

The weather was clear as the Tomcats and Hornets flew north up the legendary Khyber Pass, with the moon breaking the horizon to the north-east of the strike aircraft. The Hornets, positioned outside the lead section of Tomcats, were to launch their solitary SLAM-ER when 50 nautical miles (93km) from their target – located near to the SA-3 site that the VF-213 jets were tasked with destroying – and guide it from a safe distance using a datalink pod on the wing aircraft. Lt Cdr Peterson explained that moments after the Hornets executed their turn outbound, the horizon again lit up as the TLAMs began raining down on their targets:

Even though we were more than 100 nautical miles [185km] away from the target when the Tomahawks impacted, you could see the flashes right on time through our NVGs. A few minutes after the impacts, something else was quite evident on the NVGs – the entire capital had erupted

Left: "Black Lion 111" (BuNo 161159) is marshaled towards one of *Carl Vinson*'s waist catapults, the jet being armed with a single GBU-24A/B Paveway III LGB. Only a handful of these "bunker-busting" weapons were expended in OEF, VF-213 being the leading dropper with four. VF-41 made two GBU-24 attacks, with one being delivered by Lt Peter Gendreau and his RIO, Lt Cdr Scott Butler. "We flew a daylight strike on a weapons storage facility on the outskirts of Kabul. Our target was a series of closely grouped bunkers in an area that had a high potential for collateral damage, and because of this, Lt Gendreau flew several runs at it to make sure that we had our laser angled just right to achieve maximum weapon penetration. Our bomb definitely penetrated the bunker complex because we saw huge secondary explosions once it had detonated. That was the one drop I did in OEF that made the nightly news back home." (US Navy)

Right: Kabul Firework Display (2001)

with AAA. The Taliban may not have had more than one radar-guided SAM guarding Kabul, but these guys had a shitload of AAA! It looked like you could get out of the jet and walk across it.

I could see tracers from rapid-fire smaller-caliber guns [both 23mm and 58mm] burning out below us. They also had bigger weapons in the capital as well, with some of these guns having been moved up into the mountains that ringed the city – Kabul itself is located at more than 6,000ft AGL, with surrounding mountains that have peaks exceeding 12,000ft. Shells from these weapons exploded at far higher altitudes, and from where we were sitting, they looked like an Independence Day fireworks show. You could see a faint sparkling trail from these single round heavy caliber shells as they penetrated the mass of smaller fires and sailed on up to our altitude, before exploding with a brilliant bang.

Fortunately, the navy had upgraded the laser in our LTS pods to allow it to fire at altitudes up to 40,000ft AGL, instead of the old maximum altitude of 25,000ft. Descending to below 25,000ft AGL to fire the laser for weapon guidance would have put us well into the heart of the AAA. Some of the bigger guns could still reach up and touch us at 40,000ft, but the fires up there weren't nearly as dense as the concentration below us.

Identified by veteran Tomcat RIOs as the "weakest link" in the original AN/AAQ-25 LANTIRN pod as delivered to the US Navy, the laser receiver – not the transmitter – had the potential to "arc and spark" above 25,000ft AGL. This fault could cause a fire in the pod, and with the LANTIRN being a non-jettisonable store, that eventually meant an aircraft fire too. Once this problem was discovered, a software restriction was placed into the pod that inhibited laser fire above 25,000ft AGL.

In the new "40K" laser-modified pods, which reached the fleet in early 2001, Lockheed Martin replaced the old laser receiver (and a few other pieces of internal hardware) with an improved, more powerful one, and also modified and upgraded the software to allow laser system functionality up to 40,000ft AGL. Old pods were progressively reworked to

Surrounded by high peaks, Kabul proved to be a challenging target for Tomcats and Hornets throughout OEF. The SA-3 SAM site attacked by CVW-11 on night one of the war was situated near the international airport seen at the extreme left in this photograph. (Lt Cdr Jim Muse)

this standard, and a number of new ones procured. However, only VF-213 and VF-102 had received the "40K" pods by the time OEF commenced, thus severely restricting the attack profiles that VF-14 and VF-41 could fly when attacking certain fixed targets in Afghanistan.

With the mountains surrounding Kabul blocking the view of the city lights from the Tomcat crews until they had banked to the east around a range of peaks, Cdr King had to wait until the last minute to set up his predetermined attack run. This meant that although they were flying the trailing F-14D, Lt Saccomando and Lt Cdr Peterson actually got to drop the first bombs of OEF. Peterson recalled:

As we turned to the east over the capital to line up for our attack run, we were able to maintain our altitude well above 30,000ft AGL, but being up this high also had its drawbacks. The release range for our two GBU-12s was much further away from the target, and that made it considerably harder to identify the SAM site through the LTS. Nevertheless, we located our target and released on timeline. Then my NVGs lit up as I saw our lead engage afterburner and spin a quick 360-degree turn over the target area. They had located their target late and were inside release range, so they decided to make a quick re-attack.

We continued to guide our weapons on target, but they went slightly long. We later found out that they hit a barracks just beyond the intended point of impact. We continued out ahead of our lead towards Jalalabad and cleared the airspace for the follow-on attack by the B-1Bs, who were 20+ nautical miles [37km] in trail.

In "Black Lion 101," Cdr King and his RIO, Lt Cdr Paul Gronemeyer, were working hard to get their GBU-16s off on target:

> We made a hard right turn, continually doing belly checks, while trying to maintain our energy package – difficult to do when flying at 30,000ft AGL. By the time we came out of the turn, we had lost nearly 8,000ft and 50 knots [93km/h]! We acquired the target and released our weapons, and I unloaded the jet so as to accelerate to 500 knots [926km/h] and begin a gentle climb back above 30,000ft. We picked up our wingman, who was already heading south, on JTIDS, and it was then that we realized we were well behind our established timeline. Our wingman throttled back to save gas and allow us to close on him. Having re-established section integrity within ten minutes, the hunt for a tanker now commenced in earnest.

By the time the two jets had joined up, Lt Cdr Peterson had already made several attempts at contacting their AWACS controller to determine the location of their backside tanker. The section of VF-213 jets had flown 100 nautical miles (185km) south of the target area before they finally made contact with their controller. Cdr King's first job was to confirm that a tanker was on station on the north-eastern "driveway" as planned. However, according to Peterson:

> The controller calmly told us that the leaky tanker had returned to base and a new tanker had taken its place on another track. He then passed the tanker's call-sign and track location. We pulled out the chart and looked for the new tanker track, and to our horror found out that it was located not in the north-eastern but central part of Pakistani airspace more than 400 nautical miles [741km] away! We immediately went to a maximum fuel conservation profile, climbing above 40,000ft and slowing down significantly, even though we were still deep in enemy territory.
>
> We tried to get the tanker moved to the original station to meet us, but the controller said that he had other customers that meant the KC-10 had to remain on its current track. We ran the numbers and predicted that we would arrive on the tanker track with around 3,000lb [1,360kg] of gas each. This was probably not enough fuel to get us to the emergency divert airfield at Jacobabad, farther south in Pakistan, if there was a problem on the tanker.
>
> As we approached the tanker track in an idle power descent to save gas, we made contact with the KC-10 and explained the situation. He started dragging his receivers 100 miles [160km] towards us into Afghanistan, which he wasn't supposed to do, and had us call his turn so that we would wind up right behind him ready to refuel without wasting gas having to join up and get into position on our own. As we gained a visual on the tanker, with a division of four F/A-18 "chicks" in tow, he cleared out the rest of the aircraft waiting to refuel and allowed us to join right on the boom. The KC-10 crew had performed flawlessly in getting us into the basket as expeditiously as possible, saving our bacon that night.
>
> Our lead's fuel state was lower than ours, with "Black Lion 101" having only 3,200lb [1,450kg] of fuel left in its tanks, and Cdr King quickly plugged in. Having taken on just 2,000lb [907kg] of "comfort fuel," he then disengaged to the starboard side and allowed us to get all the gas that we required. As we plugged in and the fuel began to flow, I noted that we had only 2,000lb [907kg] of gas left – not enough for us to have made it to the emergency divert. After getting a few thousand pounds, our lead cycled back through, and we were able to return to the ship for an uneventful recovery.

Flying over solid cloud heading into Afghanistan, the crew of "Black Lion 103" (BuNo 163899) have unfurled a flag sent to them by officers serving in the New York City Police Department. Photographed in late October 2001, the aircraft already boasts a lengthy bomb tally. Having been assigned to VF-213 in May 1998, BuNo 163899 would also participate in OIF I with the unit in 2003 prior to being stricken and stripped for parts in October of that same year following just 13 years of service. (Lt Tony Toma)

While Cdr Chip King led his strike package to Kabul, VF-213's XO, Cdr Anthony Gaiani, and his wingman were tasked with protecting B-1Bs bombing the air base at Herat, before heading to Farah and attacking a Taliban communications facility with their own GBU-12s. Both targets were in western Afghanistan, not far from the country's border with Iran. The Tomcat was the only carrier-based aircraft that could perform a mission such as this, as Cdr Gaiani explained:

The strike I led on the first night of the war reflected both the realities imposed on us by the limited number of assets we could support (particularly with tankers), the ranges involved and the kind of enemy we faced. Unless and until we could eliminate the air defenses in Afghanistan, the tankers would not be going in-country. The result was that the Tomcats were tasked with the far targets in Kabul and Herat, while the Hornets were assigned the nearer targets in and around Kandahar.

Cdr Chip King (left) and Lt Cdr Kevin Claffy compare notes on the flightdeck at the end of an OEF mission in early November 2001. Behind them, one of VF-213's hard-working armorer teams has already started uploading GBU-12s on to the BRU-32 bomb racks attached to "Black Lion 103." "The F-14's belly-mounted ordnance pallets made rearming the jet more difficult than a Hornet, and the heat generated by the aircraft's engines during a six-hour flight proved hard to take," recalled VF-213's CWO3 Michael Lavoie. "Things would get so hot back there that a four- to six-man team working in close proximity to one another would be showering in each other's sweat." (Lt Tony Toma)

In reality, all three locations were a real stretch, requiring precise planning, absolute discipline and jets kept in top shape. We were blessed to have all three. In fact, I would go so far as to say that range and endurance were as much a weapon as the LGBs we carried – without the Tomcat's ability to reach the distant targets and return to the tankers unrefueled, the campaign might have lasted much longer. And I believe it is safe to say that other missions – particularly SCAR and CAS – would have had a significantly different flavor without the Tomcat's qualities.

CVW-8 enters the fray

As the night carrier, *Enterprise* had to wait until the early hours of October 8 before its strike aircraft could bomb their targets. Like CVW-11, CVW-8 would send two F-14As from VF-14 to strike the "Spoon Rest" A-band warning and target acquisition radar that provided guidance for the SA-3 SAM battery at Kabul. The latter had, of course, been struck earlier that night by TLAMs and two Tomcats from VF-213. Elsewhere, a pair of F-14As from VF-41 would target the cave entrance to an al-Qaeda training camp in mountains near Kandahar.

Leading the Kabul radar site attack was VF-14 CO Cdr Bruce Fecht:

Initially, a section of Hornets, a section of Tomcats and some support assets, including a Prowler, were fragged to carry out this mission. As the detailed planning got underway in the days leading up to A-Day, it quickly became clear that there would not be enough fuel available for the F/A-18s to make it from the last tanker position to the target area. The strike package quickly degraded to just two Tomcats from VF-14 and a solitary Prowler from VAQ-141 – not exactly the force we would have planned to put up for a first night strike had we been conducting an air wing training evolution from Fallon!

The instructors at NSAWC would have almost certainly castigated us for being so cavalier and non-conformist in our tactics and plans. However, logistics and operational necessity dictated that this would indeed be the first strike package to launch from *Enterprise* during OEF.

Other elements that were not optimal affected the planning and execution of that particular mission. The high altitude of the targets and the ceiling restrictions of the old LANTIRN pods fitted to VF-14's jets shaped the bomb delivery plan devised by Cdr Fecht and his operations department, as he explained:

The performance of our old LTS pods, the close proximity of the target impact points, considerations about the timing of our attack runs due to possible smoke/dust clouds over the target and defensive measures to counter enemy SAMs and AAA drove the route tactics that we adopted.

In effect, we determined it was best for Pk [Probability of kill] to violate another basic tenet in strike planning and actually plan for a re-attack along a similar, but reciprocal, heading to allow for the second target to be hit with the needed lethality. When you examined the plan overall, it looked pretty shaky – a first night attack on a "center-of-gravity" Taliban target, with limited altitude sanctuary, a minimal strike package and a planned re-attack on the initial, but reciprocal, heading. I don't think many folks would have signed up to this in training.

Having launched, the small CVW-8 strike package headed north and reached Pakistan without any drama. However, shortly after going "feet dry," Cdr Fecht's "Dash-2" jet, flown by Lt Cdr Marcell Padilla (and with Lt Cdr Art Delacruz as his RIO), reported smoke in the cockpit. "We had a power spike that caused the loss of our INS and saw the cockpit fill with smoke and fumes," recalled Padilla. "I told 'Cutlass 43' [Cdr Fecht – VF-14's crews were using ATO-derived call-signs] that I had seen 'arcs and sparks' and had shut down the Weapon Control System in order to avoid the possibility of a fire. After careful consideration, the system was brought back online and my RIO and I assessed our ability to continue the mission. It appeared that the AN/AWG-15 [weapons control panel] and LTS were unaffected, so we decided to press on without an operable radar."

According to Cdr Fecht:

Typically, smoke or fire in the cockpit is a definite mission terminator and an immediate return trip to the boat for a landing as soon as possible. I didn't want to make that decision until I knew for sure that there had indeed been a fire. Sometimes perceived smoke in the cockpit can have other sources, or be misidentified. I also knew that launching the spare would delay the mission, along with the issue of not having the aircrew nearly as well briefed and fully prepared as our wingman. Fortunately, they were able to isolate the problem and prevent any complications, along with determining that they could indeed continue with the mission.

The next big hurdle was tanking. Lights out, higher altitude than normal and with little room for error, this was a critical node for the sortie. The use of the FLIR pod and NVGs eased the concerns surrounding the "lights out" rendezvous with the tanker. Nevertheless, whenever we worked with the KC-135 "Iron Maiden" [a nicknamed derived from the 16th-century torture device that Naval Aviators felt aptly summed up their experiences refueling from the probe and rigid drogue-equipped boom of the veteran Boeing tanker], we expected to use lots of concentration just to stay in the basket.

The way we were configured, and at the altitude [30,000ft] the tanker was flying in order to remain out of any SAM envelopes, we were having to use a little afterburner on one engine to stay in the basket after reaching about 16,000lb [7,257kg] of fuel onboard. My technique was a little tap of afterburner – around Zone 2 – on the port engine and then use the right engine to modulate any

VF-41 OEF (2001)

further power needs. This worked for me, and it was a big relief to meet the tanker and complete that part of the mission, knowing that we now had enough fuel to get to the target and make the divert airfield if there was no tanker available on the backside.

After all three aircraft had cleared the tanker, it wasn't long before we crossed the border and were eerily on our way up the valley from Kandahar to Kabul. My first impressions of the landscape below was that it very much resembled the various mountain ranges that dominated Nevada, which we had flown over on numerous occasions when performing the strike missions that all Naval Aviators and NFOs have undertaken during Fallon air wing detachments.

Kabul was dark and seemingly quiet. It wasn't like that on the computer simulation that I had trained on in the days leading up to this moment, and not what I had expected. We knew that the TLAMs had hit their targets in the area, and I assumed they had taken down the electrical grid. This lack of lighting had no effect whatsoever on the LTS pod, however, which soon started to break out the target so that it clearly resembled the intel photographs we had pored over pre-mission.

After my RIO and I had discussed and verified the target, we got ready to drop our GBU-16s. My next step was to lift the arming guard and actually set the switch to "ARM" – something I had done numerous times in training, but this was the first time I had ever done it "in anger." I remember that the red light indicating the systems were good was particularly bright in the blackness of the cockpit. I then "pickled" the thousand pounders and felt the "thunk" of the release as they fell away. We then made a 2–3G left turn to set ourselves up for the lasing portion of the drop.

After attaining the angles I wanted, we turned back to the right and concentrated on the final portion of the strike, which was guiding the weapon through the laser spot via the RIO's weapon station. During this element of the mission, I was the only one looking out of the cockpit, since the RIOs in both jets were concentrating on the sensor displays and backing each other up on the delivery tactics. My wing pilot was concentrating on formation flying and trying to stay in position. Therefore, when the bombs hit the target, I was the only one to initially see the returning ground fire.

"Hey, look at that, they're shooting at us. Low, right, five o'clock." The tone of surprise in my voice was, indeed, surprising! I saw the tracers firing from the dark world below. Although they were heading in our general direction, I felt confident that they were behind us, and not aimed with the proper lead needed.

As we headed north, we saw the airport to the north-east of us start to light up with ground fire. It appeared that the enemy gunners had a bead on us, but it was hard to tell if they were just following the noise or actually had a sensor on us. There were at least six different centers of fire I could determine, and we knew prior to flying this mission that the airport would be heavily fortified. However, we had to fly around it clockwise from the south-west to the south-east so as to line up for our wingman's bombing run.

As we started to make our turn back to the south, we knew we were bleeding knots too quickly, and that we would require afterburner to restore our airspeed – this would highlight us in the night sky. We didn't want to help the Taliban improve their targeting solution, but we had a certain minimum airspeed we had to maintain for maneuverability and threat avoidance. Somewhere in that turn I also passed the lead to my wingman and then tried to hold on for the ride.

About the time we were getting on to a southerly heading, I was now flying sucked and left of the lead, looking around for the greatest threat. There was plenty of stuff flying through the air, with lots of bullets and an occasional missile being sighted. In fact, the view in my NVGs got so concentrated with glowing green dots that I decided it was time to take them off and just follow the afterburner of the lead. I could see bullets to the left and right of us, and I was praying that they were going to miss and fall down on each other. It was better with the goggles off and the false illusion of the black night and occasional red tracer round. With the goggles on it was just too much, giving me sensory overload.

At some point in time during that run I collected a sense of overwhelming fate. It came down to training and doing what we were supposed to, and there was really no way to jink from everything being thrown at us. I just concentrated on the run and got into the "whatever happens, happens" mode.

In "Cutlass 44," Lt Cdrs Padilla and Delacruz were working hard to emulate the success of their strike lead. Padilla recalled:

By the time we had started our attack run, there were only 15–20 seconds left before weapons release. I noticed AAA flashes in front of us and at "one o'clock" five seconds prior to me pickling the LGBs. I thought to myself that this was "G-time" [government time], and maintained delivery parameters. After delivery, I noticed that only one of our bombs had left the jet, so I did a post-release maneuver. I asked my RIO if the correct settings had been punched into the AN/AWG-15, and he answered in the affirmative.

After completing my post-release maneuver, I noticed, and reported, AAA at "one o'clock," and very quickly assessed that it would not be a factor for our section. "Cutlass 43" then called out that they had seen more AAA from the airfield, as well as the launch of ballistic SAMs. I too had spotted the AAA, but not the SAMs. I then quickly came back into the cockpit to check my flight parameters to ensure full support of the solitary LGB that we had dropped, and made the

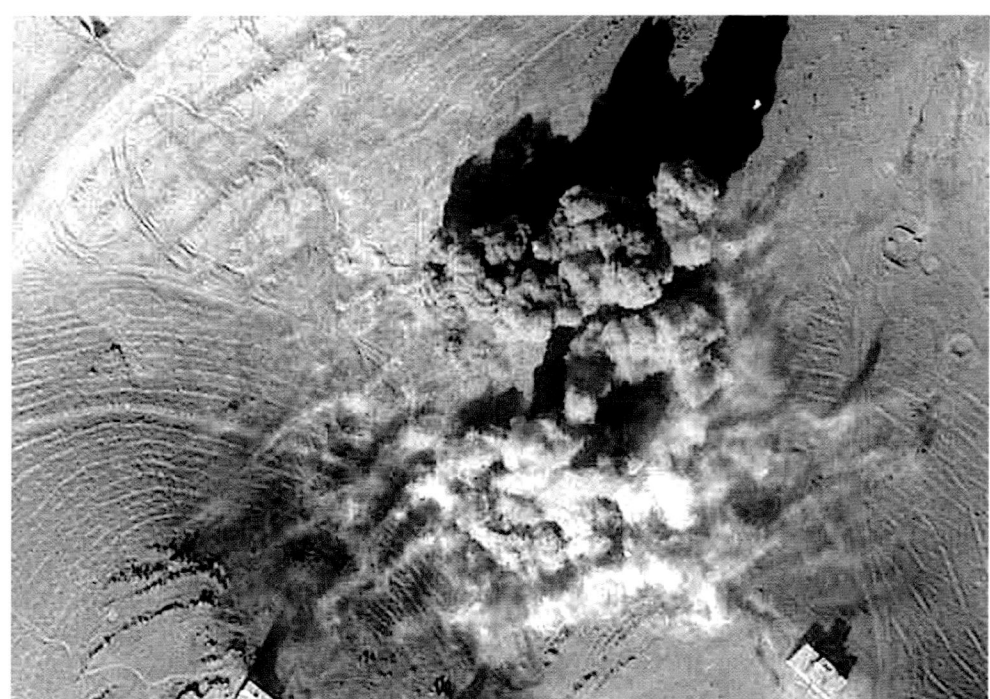

"Before and after" shots of two Taliban T-62M Main Battle Tanks caught out in the open by a TARPS-equipped VF-213 jet on November 11, 2001. These photographs were taken by the pod's KS-153B camera, the tanks having been struck by GBU-12s. (VF-213)

appropriate adjustments and stayed on parameters for the next 17 seconds until the target was hit. "Good impact" was called, and "Cutlass 43" then told us to flow on to a 190-degree heading.

It was at this time that the complete gravity of the situation set in, as there was AAA around both aircraft from all directions. In fact, there was enough light for us to plainly see "Cutlass 43" without the use of NVGs. Cdr Fecht called for a defensive maneuver against a SAM that was not seen by us until post-maneuver as "Cutlass 43" put out flares and acquired a streak of light drifting aft of their "five o'clock". At this time, I engaged afterburner and accelerated to the south, continuously checking my belly and deploying countermeasures.

Although the VF-14 crews had managed to drop three of their four LGBs, successfully hitting the target, they still faced a long flight home as Cdr Fecht explained:

Despite the heavy adrenaline, big danger, portion of the mission being over, we couldn't yet relax. We still had to find our KC-135 and conduct a lights-out tanking evolution, and then finish the whole thing off with a night trap back aboard *Enterprise* – oh what joy, all that and a night trap as well. The strike run had lasted less than five minutes, but the night trap anxiety dragged on for much longer. It usually started in marshal as you circled awaiting your "push" instructions near the carrier, and it didn't end until you felt the tug of that wire.

Thankfully, due to the experience of the Naval Aviators flying that night, the recovery ended up being very much routine – unlike the debrief, which was a lengthy affair. We had plenty of details for the intel folks on threat systems and target acquisition, along with analysis on the tactics, techniques and procedures of the mission. It was good to resolve all that immediately before we lost the detail, and with so much adrenaline pumping there was no way I could sleep till this "high" wore off anyway. I was on a cloud for hours.

Cave Busting

As noted earlier in this chapter, VF-41's target on the opening night of the war was the entrance to a suspected al-Qaeda terrorist training camp in a mountain range south-east of Kandahar. A section of jets from the unit would be involved, with the lead machine crewed by squadron XO, Cdr Pat Cleary, and NFO Lt Cdr Ed Meyle. The specialist GBU-24 Paveway III Penetrator LGB was chosen as the best weapon to achieve the destruction of this challenging target, the 2,400lb (1,088kg) "bunker buster" boasting a hardened front casing that allowed the bomb to bore through five feet of concrete prior to detonating.

Just nine GBU-24s would be dropped by F-14s during OEF – four by VF-213, three by VF-14 and two by VF-41. The latter two units had used the weapon with limited success in Operation *Allied Force*, where the precise and lengthy flight path needed for the bomb to attain the steep, almost vertical, descent to the target the GBU-24 required to be effective had proven difficult to achieve in the hostile Balkan skies. With a conventional, more flexible, LGB such as the 500lb (227kg) GBU-12, the RIO would typically have to lase the target for less than 30 seconds. However, in order to attain maximum penetration speed, the GBU-24 had to be released earlier and the target lased for 60 seconds.

VF-41's "Fast Eagle 103" (BuNo 158612) flies in close formation with a jet from VF-14 during an OEF mission on October 16, 2001. The oldest F-14 among the 22 Tomcats embarked in *Enterprise* for CVW-8's 2001 deployment, this aircraft was one of five A-models assigned to VF-14 and VF-41 that had been delivered to the US Navy in 1972. The jet, which dropped 17 LGBs on targets in Afghanistan, featured the name *Dolores* on its radome. The significance of this was explained to the Author by VF-41 CO, Cdr Brian Gawne. "Some of our personnel wanted to paint nose art on our aircraft, but we didn't have a suitable artist. In a bar in Palma, I found out that our Maintenance Master Chief, AVCM Randy Bradley, had nicknamed the jets. He joked that he had had 11 girlfriends, and had given each of the jets one of their names. Once I heard this, I told him to go ahead and paint them on the jets." (Lt Cdr Van Kizer)

The GBU-24 had a poor reputation within the Tomcat community, being dubbed "pretty unreliable and a non-user-friendly weapon that doesn't have a high hit percentage" by VF-14 *Allied Force* and OEF veteran Lt Cdr Van Kizer – he would get to drop a GBU-24 on a fuel/ammunition storage dump in Kabul on October 17, 2001. VF-41's Lt Cdr Scott Butler remembered that crews assigned GBU-24 missions had to spend "hours weaponeering and target planning in order to ensure the accurate delivery of the 'bunker buster.' It's a labour-intensive weapon tailored exclusively for use against hardened targets, and fortunately for us, there weren't too many of those in Afghanistan."

VF-41 had its bunker-busting "A team" leading the October 8 strike, with squadron CO Cdr Brian Gawne identifying Lt Cdr Ed Meyle as "our GBU-24 expert." Cdr Cleary described the mission:

> As the strike lead, I had to plan all the fuel for the jets involved, the tanker rendezvous points, the attack profiles for the target and the tactics we were to employ. Each of the VF-41 jets allocated to the mission was armed with just a single GBU-24 apiece, although the Tomcat was cleared to carry two. However, the pilot would have had to jettison one in order to get the aircraft down to landing weight should no bombs have been dropped during the course of the sortie. GBU-24s were simply too expensive to waste in this way.
>
> Aside from our two Tomcats, there was also a pair of Hornets armed with JDAM sent to hit the target, which was located 400 [644km] miles north of the carrier.
>
> Relying exclusively on S-3 tanker support, we met a section of VS-24 jets that had launched ahead of us over Pakistan in order to top off our tanks, before pressing on into southern Afghanistan. With the target – a limestone cave dug into a mountain ravine – being 40 miles [64km] to the south-east of Kandahar, we were out of SA-2 range. There were also CVW-8 strikes hitting these missile batteries at exactly the same time that we were attacking our target.
>
> Unlike the SAM sites, which were defending key military bases in Kandahar, the terrorist camp was literally in the middle of nowhere. There were no lights to be seen anywhere on the ground, so we were dropping exclusively using the imagery generated by the FLIR. When I looked out of the cockpit off NVGs, all around me was a sea of inky blackness. Although our GBU-24 guided well – this was the first bomb I had ever dropped in anger in 18 years of frontline flying – we never saw the weapon explode. This was almost certainly because it had worked as advertised and penetrated deep into the side of the mountain before detonating.
>
> Having climbed up and away to the right in order to clear the target area to allow my wingman to make his attack run, I levelled off to see if anybody was shooting at us. Literally the whole mountainside was lit up with small-arms fire!
>
> During the return leg of the mission we again refueled from two S-3s – this was the only organic strike flown by CVW-8 throughout OEF, with the USAF's sole input being its E-3 AWACS control.

All three Tomcat units continued to strike fixed military targets over the next few days, as they worked through the CAOC-driven target sets supplied to them via the daily ATO. Aside from bombing airfields, SAM sites, troop barracks and terrorist training camps, the F-14 squadrons also provided escort for the USAF C-17s that flew humanitarian airdrop missions over Afghanistan from night one of the war.

With no coalition troops in-theater to support during the early phase of OEF, the Tomcat crews worked instead with two-man SOF teams, who sought targets for them to attack – they would also provide crews with target "talk-ons." US Navy strike aircraft relied heavily on "big wing" tanker support throughout OEF, with F-14 FAC(A) crews in particular refueling at least three times from KC-10s, KC-135s, VC10s or TriStars during the course of their marathon missions.

Exclusively employing LGBs, VF-14 and VF-41 expended some 380,000lb (172,365kg) of ordnance between October 8 and 23, when *Enterprise* was relieved by *Theodore Roosevelt* (with the F-14Bs of VF-102 embarked). "VF-41 had an 82 percent hit rate, which was a level of accuracy that had never previously been achieved by the US Navy," recalled Cdr Brian Gawne. Both Tomcat squadrons in CVW-8 also buddy-lased Maverick missiles and LGBs for Hornet units.

A number of the aircrew assigned to both VF-14 and VF-41 were veterans of *Allied Force*, including FAC(A) pilot Lt Marcus Lopez. He was involved in the pursuance of Time-Sensitive Targets (TSTs) such as al-Qaeda and Taliban leaders – missions that assumed great importance once all the pre-planned fixed targets had been destroyed in the first week of OEF:

> I soon re-rolled into the FAC(A) mission, going in search of emerging targets, whilst those guys in the unit lacking this qualification stuck to hitting fixed targets on the ATO. My first FAC(A)-designated mission – although this ultimately evolved into a TST strike – was flown three days after OEF started when I got to work with one of the CIA FACs in-theater. These guys really weren't sufficiently trained to control TACAIR assets, and they were very thin on the ground. Those in Afghanistan predominantly operated around Kabul, seeking out TSTs. The CAOC specifically told CVW-8 that it wanted FAC(A) crews to go and work with the CIA "Spooks," as the latter lacked the experience necessary to provide TACAIR assets with the kind of precise target co-ordinates required to attack key "pop up" targets in urban areas.
>
> The CIA-controlled TST operation in Kabul that I became involved in on October 10 proved to be one of my longest missions in OEF – I would subsequently discover that this was often the case

VF-41 OEF (2001)

The officer cadre of VF-41 pose for their end-of-cruise photograph during *Enterprise*'s brief three-day port call to Souda Bay, Crete, in late October 2001. Following the "Black Aces'" history-making Operation *Allied Force*/OSW cruise in 1999, when the unit dropped more than 200,000lb (90,718kg) of ordnance, the 2001 deployment could have been something of an anti-climax for the crews involved. However, brief action in OSW and 16 breathless days of combat in OEF ensured that VF-41 once again led the way as the squadron became embroiled in the Global War on Terror. Despite the unit being equipped with aircraft that were, in the main, destined to be scrapped once the cruise was over, the Naval Aviators and NFOs seen here again dropped in excess of 200,000lb of ordnance (all LGBs), with most of this being expended between October 8 and 23. (via Cdr Brian Gawne)

with these types of sorties. Things really dragged on as we waited for the CAOC and the FAC to give us the green light to drop our bombs. By the time my RIO and I landed back onboard the carrier, we had been airborne for almost nine hours.

It turned out that the FAC had been waiting for key Taliban personnel to enter a tenement-style house that he was watching. He kept telling us to "wait a few more minutes," before having to go off-radio when people approached his position. He would then tell us where he was, and we could see him on the roofs of houses through the LANTIRN pod as he moved from one location to another. We ended up having to tank twice while waiting for clearance to drop, until he was finally happy that all of the personnel he was targeting were in the house for a meeting that he had received intel on. Only then we were passed target co-ordinates and cleared by him, and the CAOC via "Bossman" [AWACS], to drop a single GBU-12. The bomb went straight through the front door of the house.

The FAC was very close to the target, his building being separated from it by a small park. Our bombing run had to be flown in such a way that we passed over this park and then aimed our LGB at the front of the house so as to avoid inflicting collateral damage on surrounding dwellings.

The LANTIRN pod was critical in a mission such as this, as the Hornet's Nite Hawk pod was not powerful enough to break out urban targets with the required accuracy. The latter pod, from the height that we were operating at, would have been unable to distinguish one tenement house from another, let alone allow the pilot to count down the number of doors from the FAC's location to the target building, as we had to do.

I had a wingman with me on this mission, and although he, too, worked up the target as we circled over Kabul, his primary role was to support us in any way he could. We always had a non-FAC(A) wingman for these sorties, and it was his job to watch over us should we have to descend below the 15,000ft minimum altitude hard deck [introduced by CENTAF for OEF] in order to get our bombs, or someone else's, guided accurately to the target. Our wingman would remain above us in a position that allowed him to keep an eye on what was going on around us while we focused more closely on the target area.

Lt Marcus Lopez of VF-41 drew heavily on his Operation *Allied Force* experiences in OEF. "I was the senior JO in VF-41 going into OEF, being a FAC(A) as well as a Division Lead and a Strike Lead," he recalled. "I did not get to fly on the first night of the war because I was duty Landing Signal Officer, but I did fly once a day for the remainder of our time on station in the Arabian Sea." (via Lt Marcus Lopez)

The increasing importance of FAC(A) crews in OEF signified a shift in the way the air war was now being prosecuted. VF-14 pilot Lt Mike Bradley chronicled this change as he saw it from a junior officer's perspective:

The focus of the initial attacks was to secure air superiority for all friendly forces, hence the targets we initially went after were enemy air defense systems, strategic SAM sites, airfield complexes and aircraft. Kabul and Kandahar were common, popular targets. We also destroyed command and communication centers and terrorist training camps. Within a matter of days, however, we had expanded our scope to include emerging targets such as Taliban ground forces, tanks, vehicles and artillery, as well as the occasional TST.

Most of my flights in OEF involved targeting tanks or armored personnel carriers located in assigned "kill boxes" north-west of Khandahar. I also destroyed a SAM support truck and acquisition radar and a building complex being used for command/communication.

From my perspective as a very junior officer, mission tasking was easy. Our flight schedule would be written per the ATO, with initial targets assigned. We would gather intel and imagery of our targets, along with that of other possible flex targets, in CVIC. We would then brief, man up, launch and proceed in-country. After topping off on the tanker, we would check in with whoever happened to be the controlling agency that day [usually a USAF AWACS]. They would then clear us to go after our assigned target, or give us a new one. We then conducted our assignment, returned to the tanker and either flew straight back to the ship or manned a DCA CAP for a few hours, prior to heading home. A typical mission could last anywhere between 4.5 and 8.5 hours.

FAC(A) or SCAR?

With the emphasis shifting to mission flexibility to the extent where most TACAIR crews were now launching without knowing exactly what targets they were being sent against, those attributes that made the F-14 the perfect FAC(A) platform also meant that the jet could perform the less specialized, but no less important, SCAR role too. FAC(A)-qualified RIO Lt John Kelly of VF-41 said he felt that most of the officially designated FAC(A) missions he flew were, in reality, SCAR sorties:

The Navy has never been very good at handling the FAC(A) mission from an admin standpoint. No less than 95 percent of my ATO-fragged FAC(A) missions could have been labelled as SCAR, and the latter did not require a FAC(A)-qualified crew to perform them.

When tasked with flying a FAC(A) mission, I would walk to the jet carrying a helmet bag filled with 15 to 20 target packs. We would have one or two targets pre-planned, and we knew that we would be hitting these early in the sortie with the first strikers that would check in with us to service targets in our kill box. Such mission control saw us operating in the SCAR role, not as FAC(A)s.

Typically, there were only four FAC(A) crews per unit in OEF, and we became so important that when we had to break off for fuel, the whole kill box would be declared shut by "Bossman" until we returned. Our wingman, who would remain on station, would not be allowed to handle incoming strikers because he and his RIO weren't FAC(A)-qualified, yet most of the targets we were hitting were fixed in nature and could have been lased by a competent crew with a serviceable LTS pod.

Due to the emphasis placed on minimizing collateral damage, an F-14 equipped with an LTS pod and a PTID was the king in OEF when it came to providing target identification. This allowed us to pick out the target building, which could be as small as an outhouse, whereas our Hornet brethren could not repeat this performance without breaking the minimum altitude hard deck due to the inadequacies of their shitty Nite Hawk pod. Providing visual and laser guidance to the Hornet pilot so that he could then hit a target with an LGB or LMAV was effectively a SCAR mission.

I can't blame the F/A-18C guys for being upset with me for lasing in their bombs, as I would have been upset too if I had been in their position. It simply boiled down to us having a better pod and associated displays when it came to satisfying the ROE for identifying targets in OEF.

Technically, a mission should only be labelled FAC(A) when you are controlling air support for troops in contact, rather than hitting stationary targets with somebody else's ordnance. SCAR is more of a kill box mentality, where you are servicing fixed targets with assets that are cycled through under your control. Such missions are directed by AWACS controllers telling you which kill boxes to go and work in.

When flying a SCAR mission – even though it might have been listed as a FAC(A) sortie – we would launch in a section along with a dedicated Prowler, and then pick up two or four strikers when we arrived in-country. We would then drag them to the target area that we had been assigned either on the boat or by "Bossman." These strikers were usually other CVW-8 F-14s and F/A-18s, although they could just as easily have been jets from CVW-11 or USAF F-15Es or F-16s. We would co-ordinate fuel for all these assets once we arrived in-theater, and then drag them to the target area. When the crews in this little strike package were all following the same game plan, and there were no conflicts within the jets as to who would hit what target and who would be lasing ordnance for whom, things went well.

All the SCAR missions we flew that had been designated as FAC(A) sorties were purposely mis-labeled on the ATO by the USAF-driven CAOC so that it would have a FAC(A) crew on station at all times. This was because the latter were used to talking, dropping and recording simultaneously in a combat situation.

Given the importance placed by the CAOC on FAC(A) crews, all three Tomcat units had to work hard to make sure that this requirement was satisfied without having a detrimental effect on the flight schedule for non-qualified Naval Aviators and NFOs. VF-14's 2001 command history included the following entry, which explained how the unit dealt with the FAC(A) issue in OEF:

All three Tomcat units involved in the opening phase of OEF adopted the four GBU-12 "quad bomber" loadout within days of the campaign commencing. "Our teams could upload four GBU-12s in ten to 15 minutes," noted VF-213's "gunner," CWO3 Michael Lavoie. "Pre-war, I had planned on such a task taking 20–25 minutes to complete. Indeed, on previous cruises, I had seen teams take 25 minutes to load a Phoenix, a Sparrow and a Sidewinder. During OEF, my guys were loading four bombs and chaff and flares in half that time." (US Navy)

The squadron creatively managed the daily combat flight schedule to increase the availability of critical FAC(A) qualified aircrews. Thoughtful scheduling enabled VF-14 to provide the maximum number of FAC(A)-qualified sections, and resulted in significantly enhanced effectiveness. In addition to aircrew management, aircraft were also configured to maximize lethality. Five aircraft were designated FAC(A) platforms and configured as "quad bombers." These aircraft provided a total of four LGBs to be delivered as "marks" by the FAC(A), and maximized the bring-back capability. The remaining squadron aircraft were configured as "dual bombers." The combination of the FAC(A) with four GBU-12s and the escort with two GBU-16s resulted in unparalleled airborne flexibility.

VF-102 Enters the Fray

The *Theodore Roosevelt* battle group arrived in the Northern Arabian Sea on October 15, ostensibly to relieve the *Enterprise* battle group, which had had its deployment extended in the wake of the September 11 attacks. *Enterprise* would remain on station for a further week, ensuring that there was a seamless turnover between CVW-8 and CVW-1. VF-41 sent a Naval Aviator and an NFO to visit VF-102 as part of this turnover, and they explained what their unit had been seeing and doing in-country, as well as administrative aspects associated with getting to and from Afghanistan. They showed aircrew the driveways that they had been using when heading through Pakistan and briefed them on what tanker tracks they had found most effective.

With this information to hand, VF-102's training officer, Lt Cdr John Cummings, set himself the task of single-handedly ensuring that no Tomcats from his unit flamed out during the ultra-long missions that the "Diamondbacks" were about to fly. According to squadron-mate Lt Chad Mingo, "Lt Cdr Cummings' secret weapon was a kneeboard card given to him by CVW-8 that contained all the tanker frequencies, the various tanker tracks and how much fuel it would take us to reach those tracks from known target areas in Afghanistan. The combination of the CVW-8 brief and the customized kneeboard cards meant that we were armed with some really sound information right from the start of our time in OEF."

In the week prior to *Theodore Roosevelt* arriving off the coast of Pakistan, Lt Cdr Cummings also met with his counterparts in the three Hornet units within CVW-1 (VMFA-251, VFA-82 and VFA-86) to discuss the future missions that they would be flying:

We spoke about things like guiding LMAVs, which none of us had done before, the tactics and techniques we would employ when undertaking CAS in-theater, and working with SOF teams – we had no idea where they would be, what kinds of radios they would be working with or the frequencies they would be talking on. It was effectively trial by fire, and we just made it happen when we got into combat – we did not have a lot of guidance from on high in OEF. CAG [Capt Stephen S Voetsch] placed few restraints on us when it came to getting the mission done. I have not experienced this level of freedom on subsequent war cruises.

CVW-1 flew its first combat mission of OEF on the evening of October 15 when VF-102 was sent to attack a Taliban barracks complex north of Kandahar. Although both CVW-8 and CVW-11 had by now started to perform far fewer strikes on fixed, pre-briefed targets, the CAOC decided that CVW-1 needed to be introduced to OEF with a conventional pre-planned attack. Two F-14Bs from VF-102 led the first mission, with the RIO in the wingman's jet being squadron operations officer Lt Cdr Tom Eberhard. He recalled that things did not go well on this strike right from the start:

A pilot from VF-102 keeps a close eye on a red-shirted "ordie" who is making last-minute adjustments to the all-important guidance unit fitted to a GBU-16. "VF-102 dropped 680 LGBs in OEF, and we only had five duds," explained the unit's "gunner," CWO3 Carleton Roe. "Three were due to failures in the AN/AWW-4 Fuse Function Control Set box fitted in the F-14, which meant that the bombs were dropped without being armed." (US Navy)

Strapped in and ready to go as part of CVW-1's next launch cycle from *Theodore Roosevelt*, Lt(jg) John Lynch (flying his named aircraft, BuNo 161422) and his NFO watch a bombless F/A-18C from VMFA-251 return to the carrier at the end of yet another six-hour-long Afghanistan mission on November 15, 2001. VF-102 struck up a solid working relationship with the US Marine Corps strike fighter unit, as VF-102's maintenance officer, Lt Cdr Scott Guimond, explained. "The Marine Corps pilots tended to be more CAS savvy due to the emphasis they placed on supporting their brethren on the ground during training. They also fitted LSTs to all their jets as a matter of course, rather than as an optional extra. This store proved greatly beneficial to the Hornet pilot when he buddy-lased LGBs or LMAVs with a LANTIRN-equipped Tomcat, as the LST allowed him to pick out laser energy from the F-14's targeting pod." (US Navy)

OPERATION *ENDURING FREEDOM*

The wing spoilers on our jet would not cycle in and out, but because it was so dark up on the flightdeck, both the final checkers on the catapult and I had failed to spot it. My pilot, who was also the squadron maintenance officer, immediately knew that they were inoperable, however, but he was determined not to miss this opening mission.

Aside from no spoilers, when I checked my LTS pod after launching, I discovered that I was not getting any information from it in respect to weaponeering, including the provision of release times for our two 2,000lb [907kg] GBU-10s – I was getting a terrain picture, however. There were certain pod functions that could only be checked once the jet was in the air, and it was then that we discovered the problem. I was now going to have to manually calculate the weaponeering for our LGBs, taking into account our altitude and airspeed so as to ensure that I released the bombs no earlier than X range and no later than Y range. That way the weapons would remain in the laser energy "basket" all the way to the target.

An LGB uses "bang-bang" technology to hit its target, being guided through the acquisition of laser beam energy by its seeker head. The bomb travels up until it "bangs" into the top edge of the beam rider and then the guidance vanes steer it back down until it strikes the bottom of the beam rider – this "bang-bang" routine continues until the target is hit. Having corrected fully up or down, should the bomb then lose the beam because the designator has either been masked or the pod has gone out of range, it will follow the final input correction and miss the target by some considerable distance. An LGB is little more than a big glider, and it needs steering inputs to ensure that it hits the target.

We had been tasked with attacking a previously bombed barracks just north of Kandahar airport, this target, I am certain, having been chosen for us by the CAOC in an effort to measure our combat capabilities. This first mission was, in fact, very similar to the whole Fallon training experience in work-ups, as I was given a target pack that provided me with photos and descriptions of the barracks.

The lead jet in our section was flown by the skipper [Cdr Roy J Kelley], with the CAG in the back seat. Although he ended up doing fine on subsequent missions, CAG was not quite up to speed tactically with the LTS at the beginning of OEF, and he therefore failed to find the target on this first strike. We made four passes over it from the north to the south, during which time I tried my best to get his eyes on to the target with my degraded FLIR, but to no avail. CAG and the CO ended up having to jettison a bomb on their way back to the ship and recover with the other one still attached to the jet.

The large 8 x 8in PTID screen in the rear cockpit of an F-14B from VF-102 reveals the rugged Afghan landscape near Tarin Kowt on November 17, 2001. This view, obtained via the LANTIRN pod while the jet was circling the target area at 31,980ft, could be panned in or out, depending on the mission requirements. No other TACAIR platform in OEF – US Navy or USAF – had a tactical display of this size, and when combined with the LANTIRN pod, it allowed the Tomcat to become the precision bomber and SCAR/FAC(A) platform of choice in-theater. (VF-102)

As we ran in on our fourth, and last, pass, I was working hard trying to get CAG locked on to the target and our bombs off the jet too, as by now we were running critically low on fuel. I said to my pilot, who had flown Intruders in *Desert Storm*, to stand by to "pickle," and he questioned whether we had the barracks targeted correctly. This split-second delay in him hitting the release button meant that the bombs left the jet a tenth-of-a-mile too late. Lacking any laser energy to ensure accurate guidance, the GBU-10s missed the target by about a mile according to the BHA tapes we viewed back on the ship.

Undoubtedly my luckiest moment in OEF came when those weapons exploded harmlessly in open ground, rather than among civilian dwellings. These were the only bombs of the 44 I dropped in OEF that failed to hit their target.

The second section of jets that launched shortly after us enjoyed no more success than we had done, being let down by a tanker no-show and having to divert to the air base at Jacobabad, in Pakistan – there were really only enough big wing tankers in-theater for two air wings at that time.

Its rattlesnake insignia suitably adorned with a Christmas hat, "Diamondback 115" (BuNo 161608) joins the landing pattern on December 19, 2001 – the Tora Bora offensive ended that day. "We all found Tora Bora a challenge simply because we often didn't know exactly what, or who, we were bombing," recalled Lt Cdr Tom Eberhard of VF-102. "It was nerve-wracking dropping ordnance on targets that you could not positively identify before pickling your bomb. FACs could mark the target with hand-held infrared pointers or a smoke round from a mortar, but there was nothing to break that spot out as being an obvious target to us while we circled at 20,000ft." (VF-102)

OPERATION ENDURING FREEDOM

Both Tomcats had just 4,000lb [8,184kg] of gas apiece left in their tanks when they landed. We were all "shining our arses" as a squadron during the first couple of nights in OEF.

After this shaky start, VF-102 soon dialed itself into the mission. Four days after the unit's combat debut, the first SOF teams finally began to be inserted into Afghanistan. The ground war was about to begin in earnest.

Tora Bora

Although VF-14 and VF-41 left OEF on October 23 following the arrival of VF-102 in-theater, VF-213 remained in the thick of the action until *Carl Vinson* was relieved by *John C. Stennis* on December 15, 2001. By then the unit had expended 452 LGBs and 470 20mm cannon rounds. Both VF-213 and VF-102 flew some of the most difficult missions of OEF as Taliban and al-Qaeda fighters fled east towards the Tora Bora cave complex and the Pakistani border.

An EA-6B Prowler accompanied most of the strike packages sent into Afghanistan in the early days of OEF. This jet, from CVW-11's VAQ-135, is refueling from a KC-10A heading north over Pakistan on November 9, 2001, while a bombed-up F-14D from VF-213 waits its turn to tank. (Lt Tony Toma)

F-14 "BOMBCAT": THE US NAVY'S ULTIMATE PRECISION BOMBER

The bulk of the naval air power committed to targeting Tora Bora was provided by CVW-1, with VF-102 seeing particularly heavy mission tasking. The unit's maintenance officer, Lt Cdr Scott Guimond, summed up this period of the "Diamondbacks'" campaign as follows:

> The Tora Bora sorties proved challenging for us, as we were essentially trying to hit little more than a rock blocking a cave entrance in very rugged terrain at high altitude right on the Afghanistan–Pakistan border. The close proximity of the latter meant that we could not spill out into Pakistani airspace after making our bombing runs.
>
> Breaking out the key rock that needed to be bombed through the FLIR when it was the same colour as its surroundings proved a virtually impossible task. Things got even worse at night, for at least during the day you could talk to your RIO and the ground controller about what they were seeing. The location of friendly troops on the ground was not ideal either, as they tended to be further away from the cave entrances that needed bombing than we would have liked – this was especially the case at night. Occasionally, you would see enemy troops moving in the Tora Bora area, and the LANTIRN pod also picked up hotpots of activity.

Lt Cdr Tom Eberhard, also expressed his frustrations about this phase of VF-102's OEF deployment:

> We all found Tora Bora a challenge simply because we often didn't know exactly what, or who, we were bombing. There was also a lot more interest shown in what we were doing by the higher ups in Washington, D.C. than had previously been the case.
>
> It was nerve-wracking dropping ordnance on targets that you could not positively identify before pickling your bomb. FACs could mark the target with hand-held infrared pointers or a smoke round from a mortar, but there was nothing to break that spot out as being an obvious target to us while we circled at 20,000ft. We would drop our ordnance, nevertheless, and in turn be told, "Good effects, thanks," by our controllers.
>
> I was unsure about just how effective we had been during this offensive until several months after Tora Bora, when, on one of the rare no-fly days that we observed on our trip back home, a huge warrant officer SEAL strode into our ready room and demanded to see me. I stood up and asked him what he wanted. "Sir, I just wanted you to know that we were conducting a patrol in Tora

John C. Stennis turns into wind as it prepares to launch one of the last strike missions sent to Tora Bora. The carrier, with CVW-9 embarked, arrived in-theater on December 15, 2001, and 36 hours later VF-211 led the air wing's first OEF strike. (US Navy)

VF-102's CO jet (BuNo 163225) heads for Tora Bora in early December 2001. The aircraft carries a mixed load of GBU-12s and Mk 83 dumb bombs fitted with Target-Detecting Device fuses. According to Lt Chad Mingo, "We were forced to drop a quantity of LGBs in the sea aft of the ship at the end of some of the early December missions when we could not find targets to bomb. We were typically launching with four GBU-12s, as we had been during the previous big months of October and November, but now we were having to jettison two of them in order to make our minimum landing weight when coming back aboard the carrier. It was decided that we should start flying with mixed loads that included dumb bombs on the rear pallets, as it would be cheaper to jettison these rather than two LGBs. Once the Tora Bora offensive commenced, we no longer had to worry about bomb bring-back, but we stuck with mixed loads nevertheless." (VF-102)

A SEAL team use a grenade to mark a cave entrance for bombing in the mountainous Shah-i-Kot region south of Gardez, in eastern Afghanistan's Paktia province. By February 2002, CENTCOM believed that most al-Qaeda fighters still in-country were holed up here, resulting in the area's constant pounding by USAF and US Navy aircraft (the latter including F-14s). (US Navy)

Bora when you helped us out after we had come under heavy fire. I got your name from the CAG, as I wanted to give you these." He handed me a small bottle that was filled with Afghan sand and an optical sight that he had broken off from an SA-7 shoulder-launched SAM. "Thanks for saving our arses, Sir." He then left the ready room. I derived tremendous gratification from such brief encounters.

On December 19 – four days after the F-14As of VF-211 embarked in *John C. Stennis* had arrived in-theater – the Tora Bora offensive ended and the fighting in Afghanistan drastically reduced in its intensity. Things did not flare up again until early March 2002, when the US Army's Task Force (TF) Mountain launched Operation *Anaconda* in the mountains of eastern Afghanistan. Targeting more than 1,000 hardcore al-Qaeda fighters entrenched in ridgelines and caves throughout the Shar-i-Kot Valley, the offensive got badly bogged down to the point where the survival of US troops in contact with the enemy was only ensured through the overwhelming employment of tactical air power.

Finally, on March 3, VF-211 got the chance to show its mettle in OEF, the unit's aviators having spent almost three months kicking their heels flying uneventful XCAS (On-Call exercise Close Air Support – the CAOC's moniker for immediate CAS missions) patrols over Afghanistan. Having missed the first day of *Anaconda*, the "Fighting Checkmates" soon made up for lost time. As had been the case during the ground campaign in the early months of OEF, FAC(A)s such as Lt Cdr Nick Dienna played a particularly important role in this chaotic offensive:

In *Anaconda*, we had a much larger force than had previously been seen in Afghanistan operating in a much smaller area. A standard kill box controlled by one FAC at the start of OEF was 30 nautical

The Naval Aviators and NFOs of VF-213 (led by Cdr Chip King, who is squatting third from left) get together with "Black Lion 101" and a selection of LGBs (from left to right, a GBU-16, a GBU-24 and a GBU-10) towards the end of the unit's OEF cruise. Boasting a 99.6 percent sortie-completion rate during its ten-week spell in the Northern Arabian Sea, VF-213 expended 452 bombs and 470 20mm cannon rounds on cruise. Of the ordnance dropped by the unit, only eight Mk 83 bombs and two Mk 99 CBUs were non-PGMs. The "Black Lions" were credited with expending four GBU-24s, ten GBU-10s, 157 GBU-16s and a whopping 271 GBU-12s. By comparison, in OIF I, the unit dropped just 102 LGBs and 94 JDAM due to its operations in northern Iraq being badly affected by poor weather. (VF-213)

miles by 30 nautical miles, but in *Anaconda* that area had shrunk to eight nautical miles by eight nautical miles, and it was controlled by more than 30 coalition SOF, US SOF and TF Mountain controllers! With a much larger friendly footprint on the ground, you now needed those more traditional controls that we strictly observed when conducting CAS training during our work-ups at home. Unfortunately, these did not exist in the early stages of the campaign.

Quickly realizing that there was no real airspace control plan for *Anaconda*, our FAC(A) crews took it upon themselves to organize TACAIR support in their assigned target areas. They would initially check in on the primary control frequency given to them by "Bossman" and then try and get all other TACAIR assets in the immediate area to tune into this frequency too. Having determined who was talking to which FAC, and where the FACs were located, the FAC(A)s then went about deconflicting the strikers either laterally or vertically. This worked well, reducing the number of aircraft in the Shah-i-Kot at any one time to manageable levels – typically two divisions [totaling eight jets] at a time from CVW-9.

The first division would check in, with the second division some 45 minutes behind it. These would then alternate between the target area and the tanker so that there was always a two- or four-ship formation over the target the entire time. The divisions were typically mixed, with a single section of two Tomcats being paired up with a similar number of Hornets. The latter were usually armed with JDAM and laser Mavericks, while the F-14s carried LGBs and iron bombs – a spread of weapons that CVW-9 found covered most targeting requirements during *Anaconda*. We would split up into sections once over the Shah-i-Kot due to the jets' differing tanking cycles. There were also some USAF F-15Es in the mix, but they tended to do their own thing.

VF-211's most memorable day of fighting in OEF occurred on March 4 after a SOF team was ambushed soon after dawn as they attempted to insert themselves on the ridgeline of Takur Ghar. They were heading for Objective Ginger, which had a commanding view of the entire Shah-i-Kot Valley, but al-Qaeda forces in hardened and well camouflaged bunkers shot up their MH-47E just as the helicopter landed, forcing it to hastily leave – crucially without US Navy SEAL PO Neil Roberts.

Upon word reaching Bagram that there was a soldier missing behind enemy lines, a US Army Ranger quick reaction unit scrambled in two more MH-47Es. When the first of these touched down 50m (164ft) from the top of Takur Ghar, the enemy again targeted the helicopter and shot it down through a combination of rocket-propelled grenade (RPG) and machine-gun fire. Four crew died and others were wounded, with survivors setting up defensive positions just 150m (492ft) from one of the snow-covered al-Qaeda bunkers.

The US Army Rangers were eventually rescued at around 2000hrs, having relied exclusively on CAS support to keep the enemy at bay. Their combat controller, USAF Capt Gabe Brown (call-sign "Slick 01"), later told his superiors that he had handled some 30 CAS sorties throughout the day. Flying one of those aircraft near Takur Ghar was VF-211 pilot Lt Dan Buchar:

Shortly after dawn, I launched as part of a division of four F-14s sent into Afghanistan in support of *Anaconda*, having been briefed to head to the Shah-i-Kot Valley to help troops in contact as they continued to battle with enemy forces. As we headed north, the SOF MH-47E was shot down near Objective Ginger. Shortly after that, our division lead, Lt Larry Sidbury, got a call from "Bossman"

Junior officers from VF-211 pose with a "bunker-busting" GBU-24B/B shackled to the forward belly pallet of "Nickel 100" (BuNo 159428). Squatting to the left is Lt Larry Sidbury, hero of VF-211's efforts to defend SOF and US Army Rangers surrounded on Roberts' Ridge on March 4, 2002. (Lt Shaun Swartz)

telling him that our bombs were needed straight away. We had to refuel first, however, so each jet quickly topped off its tanks and then headed independently to the target area.

Lt Sidbury and his RIO, Lt Cdr Tim Fitzpatrick, who were both FAC(A)-qualified, reached Takur Ghar first and quickly made contact with "Slick 01." The latter was pinned down near the wreckage of the MH-47 along with the survivors of the Army Ranger quick reaction unit. Lt Sidbury and his wingman, Lt Bryan Roberts, worked directly with "Slick 01," and they dropped ordnance within 500m [1,640ft] of the friendlies.

Ordered to return to the ship by the AWACS after dropping all of his bombs, Lt Sidbury told everybody – including a rear admiral and a USAF general – "No" because the guys on the ground were still taking withering fire. He got to the point where he turned his radios off, thus blocking out the distraction of the "return to base" calls. Eventually relieved on station, Lt Sidbury somehow made it to the tanker before running out of gas and finally trapped back aboard *John C. Stennis*.

His CO [Cdr O P Honors] was still flying at the time, but his XO [Cdr Kevin J Kovacich] started grilling him about why he had ignored the calls to return. At this point, CAG [Capt Donald

VF-211's oldest aircraft during the unit's first OEF deployment was "Nickel 104" (BuNo 158618), which had been delivered to the US Navy in October 1972. Seen here accelerating down waist catapult four in Zone Five afterburner during an early *Anaconda* mission, the aircraft was lost on March 8, 2002 when its tailhook separated on landing at the completion of an OEF mission. The aircraft was the second Tomcat lost in less than a week, for on March 2, VF-143's "Dog 101" (F-14B BuNo 162923) had suffered a nose gear strut failure launching from *John F. Kennedy* off the coast of Crete while the carrier was en route to the Northern Arabian Sea. Although both crew from the VF-211 jet successfully ejected and were rescued from the sea, only the RIO of the VF-143 aircraft survived. (US Navy)

P Quinn] stormed into the ready room and started to tear strips off Lt Sidbury for disobeying a direct order. A few minutes later, the admiral [Rear Admiral James M Zortman] also walked in and everybody immediately stood to attention. His first words to Lt Sidbury were, "That's the best thing I ever saw. Don't you ever do anything different."

While Lt Sidbury and his wingman had been controlling the airspace over Takur Ghar, my RIO, Lt Cdr Ed Galvin, and I, along with Lt Mark Bruington and his RIO, Lt Shaun Swartz, had been sent to attack some mortar positions that were firing on our troops in the south-eastern corner of the Shah-i-Kot Valley.

We got to work with an EP-3E rather than a JTAC during this mission, with targeting information being fed to us by the aircraft as it circled high over the valley. The EP-3E had a SOF guy in the back picking out the mortar positions with the Orion's various sensors. The descriptions that he gave us as to the location of the mortars were eye-watering in detail! He would say, "Do you see the mud hut with the courtyard?" as he tried to talk us on to a building while we were circling at 20,000ft! The enemy's firing positions were actually within mud huts on the edge of the town of Shah-i-Kot itself.

After we failed to locate the target, the EP-3E had a B-1B that was on station throw a JDAM down to act as a marker for us. My RIO quickly spotted the impact point on the FLIR, and then the SOF guy also talked his eyes on to it, and we hit the mortar pit and two vehicles with a pair of LGBs. Their destruction was confirmed by the EP-3E, and we then targeted another position with our remaining GBU-12s later in the mission. All four jets in our division returned to base without their bombs.

There was also a Predator UAV on station near the EP-3E, as well as the B-1B, and some F-15Es checked

A lone Tomcat leads a second jet north over mountainous terrain in eastern Afghanistan in early April 2002 during one of the last operational missions flown by VF-211 on its OEF cruise. The conditions that carrier-based squadrons faced in-theater contrasted markedly with those experienced during the decade of constant OSW deployments. Fortunately for Tomcat units committed to the campaign, the reliability of their aircraft allowed them to fly challenging six- to eight-hour missions deep into Afghanistan with confidence. (Lt(jg) Mitch McAllister)

in just as we left. It was an all-out effort to provide TACAIR coverage for the guys pinned down on Takur Ghar, as well as our troops coming under fire in the valley below.

Things had gotten a little frantic for a while there when we were struggling to locate the mortar position. We actually had the B-1B split our section at one point as Lt Bruington was working the target area and I was providing high cover for him – the bomber basically flew between us. It was after this that we decided we needed to get altitude separations between aircraft types sorted out before we bombed anything! We decided to stick to individual CAP points that were widely spaced, and only came into the valley when instructed to by the EP-3E. We never actually saw the Orion at any stage during this mission, the aircraft flying higher than us some distance away from the valley. We did, however, spot the Predator, which was down close to 10,000ft, while we stuck to a hard deck height of 20,000ft.

As we headed back to the carrier, we could hear the gunfire and calls for help over the radio from the guys stuck on Takur Ghar, and although we had done our best to help the Rangers, they still weren't out of trouble when we returned home out of ordnance. That was the worst I felt during the whole cruise, and I couldn't sleep when I got back to the boat, despite the exhaustive nature of the mission. It was a huge relief to everyone when we heard that they had gotten the survivors out that evening. We got to meet some of these guys in Bahrain during a port call two weeks later, when they told us that our actions had saved their lives.

Iron Bombs

According to Lt Cdr Nick Dienna, VF-211 had started carrying a mixed load of two LGBs and two general purpose (GP) iron bombs within days of *Anaconda* starting. As with other Tomcat units that had previously used GP ordnance in OEF, the "Fighting Checkmates" predominantly uploaded 500lb (227kg) Mk 82 or 1,000lb (454kg) Mk 83 Target Detector Device-fused bombs to ensure maximum area coverage against troops in the open or in shallow trenches. Dienna explained:

> VF-211 had not dropped any GP bombs up until then in the deployment. Iron bombs were often used for reconnaissance-by-fire purposes, marking out targets for LGBs. We always tried to get the GP bombs off as soon as we checked in, rolling in at a 45-degree angle once cleared to drop in the Shah-i-Kot Valley. We would then wait for any on-call tasking for our LGBs, which were typically aimed at high value targets, or targets where there were collateral damage issues. A number of them were also dropped in fire-for-effect situations, rather than against specific target.

One of the first Naval Aviators from the unit to drop a GP bomb during *Anaconda* also happened to be VF-211's newest pilot, Lt(jg) Kevin Robb. He recalled:

VF-211's "Nickel 102" (BuNo 161612) leads "Nickel 101" (BuNo 161603) into *John C. Stennis*'s overhead prior to joining the landing pattern. This photograph was taken shortly after Operation *Anaconda* had reached its climax in mid-March 2002. The completion of the offensive effectively brought to an end all bomb dropping in Afghanistan, which is why both aircraft are still carrying their GBU-12s. Lighter than the F-14B/D, the A-model jet could land back aboard in "quad bomber" configuration. (Lt(jg) Mitch McAllister)

> I joined the "Fighting Checkmates" with just 30 hours in the A-model Tomcat in my logbook, having been trained to fly the F-14B for much of my time in VF-101. I was scheduled to join a B-model squadron, but they needed a priority F-14A student graduate to join VF-211. I got my deck qualification in the jet with VF-101 off the Norfolk coast on a Monday and was then sent by civil airliner to Bahrain four days later. I started flying combat missions within 24 hours of my arrival on *John C. Stennis*, and less than a week after getting deck qualified in the F-14A!

My bomb-dropping mission came about a week into *Anaconda*. As with previous sorties I had flown in OEF, we launched in the late afternoon as a section of Tomcats designated as an XCAS asset without a pre-planned target. We headed north and rendezvoused with our first tanker on the Afghan border. As soon as the skipper checked in with "Bossman" at our allocated loiter time, he was told that there was urgent tasking for us. We were instructed to check in with a JTAC, who in turn gave us co-ordinates for a road just north of the Shar-i-Kot Valley that he had been watching for some time.

Enemy forces had started to retreat into the mountains in cars and trucks, and the JTAC wanted the road severed and some vehicles traveling on it hit so as to block it off as an escape route for al-Qaeda/Taliban fighters. Time was of the essence here, and we were given immediate approval to drop by "Bossman" once overhead the target area. My skipper asked the JTAC where and when he wanted the bombs, and as his wingman it was my job to listen carefully to the instructions that the CO was given, for I would attack immediately after him.

I remained in a high cover position – with a 30-degree angle of bank – for the skipper as he rolled in and dropped his ordnance, and I then worked off his hits. The JTAC told me to put both of my bombs directly on top of my CO's, so that's what I did. We came in on the road at an angle of 50 degrees, some 45 degrees off the target itself, which was pretty exciting to say the least – VF-101 hadn't really prepared me for dive-bombing targets with unguided ordnance from such a high altitude and at such a steep angle. We had to roll in from around 30,000ft simply because the terrain in the area was so high. The road, which was carved into the side of a mountain, was at about 6,000ft AGL. With bombs gone, we then pulled off to the left away from the mountain.

My compatriots had been almost exclusively dropping GBU-12s up until this mission simply because of the nature of the CAS sorties they were performing and the awesome targeting capabilities of the LANTIRN pod. I was convinced, therefore, that if I got to drop any ordnance in OEF, it would be an LGB. When the JTAC asked for Mk 82s, I quickly studied my kneeboard Z-diagram and figured out the best attack profile to employ. I also confirmed my proposed course of action with my RIO, squadron XO Cdr Kevin Kovacich – a former A-6 B/N, he was one of the most experienced back-seaters in VF-211.

With an LGB delivery, the RIO would have done most of the work, designating the target with the LANTIRN pod. However, as I had already realized, Cdr Kovacich told me that the success

Four storeless VF-211 jets join up in close formation for their run in and break over *John C. Stennis* in late March 2002. Both "Nickel 101" (BuNo 161603) and "Nickel 102" (BuNo 161612) feature subtle bomb tallies beneath their cockpits. The canopy rails of each VF-211 Tomcat bore the names of New York Police and Fire Department personnel killed on September 11, 2001. (Lt(jg) Mitch McAllister)

of this attack was almost entirely in my hands. He played his part by talking my eyes on to the bomb hits from my CO's ordnance, and that helped me get correctly lined up with the target. Cdr Kovacich also put out chaff and flares as we dived on to the target and as we pulled out as well.

Fortunately, the weather was glorious, as the skipper kept reminding me. He could not believe that I was going to get to drop ordnance for the first time operationally in an angle-dive attack in gin clear conditions. This was the first time that Cdr Kovacich had dropped Mk 82s in anger as well. My CO and I expended both of our bombs in pairs, as the JTAC wanted them as quickly as possible. He then responded with "good effects" after both runs, and we left the area. By the time I got back to the boat, we had been airborne for five-and-a-half hours, which made it my longest OEF mission. This was the one and only time I got to drop bombs on cruise.

The crew of a VF-211 jet pop flares from the 30-cartridge AN/ALE-39 chaff and flare dispenser located to the left of the tailhook on the underside of the F-14A's "beaver tail." Both expendables were regularly used during the pull-out phase of dive-bombing attacks with dumb bombs in OEF. Crews also employed chaff and flares to defeat SAMs launched at aircraft in the early days of the campaign. (US Navy)

Typically, we carried two GBU-12s and two Mk 82s on most *Anaconda* missions, as we could trap back onboard the carrier with this configuration should the bombs not have been needed. For night missions, we would switch to four GBU-12s, as it was difficult to accurately aim dumb bombs in the dark. The unit also sortied with some GBU-16s early on in the cruise, and these were also uploaded again during *Anaconda*, when crews knew that they would be definitely dropping ordnance.

We were flying some pretty old Tomcats in 2002, and our maintenance folks did a Herculean task keeping them airworthy. We lost a jet on March 8 when its tailhook separated on landing, but the crew punched out okay. During the course of that particular day, we had managed to sortie eight of our ten jets.

This was the second Tomcat lost in less than a week, as on March 2 VF-143's "Dog 101" had suffered a nose gear strut failure while launching from *John F. Kennedy* off the coast of Crete. Although the RIO managed to eject safely, the pilot of the F-14B, Lt Cdr Chris Blaschum, was killed. Squadron-mate Lt(jg) Joseph Greentree recalled:

Following the crash, all Tomcats were temporarily grounded so that their nose gear could be checked for the type of fatigue cracks that had caused the undercarriage to fail during its catapult shot. Airframers in the maintenance departments of both Tomcat units in CVW-7 worked round the clock removing nose gear legs and inspecting them. They managed to turn all the aircraft around in just 72 hours. Soon after we arrived in-theater, VF-211 then lost a jet when its tailhook separated, and again we were grounded for three days while our jets were inspected and passed fit to fly.

These failures were rather disconcerting because I remember one of my instructors in VF-101 telling me the two things that would never break on this aircraft were the nose gear and the tailhook! Both failures were put down to undetectable corrosion caused by the sheer age of the jets involved.

During the final stages of *Anaconda*, *Theodore Roosevelt* was relieved by *John F. Kennedy*, whose CVW-7 conducted a theater turnover with CVW-1 as the carriers passed each other in the Red Sea. By then, VF-102 had dropped 420,000lb (190,508kg) of ordnance and buddy-lased an additional 50,000lb (22,680kg) from other platforms. The unit had tallied 5,000+ hours of

VF-102 flew in excess of 5,000 hours in its F-14Bs during the unit's OEF cruise – more than any other TACAIR squadron in CVW-1. With the Tomcat averaging 60–70 maintenance hours per flying hour, the effort put in by the unit's support personnel (seen here in green, dark brown, red and white) was nothing short of Herculean. (VF-102)

flying time (more than any other TACAIR unit in CVW-1), with 1,184 hours in November alone and a staggering 61.8 hours on December 8 in the lead-up to the Tora Bora campaign.

CVW-7's F-14Bs from VF-11 and VF-143 made their combat debut over Afghanistan on March 11. *John F. Kennedy* had been given the job of night carrier, and the vessel launched its first wave of 11 aircraft in the early hours of the morning on the 11th. Heading for the Shah-i-Kot Valley, the aircraft were "set up for heavy FAC(A) and CAS targets" according to VF-11 CO, Cdr John Aquilino. One of these jets was also making history for the Tomcat, as it was equipped with a single 2,000lb (907kg) GBU-31 JDAM.

GPS-guided JDAM had been introduced to the fleet by the F/A-18C during *Allied Force* in the spring of 1999. By early 2001, NAVAIR had engineered the OFP321A software upgrade for the F-14B that would allow the jet to employ JDAM. VF-102 was too far progressed in its work-up cycle to have the software installed in its jets, so CVW-7's two Tomcat units were designated as the first fighter squadrons to receive the upgrade.

The VF-11 JDAM jet sortied on March 11 was flown by Lt Cdrs Scott Knapp and Chris Chope, who Cdr Aquilino described as "the most qualified crew in the unit when it came to JDAM employment, having done most of the integration work with the GBU-31 from the squadron standpoint." Once on station over the Shah-i-Kot Valley, the section of "Rippers'" Tomcats was told by a JTAC to bomb a mortar pit near the "Whaleback" – the dominant feature within the valley, and the scene of some of the fiercest fighting during *Anaconda*. According to former A-6 B/N Lt Cdr Chope:

> We set up a trail attack profile whereby our wingman would drop his GBU-12 first and we would drop the JDAM moments later through the smoke created by his weapon. Seen through our LANTIRN pod, his weapon skipped off a nearby rock and exploded right alongside the target. When it came time for us to pickle our GBU-31, the weapon hung, refusing to come off the jet. Lt Cdr Knapp and I did everything by the book, but the bomb refused to drop. After a long, silent flight back to the ship, we recovered aboard and the armorers removed the GBU-31 from the aircraft and tested it, whereupon the weapon was discovered to be faulty. The guidance unit in the tail kit, which has its own GPS receiver and navigation system, was not talking to the weapons computer in the jet.

The following evening, CVW-7 launched another wave of aircraft over the beach, with Cdr Aquilino and his RIO, Lt Cdr Kevin Protzman, leading a section from VF-11. Aquilino recalled:

> We were cleared to hit a cave complex on the "Whaleback," having already been given a rough target set aboard the ship during our pre-mission briefing. There was no guarantee that we would be cleared to expend any ordnance, however.
>
> Once on station, we received targeting support from an E-8 Joint STARS [Surveillance Target Attack Radar System] that was circling overhead the Shah-i-Kot, as well as aim point GPS co-ordinates for our JDAM from "Bossman." My RIO punched the numbers into his mission computer and then reread them back to the AWACS controller to confirm that he had typed them in correctly. Only then were we given approval to release the bomb. We made our run in on the target from high altitude and recorded good BHA video of the weapon hitting the cave entrance.
>
> The assessment post-delivery was that the bomb had hit well within the CEP of the weapon. We later heard from troops on the ground that they had been impressed by the strikes flown that night, with all the targets that they had identified being hit. The cave complex we had bombed burned for about 12 hours.
>
> Our jet was carrying a solitary GBU-31, as at that time we only had clearance to carry the 2,000lb [907kg] weapon – we were yet to receive permission to pair the JDAM with an LGB or a dumb bomb. With targets being at a premium, we decided that there was no point in taking two JDAM, as we would have had to jettison one in the water in order to get down to our minimum landing weight if we failed to find anyone who needed our bombs. Those GPS kits weren't cheap, and we didn't have too many GBU-31s in the armory, so jettisoning ordnance was not an option.

Their aircraft armed with a single GBU-31 JDAM, VF-11 CO Cdr John Aquilino and RIO Lt Cdr Kevin Protzman prepare to launch from *John F. Kennedy* on an OEF mission shortly after *Anaconda* commenced. This crew made Tomcat history by dropping the first JDAM in anger from "Ripper 201" (BuNo 162912) on the night of March 11–12, 2002. (US Navy)

OPERATION ENDURING FREEDOM

The pilot of VF-11's "Ripper 210" (BuNo 162911) jousts with a KC-135 over a solid undercast during *Anaconda* while trying to keep his F-14B plugged into the drogue basket. The fighter has a solitary GBU-31 JDAM attached to its forward port BRU-32 bomb rack, CVW-7's two F-14B squadrons lacking clearance at the time to pair the PGM with an LGB or dumb bomb. (Lt Brian Vanyo)

The GPS-guided weapon would subsequently see widespread use with F-14B/D-equipped units (the A-model lacked the software to employ JDAM) in OIF.

Following the successful conclusion of Operation *Anaconda*, and the apparent disintegration of the Taliban as a credible fighting force in Afghanistan, CENTCOM scaled back its operations in OEF. As part of this drawdown, CTF-50 was reduced in strength to a single carrier strike group (the term battle group was abandoned in 2002) from April 18, 2002, when *John C. Stennis* and CVW-9 left the Northern Arabian Sea and headed home. VF-211 had expended almost 100,000lb (45,359kg) of ordnance in OEF, with most of it being dropped during *Anaconda*. The unit's veteran F-14As had flown 1,250 missions (CVW-9 totaled 3,242) and logged 4,200 hours in combat over Afghanistan (the air wing tallied 13,500 combat hours), with VF-211's maintenance department coaxing a 99.7 percent sortie completion rate from its ageing charges.

Armed with four AGM-114 Hellfire missiles, the plane guard HH-60H BuNo 165116 of HS-5 flies across the bow of *John F. Kennedy* as it sails at speed in the Northern Arabian Sea in the early summer of 2002. The carrier's preparation for its OEF deployment had not been easy. In an unprecedented move, the ship's captain had been relieved of his command when *John F. Kennedy* failed its Board of Inspection Survey in December 2001 – a decision that delayed the ship's departure on cruise. The carrier's battle group also initially failed its Joint Task Force Exercise, in which CVW-7 had participated "from the beach" because *John F. Kennedy* was still stuck in Mayport, Florida, being hastily prepared for deployment. The carrier finally left on its OEF cruise on February 16, 2002, much to the relief of *Theodore Roosevelt*'s crew – the latter vessel's time on station in the Northern Arabian Sea had been extended due to *John F. Kennedy*'s delayed departure. (US Navy)

With the departure of *John C. Stennis*, *John F. Kennedy* and CVW-7 now provided the primary TACAIR presence in-theater. Afghanistan was very quiet for the remainder of 2002, however, as VF-11's Lt Cdr Chris Chope explained:

When things petered out after March 16, it was tough for CVW-7 just to sit on the line for the next four months and see next to no action. I was FAC(A)-qualified, which meant that I spent a lot of time working with troops in the field, as well as other TACAIR assets such as B-52Hs and B-1Bs. The TACPs on the ground were clearly working hard while we were in-country, as you could hear these guys panting as they ran up and down mountains in an effort to remain in radio communication with us.

Keeping in contact with troops on the ground was tough there due to the size of the mountains over which we were often operating. If you got on the wrong side of the peak from where the JTAC was situated, you would lose contact. Trying to visually acquire a team of six or seven guys operating in mountainous terrain was also a challenge, and I only saw them very occasionally through binoculars or with the LANTIRN. By definition of their mission, they tended to blend in well with the terrain in which they were operating. Despite having no visual reference on their position, we usually had a good feel for where the SOF teams were, and we would help them out by scouting the area ahead of them in advance of their movement on the ground.

In our capacity as a FAC(A) crew, my pilot and I would routinely operate with US Marine Corps F/A-18Ds and USAF A-10s from Manas, in Kyrgyzstan, P-3C/EP-3Es from Masirah Island, off Oman, F-15Es and F-16Cs from Al Udeid, in Qatar, B-1Bs from Seeb, in Oman, and B-52Hs from Diego Garcia and Guam. These guys would check in once on station, and we would occasionally ask them to go and investigate a contact raised by a JTAC, after which they would hit the tanker and head home. In the main, things were very quiet while we were in-theater.

With the need for LGBs and JDAM having gone, at least for now, high-speed show-of-force flybys soon became the order of the day. According to Cdr John Aquilino:

This was not a mission that we had been trained to fly pre-cruise, as it had never appeared in our work-up syllabus. We had been told by VF-102 during our OEF turnover that flybys would soon

VF-11 NFO Lt Cdr Chris Chope keeps a watchful eye on his pilot as their F-14B takes on mid-mission fuel from a KC-10 during an OEF patrol in May 2002. "You would raise the tanker on the radio to find out where he was and then you would lock it up with your radar," Chope explained. "Having rendezvoused with the tanker, you would ask him to reel out the drogue, and once this was in position, you would extend your probe and plug in. If you were a little short on fuel, or you needed to be back on station ASAP, you could ask the pilot of the tanker to roll out on to a certain heading so as to drag you closer to your objective while you were still plugged in replenishing your tanks. Most crews were pretty good at accommodating such requests, although the USAF guys did not like to stray too far off their designated tanker tracks. Some crews were so obliging that they would even do the leg work with the AWACS in respect to letting the controller know that they were moving from one track to another." (Lt Cdr Chris Chope)

feature regularly in our mission tasking. VF-211 emphasized this point too, explaining to us that although such flybys may not have seemed to be too big a deal from the crew's perspective, they meant a lot to the soldiers on the ground. When crowds were gathering near troops on patrol, and their intention was unknown, the fact that there were US jets in the air overhead that could have a direct impact on the situation was greatly appreciated by coalition forces.

With months of experience in-country, our troops had supplied the CAOC with feedback via "Bossman" on how they could get the most value out of our flybys. We were told that it was crucial to find an area near to the disturbance where the jet's engines could clearly be heard – it didn't matter if the aircraft could not actually be seen by friendly troops. A successful flyby was one that dispersed a crowd or helped buy the guys on the ground time to get a better understanding of what the crowd was going to do.

I was requested to perform just one show-of-force flyby in OEF, and this was during the course of a marathon ten-hour daylight mission. We launched from the ship and flew up the eastern part of the country until we reached central northern Afghanistan. Assigned a CAP station and a frequency for a ground FAC who we were to talk to, we checked in and then waited during our vulnerability

VF-11's "Ripper 207" (BuNo 161437) loses altitude over the Northern Arabian Sea as the crew prepare to recover aboard *John F. Kennedy* following a late afternoon Unit-Level Training (ULT) flight. With OEF missions being a priority for TACAIR units, ULT hops rarely featured on the daily flight plan once the air wing was committed to CTF-50 operations. When time allowed, sorties to a weapons range in Oman were undertaken so as to keep crews current on bomb delivery methods. These flights became increasingly more important after March 2002, as very little ordnance was being dropped in Afghanistan following the completion of *Anaconda*. It was a 200-mile (322km) transit flight from the carrier to the range, where crews dropped a mix of blue Mk 76 Mod 5 practice bombs and BDU-59 LGTRs. (Lt Brian Vanyo)

OPERATION *ENDURING FREEDOM*

window to provide this guy with whatever he might have needed, be that weapons on target, a show-of-force, or simply a presence in the sky above him.

After 20 minutes of silence, he got back on the radio and asked us to fly a show-of-force. He passed us the latitude and longitude for the flyby, what direction he wanted us to come in from, the height we were cleared down to and any known threats in the area. We set up our jets in a two-aircraft defensive spread for mutual support and optimum threat coverage, and then each of us took it in turns to make a single high-speed, subsonic pass with the afterburners "cooking" so as to make plenty of noise. We also pumped out lots of flares during the runs just in case there were any MANPADS in the immediate area.

Very occasionally, CVW-7 would be called on to back-up all this noise with a little muscle, and VF-143 pilot Lt Bill Mallory was one of only a handful of Naval Aviators to drop ordnance from a Tomcat post-*Anaconda*:

I was the wingman in a section of "Dog" jets patrolling near Kabul during a seemingly routine night XCAS mission at the midway point in our OEF cruise. Our jet (my RIO was squadron XO, Cdr Christopher Murray) was armed with two GBU-16s, and the lead aircraft, flown by Lt Cdrs Jonathan Stevenson and Matt Leahey, carried a single GBU-31 JDAM.

VF-143's CAG jet (BuNo 163220) flies over Afghanistan carrying a GBU-31 JDAM and a GBU-12 LGB, clearance to carry such a mixed load only reaching CVW-7 from NAVAIR mid-cruise. This aircraft (which had previously seen combat in *Desert Fox* with VF-32) also served as VF-143's CAG jet during the squadron's final Tomcat deployment in 2004, when it was heavily involved in OIF II. (Lt Joseph Greentree)

The launch, rendezvous and transit through Pakistan into Afghanistan went as it had done a dozen times before, and it was becoming a real challenge not to lose interest in these missions. Therefore, when the AWACS controller told us that he wanted us to head to a particular kill box, we were ecstatic. It was readily apparent from their radio comms that Lt Cdrs Stevenson and Leahey really wanted to drop their first JDAM, but the controller didn't want any more than 1,000lb [454kg] of ordnance, so Cdr Murray and I were in luck and we were passed the lead.

Unfortunately, either incorrect co-ordinates or a well-hidden target, coupled with the functionality of the new "40K" LTS pods that were fitted to our jets, gave us a hell of a time when it came to finding the aim point for our weapon. This led to too much "FLIR fishing," which was disorienting for both my RIO and I. We were quickly running out of our allotted vul time (which was just 15 minutes), so when Lt Cdr Leahey thought he had acquired the target in question, we passed the lead back to him and executed a "buddy bomb" attack. This particularly bothered the "tactically savvy" Cdr Murray, because he was unable to find the target. He let me know how much he was annoyed all the way home!

As far as the target goes, if it was a rock, then we got a direct hit. We accomplished all that was asked of us, which was to cover the egress of a SOF unit. Unfortunately, we ended up dropping more drop tanks than bombs on that cruise.

John F. Kennedy and CVW-7 ended their OEF tour of duty on July 19, 2002, when they were relieved by *George Washington* and CVW-17. While on station in the Northern Arabian Sea, CVW-7 had completed 2,599 combat missions and expended 64,000lb (29,030kg) of ordnance.

F-14B-equipped VF-103 was CVW-17's sole Tomcat unit, and it commenced flying XCAS, TARPS and FAC(A) OEF missions on July 20. After seeing very little in the way of action (VF-103 expended no ordnance in Afghanistan), *George Washingto*n headed west into the NAG in early September, whereupon CVW-17 commenced flying OSW missions. This was the first time that carrier aircraft had flown into southern Iraq since *Enterprise* had sent TACAIR assets from CVW-8 in-country on September 9, 2001.

Abraham Lincoln and CVW-14 then became the OEF carrier, with the air wing's F-14D-equipped VF-31 flying missions between September 11 and October 28, 2002, prior to heading into the NAG. Like VF-103, VF-31 obtained plenty of TARPS-generated imagery but expended no ordnance in Afghanistan.

VF-211, now flying with CVW-1 from *Enterprise*, became the last Tomcat unit to undertake OEF missions when the carrier got the call to leave the NAG and participate in Operation *Mountain Resolve* in early November 2003. This coalition offensive, launched in the Nuristan and Kunar provinces of Afghanistan, involved an airdrop into the Hindu Kush mountains by the US Army's 10th Mountain Division, and resulted in the killing of Hezbi commander Ghulam Sakhee.

A number of the Naval Aviators and NFOs serving with VF-211 at the time had seen combat during the "Fighting Checkmates'" previous deployment in 2001–02. One of those was pilot Lt Dan Buchar:

We were hastily pulled out of a port call to Dubai and rushed to the Northern Arabian Sea, where the unit was tasked with flying night missions from the get go. The CAOC wanted at least one FAC(A) crew per section over the beach, and that really hit a small number of us hard, for VF-211 only had four suitably qualified crews. I pulled the 0200hrs to 0800hrs FAC(A) watch with my RIO, which was not a lot of fun – CVW-1 ran the FAC window at night and Bagram-based Air Force A-10s ran it during the day.

I got to fly a couple of show-of-force passes during this period, where we were cleared down to 3,000ft – these were the standout missions for me during the 12 days that we supported *Mountain Resolve*. We flew some very long sorties lasting more than nine hours at a time, and I undertook

VF-143 pilot Lt Bill Mallory took this close-up photograph of the mixed-loadout bomb configuration of his F-14B during the final weeks of CVW-7's commitment to OEF in 2002. The unit had few opportunities to use JDAM following the cessation of *Anaconda* just as *John F. Kennedy* arrived in-theater. According to Mallory (who did, in fact, drop a single GBU-31 in anger), "As it turned out, the most ferocious thing in 'Peace Wing 7' [as CVW-7 was disparagingly referred to by aircrew, who were convinced peace usually broke out just as the air wing arrived in-theater] was the 'Dog's' mascot, a Siamese Fighting Fish. It flew in every Tomcat embarked by the unit, had more than 100 traps and was awarded an Air Medal for 20 flights in a combat zone." (Lt Bill Mallory)

"Victory 112" (BuNo 161855) of VF-103 takes on fuel from RAAF Boeing B707-338C A20-623 of No 33 Sqn, operating with the call-sign "Fosters." Two Australian tankers were committed to OEF as part of Operation *Slipper*, the aircraft flying from Manas Air Base in Kyrgyzstan. US Navy aircraft routinely refueled from the RAAF B707s in 2001–02, making full use of the aircrafts' twin Flight Refueling Ltd Mk 32B underwing pods. (Capt Dana Potts)

three of these in a 72-hour period, all at night, as the CAOC wanted round-the-clock FAC(A) cover in-theater. The air wing set up E-2C and EA-6B dets at Bagram in support of *Mountain Resolve*, but had the TACAIR assets fly in directly from the boat. The whole operation was well executed by CVW-1, but the folks on the ground failed to find any worthwhile targets for us to bomb.

RIO Lt Mario Duarte also flew a number of OEF missions during *Mountain Resolve*, stating that the jet's "LTS pod made the F-14 a critical tool for the CAOC during this operation, as it was the best infrared sensor in-theater when it came to searching for vehicles or people moving at night. If something happened to be detected, we would remain on station overhead and use different fields of view and polarities to work out who was moving on the roads below. The coalition was keen to stop people traveling at will across the central border area with Pakistan."

The last section of VF-211 jets to patrol over Afghanistan recovered back aboard *Enterprise* soon after dawn on November 14, 2003, thus bringing to an end the Tomcat's considerable contribution to OEF.

F-14 "BOMBCAT": THE US NAVY'S ULTIMATE PRECISION BOMBER

Above: VF-211 was the only Tomcat unit to make a return visit to Afghan skies, the "Fighting Checkmates" participating in Operation *Mountain Resolve* in early November 2003. By then, the squadron had transferred from CVW-9 to CVW-1, which was embarked in *Enterprise*. VF-211's commitment to the operation was to last just 12 days, with the Tomcat bidding OEF farewell for the last time on November 14, 2003. Here, "Nickel 111" (BuNo 161295) and "Nickel 114" (BuNo 161275) hold station behind a KC-135 while flying over a solid undercast. The lead jet in this section was subsequently declared written off following a mid-air collision over the Red Sea with "Nickel 104" (BuNo 158635) on February 2, 2004. Both jets managed to land safely on board *Enterprise* after the incident. (USAF)

Opposite: VF-31's "Bandwagon 101" (BuNo 164600) prepares to land aboard *Abraham Lincoln* following a routine TARPS mission over Afghanistan on October 18, 2002. Although the unit spent many hours photographing the key "hot spots" in-country, VF-31 did not drop a single bomb during the 47 days CVW-14 was committed to OEF. (Lt Jim Muse)

CHAPTER 10

STRAFING

From the outset, the Tomcat had been designed and fielded with an internal 20mm M61A1 cannon mounted in the lower left side of the fuselage beneath the cockpit. Initially seen as a close-in air-to-air weapon, the cannon came into its own in the air-to-ground role during the "Bombcat" phase of the F-14's career.

Although the F-4 Phantom II had proven itself in aerial combat during the Vietnam War, the fighter had been frequently criticized by frustrated crews who had missed out on a certain kill because the aircraft lacked an internal cannon. The USAF rectified this omission with the F-4E, but the US Navy stuck with missile-only Phantom IIs. Keen to avoid such issues with its new fleet fighter, Grumman made sure that the aircraft was designed with a single cannon to complement its formidable missile armament.

The tried-and-tested General Electric M61A1 Vulcan cannon was the weapon chosen for fitment into the F-14, the 20mm "Gatling gun" having been fielded in US fighters since 1959. Some 74in (188cm) long and weighing 265lb (120kg), the cannon had a magazine capacity of 678 rounds when installed in the Tomcat. However, no more than 600 rounds were usually carried to ease loading and unloading evolutions. The cannon could fire 20mm training, M56A3/A4 High-Explosive Incendiary (HEI) or PGU-28 Semi-Armor-Piercing High-Explosive Incendiary (SAPHEI) shells using a linkless feed system, emptying the ammunition drum in just seven seconds. Operating as a "closed system," the spent shell cases were retained in the belt and returned to the ammunition drum, rather than being expelled from the aircraft and possibly sucked into the left intake. Such a system also minimized any shift in the aircraft's center of gravity once the weapon had been fired.

The gun was capable of being fired in "Low" 4,000 rounds per minute or "High" 6,000 rounds per minute modes. The lower rate was usually used for air-to-ground engagements and the higher rate for air-to-air engagements. The maximum number of rounds fired by each trigger pull was set by ordnance personnel to be either 50, 100, 200 or unlimited. Fifty rounds was often set for air-to-air applications, while 100 rounds was preferred for planned air-to-ground engagements.

The six rifled barrels of the M61A1 rotated counter-clockwise when fired. General Electric had purposely gone with a six-barrel configuration in order to lessen wear and help dissipate heat, thus allowing the weapon to achieve its high rate of fire. A limiter was also fitted to reduce the risk of overheating or jams. A muzzle gas diffuser was employed to avoid structural damage being inflicted on the airframe when the weapon was in operation, while a muzzle clamp prevented the dispersal of rounds fired. The cannon bay was cooled using ram air, with the intake opening automatically when the trigger was actuated and closing ten seconds after it was released.

The General Electric M61A1 Vulcan cannon was a snug fit in the lower forward fuselage of the Tomcat, the 20mm "Gatling gun" being 74in (188cm) in length and weighing 265lb (120kg). This example was carried by VF-31's final CAG jet, F-14D BuNo 164342, and it was photographed on the flightdeck of *Theodore Roosevelt* in the NAG in early January 2006. (Erik Hildebrandt)

Aviation ordnancemen from VF-21 use a Linkless Ammunition Loading System (LALS) to replenish the ammunition drum of an M61A1 installed in an F-14A embarked in *Independence*. The empty ammunition belt can be see entering the LALS at the bottom right-hand corner and then emerging with 20mm rounds in place in the center of the photograph as the linkless belt is carefully loaded into the ammunition drum situated immediately below the rear cockpit. The silver "flex drive" cable to the left of the belt spun the LALS. *Independence*, with CVW-14 embarked, was sailing in the Gulf of Oman during Operation *Desert Shield* when this shot was taken in the fall of 1990. (US Navy)

VF-143's "Dog 106" (F-14B BuNo 161434) has its gun bay panels hinged open or removed on the ramp at Oceana, clearly revealing the M61A1 to the left of the extended crew ladder and the 678-round drum to the right of it. The ammunition feed from the drum to the gun passed through the ladder storage bay. (Danny Coremans)

All Naval Aviators received air-to-air gunnery training with the M61A1 while undergoing their flight training with either VF-101 or VF-124, and this continued once in a fleet squadron. Typically, they would use either the Aerial Tow Target TDU-32A/B (radar reflective) or B/B (laser reflective) during such evolutions, this store (which was 40ft/12m long and 7½ft/2.3m high) being towed aloft by an F-14. A Tomcat with a TDU-32 banner attached could safely take off from either an airfield or a carrier flightdeck. Its speed was limited to 250 knots (463km/h), however, and hits would be assessed by counting holes in the banner. If multiple crews were carrying out firing passes on the tow target, wet lacquers in different colors would be applied to the 20mm shells by armorers prior to the cannon being loaded to facilitate post-mission banner assessment.

Although air-to-air gunnery was practiced by all fleet crews, the only operational aerial success enjoyed by the F-14 with the cannon was achieved by an aircraft of the Islamic Republic of Iran Air Force (IRIAF) on September 7, 1980. In the build up to the Iran–Iraq War, which commenced on September 22, 1980, skirmishes took place between both sides. On the afternoon of the 7th, five Mi-25 "Hind D" attack helicopters of the Iraqi Army Air Corps shot up several border posts in the Zain al-Qaw region. Having been detected by IRIAF radar, the helicopters were intercepted by two F-14As of the 81st Tactical Fighter Squadron flying from Khatami Air Base.

The unnamed pilot of the lead Tomcat fired two AIM-9P Sidewinders at one of the Mi-25s, but both weapons failed to lock on to their target and flew into the ground. Selecting "GUN" on his control column, the pilot placed his gunsight pipper over the rearmost helicopter and opened fire with a burst of 400 rounds. The Mi-25 was hit multiple times and exploded.

US Navy Tomcat crews also practiced strafing runs with 20mm cannon, firing the weapon against targets at ranges on the east and west coasts when flying from naval air stations during pre-cruise work-ups or, when embarked in a carrier, attacking inflated practice targets

An Aerial Tow Target TDU-32B/B lies alongside *Harry S. Truman*'s waist catapult three, its lanyard having been attached to the underside of VF-32's "Gypsy 100" (F-14B BuNo 162916). This training flight was generated while the carrier was sailing in the NAG in March 2001. (Gert Kromhout)

nicknamed "killer tomatoes" because of their red color. The latter would be dropped in the carrier's wake and aircraft then cleared to undertake strafing passes at the floating target. According to Lt Cdr Lloyd Abel, "the gun, in certain air-to-ground attack modes, was a 'death ray,' but this was not so much the case when it came to air-to-air engagements."

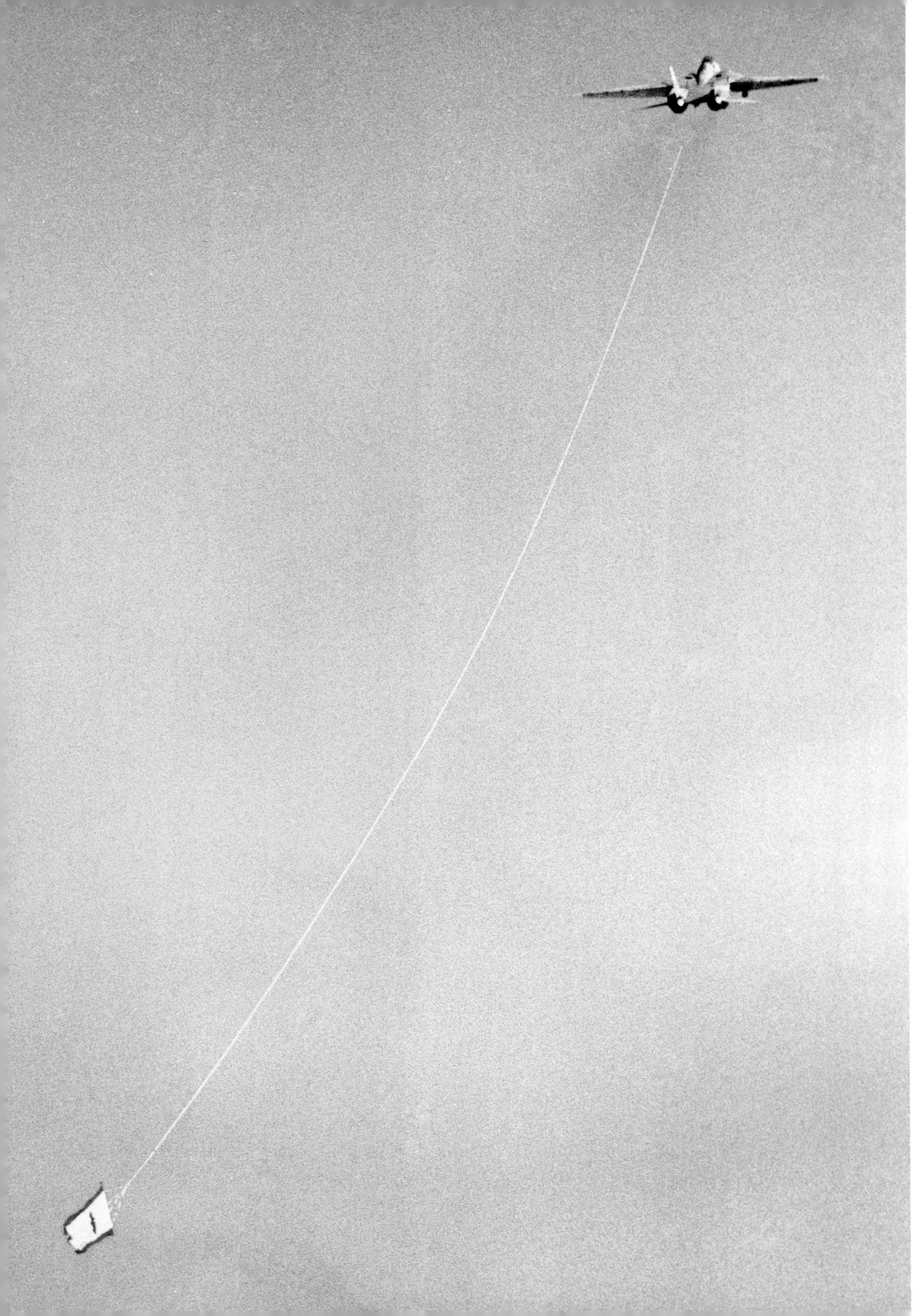

Above: A well-used target ship moored in the waters of the Bombing Target 11 range off Piney Island, North Carolina, sustains more holes during a strafing pass by an F-14A from VF-102 in 1987. This TARPS-generated photograph was taken with a KS-87 camera installed in a pod carried by a second VF-102 jet. A stationary floating target was easy for a competent Tomcat pilot to hit with 20mm rounds. An agile vessel traveling at speed proved to be an entirely different proposition, however, for Lt Cdr Bill Sizemore and Lt Rob Reed during Operation *Desert Storm*. (US Navy)

Left: With the banner trailing in its wake, an F-14D from VF-2 climbs away after launching from *Constellation* in the Pacific on April 1, 2001. (US Navy)

F-14 "BOMBCAT": THE US NAVY'S ULTIMATE PRECISION BOMBER

Operation *Desert Storm*

As previously noted, the Tomcat's performance as an air superiority fighter in Operation *Desert Storm* was disappointing. Although the units committed to the campaign undertook countless CAPs and vital TARPS missions, the jet was prevented from engaging IrAF aircraft due to its crews' inability to solve all the RoE requirements independently of AWACS support. This was because, at the time, the older F-14A/B (unlike the F-14D, which was not yet in operational service) lacked the very latest IFF systems and software fitted in the USAF's F-15Cs.

Clearly frustrated at not being able to play a more active part in the war effort, and with their jets still some years away from being effective strike aircraft, a handful of Tomcat pilots resorted to strafing ground targets with the M61A1. On one such occasion, according to a

The TDU-32B/B banner, some 40ft (12m) long and 7½ft (2.3m) high, was speed restricted to 250 knots (463km/h). Eyeing up their target before making a firing pass, the crew of F-14A BuNo 159001 have yet to engage the banner – the spotless aircraft lacks any tell-tale gunpowder residue around the muzzle. Although wearing VF-1 markings, this jet was still assigned to VF-124 at the time while the fleet unit was completing its conversion on to the Tomcat at Miramar with the FRS. The aircraft was brought down by an inflight engine fire on January 14, 1975 while flying from *Enterprise* during the first-ever operational Tomcat cruise. (US Navy)

crew from CVW-3's VAQ-130, embarked in *John F. Kennedy* sailing in the Red Sea, they were flying an EA-6B deep into Iraq and approaching their target when their F-14 escort left them in full afterburner without warning. Unperturbed, the Prowler crew pressed on and made their jamming orbit, shot a HARM and egressed safely with a pair of A-7Es from CVW-3 that had joined on them – each of the Corsair IIs were armed with Sidewinders, thus offering the unarmed Prowler a modicum of protection from IrAF fighters.

The Tomcat crew, meanwhile, had descended to strafe buildings, and they only saw the EA-6B again when they were receiving backside fuel from a tanker over Saudi Arabia. Upon returning to the carrier, the crew of the Prowler vociferously protested to the CAG (Capt A Hardin White) about being abandoned over enemy territory. He in turn spoke to the CO of the F-14 squadron about it, and the pilot and RIO of the fighter were admonished for leaving their high value unit unprotected. They subsequently visited the VAQ-130 ready room and apologized to the Prowler crew.

Referring to the same episode, a source from within CVW-3 noted "a VF-14 Tomcat descended and strafed an Iraqi observation post on egress. 'Ordies' from both fighter units in the air wing [VF-14 was paired with VF-32] asked crews to test fire the gun on every sortie, and on this particular occasion a jet that was being flown by a Flag staff member who had just joined the ship rather overdid his firing 'test', resulting in him being grounded by the admiral. Prior to the start of *Desert Storm*, all F-14 crews had been told that there was a standing policy in effect preventing jets from descending below 20,000ft over enemy territory due to the density of AAA." The policy was firmly restated to all squadrons within CVW-3 as a consequence of this isolated incident.

There was one legitimate F-14 strafing event in *Desert Storm*, however, involving single jets from VF-41 and VF-84 embarked in *Theodore Roosevelt* as part of CVW-8. Leading the mixed section was Lt Cdr Bill "Size" Sizemore (with RIO Lt(jg) Carl Klotzsche) of VF-41, while his wingman was Lt Rob "Astro" Reed (with VF-84 XO Cdr Ron Rahn as his RIO). Reed recalled:

We launched on a CAP mission on February 12, 1991, fragged for central Iraq to patrol an area just south of Baghdad. It was a double-cycle mission, meaning that we launched and stayed airborne until the second recovery after launch. We were scheduled to get gas from the USAF tankers on one of the "berry" tanker tracks – there were several, namely "strawberry," "raspberry" and "blueberry."

I was flying "Victory 211" [BuNo 161164], a TARPS aircraft. This jet was a great airplane. It would run at 1.2 IMN [Indicated Mach Number] and about 780 KIAS [Indicated Air Speed – 1,444km/h]

The dark blue waters of the NAG are churned up in a neat line as 100 20mm rounds impact the surface during a "gun-ex" undertaken by VF-31 in early January 2006. Tomcat crews would routinely attack inflated practice targets dropped in the carrier's wake when on deployment. (Erik Hildebrandt)

Lt Cdr Chris "Limp" Richard rolls in hot and squeezes off 100 rounds during a strafing pass over the NAG in early January 2006. His F-14D, "Felix 100" (BuNo 164342), is also armed with a GBU-12 LGB, a GBU-38 JDAM and a solitary AIM-9M. Aircraft from CVW-8 fired both M56A3/A4 HEI and PGU-28 SAPHEI rounds during the OIF IV deployment of 2005–06. The PGU-series of electrically-primed ammunition was introduced in the late 1980s and is now the standard 20mm round for US Navy, US Marine Corps and USAF aircraft. A "low-drag" design developed to increase muzzle velocity, each PGU-28 round combines armor-piercing characteristics with high-explosive incendiary effects. It has an effective velocity increased to 3,450ft per second, yielding greater range and more penetrative power, as well as better accuracy. (Erik Hildebrandt)

at 10,000ft in zone two AB [Afterburner]. It was tuned up, flew straight and was fast. I usually came off a TARPS run with several thousand pounds more gas than my wingman, as they needed full AB, or at least zone four, to keep up with "211." It was my favorite Tomcat, and the aircraft even had my name on the canopy rail for the front seat. How could I not love that jet? I flew it on many TARPS missions during the deployment, but on this day, it was loaded for air-to-air. We carried two AIM-9Ms, two AIM-7Ms, one AIM-54C and 500 rounds of 20mm ammunition. It was a squadron decision to carry the Phoenix. VF-41 did not, preferring the lower drag of the Sparrows, but we did, preferring the launch and leave [active mode] options that the Phoenix offered.

AMSAN Dave Allen and AO3 Jay Davis apply radar silhouettes to VF-14's "Camelot 100" (F-14A BuNo 162700) following the aircraft's impromptu strafing attack on military buildings in Iraq during a Prowler escort mission in mid-February 1991. (AO3 Jay Davis)

Lt Cdr Sizemore's wingman during the February 12, 1991 strafing attack was Lt Rob Reed (who had VF-84 XO Cdr Ron Rahn as his RIO). Following his tour with VF-84, Reed became an instructor with VF-101. (via Capt Rob Reed)

We rendezvoused overhead CVN-71 and headed north into Iraq. CAP was uneventful, with a couple of tanking events on the KC-135s and no radar contacts, and when our time on station was up, we headed home. The weather was clear, with that ever-present bit of sandy haze in the air. As we crossed from Iraq and headed for the Gulf, we switched up "Red Crown" – the surface ship that controlled air defense for the carrier forces in the Gulf – to identify ourselves and return to the boat.

"Red Crown" asked us, "Are you equipped with '20 Mike-Mike'?" to which "Size's" RIO answered "Affirmative." "Red Crown" then asked if we were "capable of engaging a surface target?" to which we answered "Affirmative." The excitement started. They were asking if we could strafe a target! For weeks we had been flying with no real opposition and no real targets for the fighters. We were vectored to an area between Kuwait City and Failaka Island – a bit of a bay about ten miles across. Failaka had been occupied by Iraqi troops when Kuwait had been invaded, but they were now mostly cut off from re-supply.

Lt Cdr Sizemore (in "Fast Eagle 104" BuNo 160387) was the first to spot their target:

As we approached, I observed a single A-6E conducting bombing runs on a small vessel approximately 40ft [12m] in length that was transiting between Kuwait and Failaka Island. The vessel appeared to be a sport fishing boat with an upper "flybridge," and it was traveling at a fairly high speed – approximately 25 knots [46k/h]. Failaka was heavily defended by AAA batteries that would shoot at you if you flew nearby. One of our air wing A-6Es from VA-36 had been shot down, with the crew lost, near Failaka ten days earlier.

The vessel was approximately halfway between the Kuwaiti mainland and Failaka Island when we arrived on the scene. We watched the lone A-6E make multiple passes to bomb the vessel. The boat would jink aggressively during the attack runs, causing all the bombs to miss their target. After the A-6 was "Winchester" [out of ordnance], we were called in to strafe.

I was the flight leader and set us up for individual standard 15-degree high-speed strafing runs in a left-hand pattern to keep us as far away from Failaka Island as possible. We made multiple passes on the vessel, which would jink aggressively as soon as we started our runs. All of my shots missed the vessel close aboard, aft and forward. I could see two individuals on the boat – one was driving from the elevated bridge and the other was near the bow, and the latter would shoot at us as we flew by during each firing run.

Lt Reed also failed to hit the boat:

The vessel was going quite fast, and given the 90-degree angle between our heading and his, the amount of lead we had to give it was both considerable and a complete guess. I lined up, pulled some lead and was all set to press into firing range, which for strafing was about a mile. My RIO called for me to shoot, and so I did, but before I was ready and before I thought we were in range. My bullets walked up to the boat but stopped just short – VF-84 had elected to set the burst limiter at "100," and the weapon ceased firing at the worst possible time.

I pulled straight up in afterburner [AB] to get away from the threats on Failaka and in Kuwait, and we set up for another run. Let me pause here to say that it's a terrible idea to roll in on the same target twice – the enemy can see you coming, and has a chance to prepare for you. I didn't even think of that. All I wanted was to hit that boat. I rolled in a second time, with a much shorter run. I got much closer before pulling the trigger, and my bullets went right through his path just a few feet behind him. Good range, not enough lead. I noticed during that run a "winking" on the front of the boat. Muzzle flash. Sure enough, the enemy knew I was coming and had prepared. Small caliber, but a single bullet will ruin your day if it finds you.

I repeated the pull up and roll in, but this time as we rolled in I saw a flak burst right near my airplane. The Iraqis on either Failaka or in Kuwait had found us as well. It's hard enough to aim

VF-84's "Victory 211" (F-14A BuNo 161164) and "Victory 202" (F-14A BuNo 160391) fly as a section over the Caribbean on January 27, 1990 while embarked in *Abraham Lincoln* with the rest of CVW-8 during the carrier's shakedown cruise. This photograph was taken by Electronic Countermeasures Officer Lt Cdr Rick Morgan of VAQ-141 from the cockpit of his EA-6B. A little over a year later, "Victory 211" was involved in the only legitimate F-14 strafing event of Operation *Desert Storm*. (Lt Cdr Rick Morgan)

a gun when you're level, but to get the pipper lined up when you're doing high-G maneuvers to jink the airplane (all a combination of lateral and vertical maneuvering to throw off the enemy's tracking) is nearly impossible. My bullets went beyond the boat, just missing him again. We rolled in for a third time – same story, with jinking throwing off the pipper. The flak bursts were now more numerous, and we were more concerned with surviving versus getting good hits.

On the last roll-in (my fourth), I asked my RIO if he could lock up the boat with his radar. A Sparrow missile, with its 85lb [38kg] warhead, is quite lethal to surface targets, and while he was busy with the radar, I was both jinking and looking for a heat signature for a Sidewinder IR [infrared] lock – that missile would be a lot better at tracking than my bullets. Any missile, in fact, would have been better in this case.

We pulled off again without success, went full AB in a climbing turn to get above those threats – flak, handheld SAMs and small-arms fire – and found "Size," who was holding overhead, amused, I think, at the number of times we had tried to strafe.

Like his wingman, Lt Cdr Sizemore had also tried to target the boat with a missile:

After expending all of our 20mm ammunition during four passes, I set up a wider pattern and tried to get a radar lock and then an IR tone to possibly expend a Sparrow or Sidewinder. Neither attempt was successful, and while doing this, I became aware that the previously clear air was now full of small grey "clouds." Upon closer examination, I saw the small "clouds" were actually AAA bursts coming from Failaka, and they were getting more numerous the closer the vessel got to the island.

After both "Astro" and I had expended all of our 20mm ammunition and failed in attempts to employ other ordnance, we climbed into the upper teens [18,000–19,000ft] and moved out of range of the AAA. The last I saw of the vessel and its determined crew, it was entering a cove on the western side of Failaka Island.

Wise after the event, Reed reflected more than three decades later:

In retrospect, we really took too many chances for a simple little boat. I know my [Navy Commendation Medal] Citation says that I engaged three times, but I came back with an empty gun, so there were at least five trigger squeezes, followed by the attempt with missiles.

When I trapped aboard "TR" and the VF-84 ordnance guys came to safe the gun, you could see their excitement at all the powder residue on the gun port and the empty belt of ammo. They knew we had fired the gun. When the maintenance folks found out about my strafing adventure, and the lack of a boat sinking, a new symbol appeared on the side of "Victory 211" – a small fish, surrounded by a red circle, with a red slash through it. Like the victory flags painted on airplanes in previous wars, I had been awarded a fish "kill" by our maintenance guys right under my name on the canopy rail.

The Tomcat strafing attack off the coast of Kuwait on February 12, 1991, was led by Lt Cdr Bill Sizemore (with RIO Lt(jg) Carl Klotzsche) of VF-41. Seen here as a captain almost two decades later, Sizemore would retire from the US Navy with the rank of rear admiral (as had his Naval Aviator father in September 1982). (US Navy)

"Fast Eagle 102" (F-14A BuNo 160918) refuels over the NAG during the midway point in a DCA mission from *Theodore Roosevelt* on February 14, 1991. Two days earlier, while returning to the carrier from an uneventful CAP over central Iraq, sister-aircraft "Fast Eagle 104" (BuNo 160387) had attempted to sink an Iraqi vessel with cannon fire as it transited between Kuwait and nearby Failaka Island. (Lt Cdr Rick Morgan)

Despite both Tomcat pilots having missed the highly agile Iraqi boat, Bill Sizemore (who later attained the rank of rear admiral prior to his retirement) was fulsome in his praise of the aircraft's M61A1 cannon:

> The Tomcat was an excellent platform for air-to-air and air-to-ground gunnery, and both were taught extensively in the F-14 training squadron and practiced, and trained to, in the fleet. The key to achieving hits was to get into the proper position and then be as smooth as possible while firing. Because the cannon was located below and to the left of the pilot, the smoke generated by the weapon when it was fired didn't obscure the canopy as it did in the F/A-18, where the gun was located directly in front of the cockpit. Having later transitioned to, and flown, the F/A-18 for several thousand hours, I found the Hornet to also be a great gun platform in both air-to-air and air-to-ground modes. In the Hornet, the key to getting hits was the same as with the Tomcat, except that you had to "hold what you had" while smoothly squeezing the trigger because as soon as the gun fired, the outside canopy area was immediately covered with gun smoke, obscuring your view.

Above: A fully armed "Victory 211" is carefully guided over *Theodore Roosevelt*'s waist catapult three prior to flying an Operation *Provide Comfort* flight on May 2, 1991. One of three TARPS-capable jets assigned to VF-84, it was kept busy during this post-*Desert Storm* period producing photographic imagery of Iraqi force dispositions in the north of the country while coalition aircraft protected the Kurds from aerial attack. "Victory 211" just happened to be the Tomcat that featured Lt Reed's name on the canopy rail, and VF-84 maintenance personnel adorned the jet with a "victory" symbol consisting of a small fish surrounded by a red circle, with a red slash through it. The "kill" marking can be seen here to the left of the diagonal fuselage band. (US Navy)

Right: CVW-8's two boat-strafing jets are seen here chained down nose-to-tail on the stern of *Theodore Roosevelt* in this May 6, 1991 photograph. Both aircraft were being prepared for an Operation *Provide Comfort* TARPS flight at the time. (US Navy)

OEF Strafing

More than a decade would pass before a US Navy Tomcat next fired its guns in anger. By then the aircraft had matured into a precision strike fighter, employing both LGBs and dumb bombs in the Balkans, Iraq and Afghanistan. It was during the early stages of the ground war in the latter country that F-14 crews were presented with the opportunity to strafe once again.

On November 5, 2001, after almost a month of air strikes against infrastructure and military targets in Afghanistan during the opening phase of OEF, the Northern Alliance at last began to take ground from Taliban and al-Qaeda forces to the south of Mazar-i-Sharif. As expected, the enemy proved difficult to shift from well-prepared defensive positions, and Hornets and Tomcats from CVW-11, embarked in *Carl Vinson*, were soon called in by SOF controllers to winkle them out. Among the jets on station performing XCAS was "Black Lion 101" (F-14D BuNo BuNo 164603), crewed by squadron CO Cdr Chip "Biff" King and Lt Cdr Michael Peterson, who recalled:

> We were one of several sections of strike fighters from the air wing that were tasked with providing on-call CAS for SOF and Northern Alliance troops due to attack Taliban positions. The Northern Alliance, supported by US SOF personnel, had planned to push towards Mazar-i-Sharif from the south and south-east, but we hadn't received any report of contact so far. We felt that the chances of actually performing CAS were pretty slim.
>
> We had been on the schedule for CAS missions previously but always wound up performing SCAR/Armed Reconnaissance instead. The two missions differ from CAS in that there are no friendly forces in close proximity to the targets that you're destroying. We would usually be handed off to an observer who was safely sitting in a remote location and had "eyes on" some enemy targets that he would talk us on to and then we would destroy them. So far, we hadn't actually performed CAS, although the ATO had scheduled us for the mission several times. On this day, both aircraft in our section were carrying two 1,000lb [454kg] GBU-16 LGBs and a full drum of 20mm SAPHEI.
>
> The launch, join up, transit across Pakistan and entry into Afghanistan were uneventful. We topped off on one of the southern tanker tracks and headed north to our assigned holding point for some additional gas. We met our next tanker north-west of Kabul. The air wing staggered the CAS sections' on-station time so that aircraft would always be available, if needed. We headed to the northern tanker track to remain topped off on fuel because we were scheduled to have the last on-station time during this cycle of CVW-11 aircraft in-country.
>
> Other sections tasked with CAS support headed directly to the holding point to provide immediate cover and would then hit the tanker before heading south. The F/A-18 Hornets usually had a much shorter on-station time when compared to the Tomcat due to fuel consumption rates and gas onboard, so we patiently waited for our on-station time to arrive. An AWACS controller eventually passed us a set of co-ordinates to head to, a FAC call-sign ["Tiger"] and a secure working frequency to contact him on.

Lt Cdr Michael Peterson and Cdr Chip King, CO of VF-213, participated in the squadron's opening night strike on Kabul. Both men were Tomcat veterans by then, with Cdr King being CVW-11's senior strike lead and Lt Cdr Peterson having seen combat as an SFTI augmentee with VF-14 and VF-41 during Operation *Allied Force*. Perhaps their most memorable mission in OEF was flown on November 5, 2001, when they became the first US Navy Tomcat crew to use the M61A1 in anger in almost 11 years. (via Cdr Michael Peterson)

> The location assigned was about 20 nautical miles [37km] south of Mazar-i-Sharif, a strategically located city in north-central Afghanistan about 30 nautical miles [55km] south of the border with Uzbekistan. There was also an airfield about five nautical miles [nine kilometers] to the east of the city.
>
> So far, everything was like a standard, low key flight. We'd probably contact the FAC, who may or may not have some targets of opportunity that he had been observing from a distance, and hopefully we'd get to "flex" to armed recce the area and then drop some bombs. As we approached our assigned co-ordinates, we had difficulty establishing radio contact with the FAC, but this was also normal due to line-of-site restrictions in mountainous terrain. Once we were within 20 nautical miles [37km] of him, we finally made contact with "Tiger," and, as a good FAC(A) myself, I was going to give him a complete and proper by-the-book check-in.

I was about two lines into my "MNPOPA" [(call-sign), Mission Number, Number/Type of aircraft, Position, Ordnance/Systems, Playtime and Abort Code] brief when he promptly told me to "shut up" and asked "did I have the friendly position that the AWACS passed?" There was a great deal of urgency in his voice and the sound of gunfire in the background. I quickly read the co-ordinates to him and bypassed the standard authentication procedures, as he was talking on the pre-assigned secure frequency in an American voice, and that was good enough for me.

The FAC relayed that he was being overrun by Taliban fighters, and that he wanted ordnance "on his position" now. I asked for an altitude so I could accurately cue the LTS and a quick description of his position. After entering the altitude and cueing the LTS with the AWACS' co-ordinates, we could see via the FLIR a hill top with a large dry wadi running from north-to-south down the western side. I described the position to the FAC, and he confirmed that his unit was on top of the hill and being overrun by enemy forces coming up the south side. He requested that we hit the base of the south side of the hill.

We set up a south-to-north run in for that attack, designated the base of the south side of the hill with the laser and delivered a single GBU-16, telling the FAC to get his head down when we released the bomb. Our wingman set up in high cover to watch for any MANPADS being shot at us, and when our weapon hit, the FAC provided a hasty target correction, along with the instruction to "Hit them again – put it half way up the hill." We quickly made the second run as requested, after which we switched places with our wingman since we had dropped both of our LGBs. The FAC asked our wingman if he had seen our hits, and he told him to put the next bomb three-quarters of the way up the hill.

Having delivered his bomb, and with only one GBU-16 remaining, the FAC gave our wingman one final correction, and told him to "Put it on the top of the hill! They're all over us!" We could still hear the firefight going on in the background over the radio, with the sound of small arms fire, and I quickly told the FAC "Understand Danger-Close," which meant that friendly forces were going to be inside the effects of the LGB. The FAC responded with "Just put it on the fucking hill top!" Our wingman rolled in and delivered his final weapon on the hill top. The whole evolution from check-in to four weapons expended took about four minutes.

After the final 1,000-pounder impacted the hill top, I called the FAC and asked him for feedback. "'Tiger,' how was that?" but we failed to make radio contact after several attempts by both aircraft. It was the lowest point I have ever had in my life. I remember feeling all the blood drain from my face and feeling sick in my stomach. I was sure that the last bomb had killed the FAC and other friendlies. We tried to re-establish radio contact a few more times, with no luck, and continued to orbit the position of the attack.

About ten minutes later (an eternity at that point), "Tiger" contacted us on the working frequency. I went immediately from the depths of depression to completely ecstatic. The FAC exclaimed, "That was fucking outstanding. I'm okay, my guys are okay, my horse is okay. You must have got 50 of those fuckers!!! What have you got left?" referring to how much ordnance and time we had remaining. We replied that we were "Winchester", with the exception of 20mm. He said that the

With its refueling probe extended, VF-213's "Black Lion 101" (carrying two GBU-16s) closes on a tanker for frontside fuel en route to Afghanistan on November 7, 2001. Two days earlier, it had been involved in a strafing attack on Taliban fighters attempting to attack Northern Alliance troops, supported by US SOF personnel, as they pushed on Mazar-i-Sharif. (USAF)

SOF "Tiger" FACs were making contact all over the place, and that he needed us to get more CAS aircraft on station ASAP.

We radioed the AWACS to relay his request, but were told that we were the end of this wave of CAS support. Other aircraft were on their way into country, but it would be 20–30 minutes before they arrived on station. We passed the information on to the FAC, who told us to "hang out and let them hear you – they don't know you're out of bombs," so we agreed to remain on station as long as possible to provide a deterrent effect.

Fifteen minutes after expending their second LGB, the crew of "Black Lion 101" received an urgent call from another "Tiger" FAC located a few valleys over. Cdr Chip King explains what happened next:

We were the last section on station, and although we had already dropped all of our ordnance and were heading for the tanker, we continued to monitor the FAC frequency. Having heard the frantic call from "Tiger," we headed back to the area, informing him that all we had remaining was 600 rounds of 20mm SAPHEI, and given the current SPINS [Special Instructions] for OEF, we were unable to descend to an altitude that would allow our attacks to be effective. The urgency in his voice was obvious, and the AWACS responded for us to standby as Lt Cdr Peterson notified it on J-Voice [Link 16] that we were going to proceed below the hard deck to provide support with 20mm runs. Although we really weren't asking for permission, the CAOC commander, who was monitoring the channel, came up on the radio and said we were "cleared to prosecute as necessary."

We located the Northern Alliance position on the ground with our FLIR and began maneuvering for an immediate attack, if cleared. You could see that they were being overrun by enemy troops, mostly either on horseback or in all-terrain vehicles. We had positioned ourselves roughly 60 degrees off the target when we finally got clearance below the hard deck.

According to Lt Cdr Peterson, "We received a quick talk-on, descended, and set up an attack on the target area – a wide wadi where we could see the enemy 'ants' running across on the FLIR." Cdr King recalled:

The terrain was very rugged around us, and I told my wingman to remain in high cover – MANPADS were our greatest threat. We armed the guns and left them hot until we were "Winchester." We made four full strafing runs, with the gun firing out on the fifth run. We set the radar altimeter for 2,500ft AGL so that we would level out by the time we reached 2,000ft AGL.

I made sure that we entered each strafing run with a speed of more than 540 knots [1,000km/h] over the nose of the jet, with most of our run-ins being made along the same approach line. You could see the SAPHEI rounds exploding on impact, and the flashes of small arms fire shot back in our direction in return. We knew that we were out of their effective range, but we never discounted the "golden BB" rule. It would have only taken one round to ruin our day. I remember thinking to myself at the time what a disparity in technology. It was like "Buck Rogers" meeting the "Arabian Nights," with the Taliban fighters on horseback. Our fire was obviously effective, allowing the heavily outnumbered Northern Alliance forces to retreat without suffering further casualties.

Once we were "Winchester", we stayed overhead until "Tiger" called "Clear." He was very grateful, and thanked us, saying that he looked forward to buying us a beer upon our return home. By then Lt Cdr Peterson and our wingman were co-ordinating relief with new strike assets that had been vectored to the fight as we made our way to the tanker.

Lt Cdr Peterson's view of the strafing passes was as follows:

Having been given a "cleared hot" call, "Biff" squeezed the trigger and put in a little back stick to string a line burst of 100 rounds of 20mm in about two seconds. You could see the SAPHEI

"Black Lion 101" (BuNo 164603) was photographed on the flightdeck of *Carl Vinson* on November 5 with its cannon muzzle and surrounding fuselage panels still covered in gunpowder residue following the weapon's firing by Cdr Chip King earlier that day. The aircraft also featured an impressive bomb tally after almost a month of near-constant operations. (VF-213)

impacting the enemy troops and the area all around them on the FLIR. It was "quite spectacular" to see personnel engaged with 20mm. The FAC asked us to make a few more runs before he was able to break contact. It was everything that they tell you not to do during training – go low and re-attack the same target multiple times from the same direction. The first run may have been a surprise, but the next two certainly weren't.

We could see automatic weapons fire from both sides of the wadi as I was pumping programs of IR low-visibility flares from our BOL missile rail launcher/expendable dispensing units – the F-14 could carry an impressive 540 expendables if outfitted with two "buckets" and three BOL rails – every few seconds in case MANPADS were launched at us. I saw at least three fired in our direction, but they would take abrupt turns in our wake when they locked in on the flares we were putting out. Even with our wingman flying high cover, the last two runs were a little uncomfortable.

The FAC was able to break contact, and we headed to the backside tanker as we heard the next wave of CAS aircraft check in. Fortunately, the Taliban fighters hadn't quite figured out the concept of leading a 540-knot [1,000km/h] target, and a quick check of the aircraft by our wingman while on the backside tanker showed that we had escaped without taking any hits. We received a "job well done" from the CAOC commander himself when we passed our inflight mission report to the carrier.

One of the best parts of the mission occurred after we trapped aboard the carrier. Word of our attack had not reached the flightdeck by the time we recovered, and as we taxied to the de-arming area in front of the carrier's island, "Biff" and I saw the wide-eyed expression of squadron gunner CWO3 Michael Lavoie when he spotted all the gunpowder residue around the gun gas exit point and realized the cannon had been fired. After he heard the story, he wasn't so upset about the large number of expendables we had "wasted" during the flight.

Shortly after this mission, a second VF-213 crew also strafed a target while supporting SOF elements working with the Northern Alliance. Lts Andrew Mrstik and (RIO) Martin Velazquez fired 83 rounds during a single strafing pass.

VF-102 was the only other F-14-equipped unit to successfully use the 20mm cannon in action during OEF. On November 17, with Northern Alliance forces now heading south towards the Taliban heartland around Kandahar, jets from the squadron were involved in one of the most memorable Tomcat actions of the campaign. Two aircraft would strafe, with one suffering a jammed gun after its first pass. However, the F-14B flown by Lt Cdr Scott Guimond and RIO Lt John Depree had no such problems. Guimond described the highly eventful mission to the author:

My RIO and I had just finished tanking, and we were on our way up to Kunduz to help support the Northern Alliance when two Marine Hornet pilots from VMFA-251 got on the radio to "Bossman" in a circling AWACS and said that they had spotted a convoy of 40+ Taliban vehicles heading north to Tarin Kowt [capital of Uruzgan province, in central Afghanistan]. The residents of this city [recognized by most Afghans as being at the heart of the Taliban movement] had ejected a small Taliban force from the area

The pilot of "Diamondback 102" (F-14B BuNo 163225) is guided on to bow catapult one by a yellow-shirted plane director onboard *Theodore Roosevelt* prior to flying an OEF mission on December 31, 2001. Lt Cdr Scott Guimond and RIO Lt John Depree had used this aircraft the previous month to strafe Taliban forces attempting to retake Tarin Kowt and seize Pashtun leader, and future Afghan president, Hamid Karzai. (US Navy)

just hours earlier. We quickly realized that the convoy was only 40 miles [64km] away from us, so we turned our section around and headed back south prior to "Bossman" telling us to do so – we knew that there was a lot going on there.

Hours earlier, Pashtun leader, and future Afghan president, Hamid Karzai had sent word to his supporters in Tarin Kowt to start a revolt, and once this was underway, he and his small force of Afghan fighters had moved in. Eleven members of US Army Special Forces Operational Detachment Alpha [A-Team] 574 had also accompanied Karzai into the city. The Taliban in Kandahar quickly found out that Tarin Kowt had fallen, and they were determined to retake the town and seize the Pashtun leader.

Guimond continued:

As we raced towards Tarin Kowt, my RIO managed to get on to the frequency that was being used by one of the combat controllers who had the enemy convoy in his sights as it neared the town. He told us that there were vehicles on the move up the valley that led to a mountain pass some two miles from Tarin Kowt. The controller wanted to know if we could see the road that the enemy force was driving along.

The Marine Hornets had already left the scene due to a shortage of fuel, but they had hit the convoy with their GBU-12s prior to departing. Despite thermal crossover problems besetting their Nite Hawk pods – the Hornet pilots had attacked at dawn – they had destroyed one vehicle and disabled a second. The smoke rising from these burning pick-ups clearly marked the convoy's position for us to the south of Tarin Kowt.

The vehicles had been hit near a bridge at the entrance to the bowl-like valley, and by the time we reached the target area, they had moved considerably closer to the mountain pass that led directly to the town. Despite having taken losses, the Taliban had pressed on regardless. This was the first time that we had seen this kind of determination by the enemy to reach their objective. When we checked in, I started panicking because I knew that the second the Taliban got into the city, we would not be able to attack them due to their proximity to friendly forces and Afghan civilians.

The controller on the ground then told us that his squad was heading back to Tarin Kowt from their scouting position on a rocky outcrop overlooking the valley. The Northern Alliance fighters with A-Team 574 had piled back into their vehicles upon seeing the sheer size of the Taliban force heading their way, and the SOF troops could not convince them that air power would deal with the threat! They too had had to head back to Tarin Kowt, as they had no other form of transport with which to evacuate the area should things have become too hot on the ground.

By the time we spotted the convoy, enemy forces were heading for the town in military trucks, rather than civilian-type pick-ups. As a FAC(A) crew, we immediately took control of the TACAIR aspect of the defense of Tarin Kowt. The first thing we did was to detach our wingman and tell him that his target was the northernmost truck. We then set ourselves up as "goalie" just in case he couldn't hit it. We needn't have worried, because the XO [Cdr Michael A Vizcarra] and his pilot did a superb job of stopping the truck with a direct hit, despite their LGB dudding. We hit another vehicle and the convoy ground to a halt, at which point Taliban fighters started to spread out into the valley on foot.

Moments later, the SOF controller requested that we check the other road passes that led into Tarin Kowt to make sure that no one was approaching the town from another direction, and to confirm that A-Team 574 also had a clear escape route. Our wingman detached and started to scour the area while we continued to work over targets in the valley.

By now "Bossman" had figured out that more assets were needed to stop the Taliban advance, and additional jets were vectored in from CVW-1. We ended up handling six sections of Hornets and Tomcats, and the AWACS even moved the duty tanker closer to us so that we had less distance to fly when it came time to replenish our fuel. We provided more than 30 controls for jets dropping LGBs and Mk 83 airburst bombs (the latter from Tomcats), and also expended 20mm SAPHEI rounds in three strafing passes.

The cannon was our weapon of last resort, as we had run out of ordnance and there was no one available to relieve us on station. We really didn't want to strafe, but we had no choice, as the nearest jet with bombs was still more than ten minutes away. We had been forced to use up all our ordnance in the early stages of the engagement when the Taliban threatened to overrun Tarin Kowt. The next two sections on station also did some great work, with their weapons being skillfully buddy-lased by my RIO against some rapidly moving targets.

By then our wingman was also out of bombs, and with no other assets close by, there was a lull in our attacks on the Taliban positions. Sensing their opportunity, a handful of troops got back into their vehicles, and a solitary tank, and made a dash for Tarin Kowt, forcing us to make strafing runs on them. Despite the main valley pass being blocked, enemy troops kept on remounting and taking to the scrub in an effort to get around the destroyed trucks. The vehicles were full of troops, who would quickly get out when they heard jet noise. Bombs would then destroy a vehicle or two, and as the jet noise decreased, the Taliban fighters would emerge from the scrub and climb back aboard the surviving trucks and drive another quarter-mile until the next bombing pass was made. Then the whole routine was repeated once again. They were determined to get to Tarin Kowt.

We had spoken about strafing within the squadron during our work-ups, and agreed that the gun would only be used as a weapon of last resort in a danger-close situation. With the tank clearly on the move, we went for a classical forward quarter attack. My wingman strafed first, and his gun jammed after he had fired just a single burst. We then came in from the same direction and made two passes, and although I am uncertain as to whether we hit the tank or not, it stopped and at least one crewman got out. A third pass was made on a truck that was also on the move, and it too was stopped in its tracks.

The SOF controller was happy to let us handle the TACAIR assets throughout this engagement once he realized that we had the training and experience to ensure that our bombs would indeed hit

"Diamondback 102" cruises over typically inhospitable terrain as its heads north into Afghanistan. This aircraft was assigned to the CO of VF-102, Cdr Roy J Kelly, and XO, Cdr Michael A Vizcarra, the latter participating in the Tarin Kowt strafing mission of November 17, 2001 while in the back seat of another Tomcat. Issued new to VF-24 in May 1989, BuNo 163225 later served with VF-101 and VF-103 prior to being assigned to VF-102 in 1998. It remained with the unit until transferred to VF-101 when VF-102 became VFA-102, equipped with F/A-18Fs, in May 2002. (VF-102)

their intended targets. We only had one danger-close call throughout this sortie when we spotted dust being kicked up right on the edge of town. Thinking that we had somehow let one of the vehicles slip through the net, it turned out that the dust was being created by dudes on donkeys!

Lt Cdr Guimond and Lt Depree also undertook strafing passes the following month, although the results on this occasion were not so obvious. According to Guimond:

> We helped a British Special Air Service/Special Boat Service squad, codenamed Task Force Sword, navigate its way through a large desert of dunes to their target in the south-western corner of Afghanistan. This mission took me further south than I had previously flown. They were raiding a large opium transloading site situated literally in the middle of nowhere in the hope that the operation would glean valuable intelligence on enemy forces in the area.
>
> A unique aspect of this mission was the fact that the crews involved received a briefing about the target prior to launching from the ship – the only time I experienced this in OEF. Our section was told that we were supporting SOF, and we were shown where they were going. We were also told that they did not want us to blow anything up. We were to simply remain in the overhead in case they needed CAS.
>
> By the time we arrived on-station to relieve the section led by our skipper, the attack on the compound was supposed to have already taken place under the cover of darkness. However, the timeline shifted sideways when the SOF guys got themselves lost in the dunes. It took us about an hour to guide them the final few miles to the target after taking over the tasking from the CO and his wingman. Much to our surprise, the soldiers then told us that they were going to set up for a direct assault on the compound, despite it now being well past sunrise. They gave us the timeframe for their attack, which had taken into account our need to go and get more fuel. With our tanks topped off, we came back on station and conducted pre-assault strikes on a series of targets – possible troop emplacements – that we had been briefed about back on the ship.
>
> The SOF guys encountered stiff resistance, being opposed by three times as many enemy troops as they had expected. Despite their precarious position, the British soldiers took some convincing to allow us to use our remaining bombs to help them. They were anxious not to have the compound destroyed, and thus lose any intelligence material that might have been on site, so they initially asked us to strafe.
>
> We tried to make our runs from a higher altitude in steeper dives, but we did not achieve the levels of accuracy that we had hoped for. We were being asked to "hit the guy with the machine gun," but from the heights we were restricted to, there was no way we could pick such small targets out with any real accuracy – we were also greatly concerned that we would hit friendlies due to their close proximity to the enemy. We eventually ended up descending closer to the target to the point where we could see people moving on the ground. There were two Hornets with us as well, and they also strafed, but they were as ineffective as we were.
>
> The British troops had a laser designator with them, and we eventually assured them that our bombs would only hit the areas that they marked. They talked us on to a series of targets and we dropped three LGBs in rapid succession that quickly put an end to the enemy's resistance. It had been crucial that the GBU-12s found their aim points, as we were delivering these in danger-close troops in contact conditions. One of the targets we hit was a machine gun mounted on the back of a truck positioned on the other side of a wall that needed to be scaled by the British troops.
>
> When working with the Brits for the first time, they used very plain language on the radios – operations involving SOF tended to be less doctrinally scripted than those where more traditional assets such as regular Army or the Marine Corps were the primary participants. I had served as a FAC(A) instructor prior to OEF and had participated in some pretty realistic training scenarios, but during this mission I heard cracks, pops and explosions through the radio for the first time. The controller was clearly close to the action, and when he started giving us a description of his position, we quickly realized that he was actually *in* the fight. He told us to standby on a number of occasions as he took cover from enemy fire, then moved his position to better observe the target he wanted bombed.

At least two other VF-102 crews were asked to strafe by FACs supporting coalition troops that were pinned down by a larger Taliban force, but they declined due to the heightened possibility of also inflicting casualties on the soldiers they were trying to defend. One of the pilots involved was Lt Chad Mingo:

> We were called in to strafe one night, and in the Tomcat our HUD gun symbology was only good at altitudes below 4,000ft. We weren't keen to get down this low, particularly at night, so my RIO and I spent some time extrapolating the charts out a little bit to see if we could come up with a manual nose setting for a gunnery pass. There were troops in contact below us at the time, and I told the FAC that without a good weapons solution I was not going to shoot rounds indiscriminately in his general direction. There were also two Hornets from VMFA-251 on station with us, and as they were able to get good gun symbology at a higher altitude, we turned the tasking over to them.

The final potential strafing event for the Tomcat in OEF involved an F-14B from VF-103 on July 20, 2002. Embarked in *George Washington*, the squadron was participating in CVW-17's first day of OEF operations when it was called upon to strafe a Taliban compound. VF-103 pilot Lt Cdr Lou Schager listened to events unfold from the carrier:

> Having just landed back aboard the ship from my first OEF mission, I was in CVIC monitoring the radio chat from our relief section on station when I heard one of the pilots say that they had been instructed to strafe an enemy compound in support of troops in contact in northern Afghanistan. We were excited because we thought that the "Jolly Rogers" were going to get some work done here. A short

VMFA-251 F/A-18Cs receive frontside fuel from a USAF KC-10A while heading north over Pakistan, the Hornets being armed with unguided Mk 83 bombs. "Diamondback 106" (BuNo 162910) holds station off the port wing of the tanker, waiting for one of the drogues to be vacated. A section of VMFA-251 jets had initially spotted, and attacked, the Taliban force heading for Tarin Kowt on November 17 prior to VF-102 arriving on the scene. Hornets from CVW-1 also strafed targets during the OEF deployment. (Lt Andrew Hayes)

The final Tomcat strafing event of OEF took place on July 20, 2002, when a section of F-14Bs from VF-103 was called upon to attack a Taliban compound on CVW-17's first day of operations in-theater. Not a single round was fired, however, when a gun breech jammed the weapon in the lead jet and the second aircraft had a HUD failure. Hornets from VFA-83 strafed the compound instead, with aircraft from VFA-81 also dropping two dumb bombs on the target. VF-103's CO jet during its 2002 deployment embarked in *George Washington* was F-14B BuNo 161435. (Lt Mike Meason)

while later, however, the same individual came back up on the radio and reported that the cannon in the lead jet – flown by Lt "Ralphie" Hencho, with the XO, Cdr David Landess, in the back seat – had suffered a gun jam when the first round got stuck in the breech. To make matters worse, the pilot in the second Tomcat – Lt Chuck Murphy – then experienced HUD failure so he couldn't strafe either.

Hornets from VFA-83 ended up doing some strafing, and a section of jets from VFA-81 then dropped two dumb bombs on the enemy position, leaving our guys to supply the light strike pilots with some good situational awareness about what was occurring on the ground.

The squadron felt great frustration at having not been able to strafe when the opportunity presented itself, and to make matters worse, this was the only chance that VF-103 would get to employ ordnance during the OEF phase of its 2002 cruise. The unit had little choice but to chalk it up as a good lesson learned, and I know that our "ordies" and aviation technicians couldn't have felt any worse about the situation.

Left: "Ordies" from VF-103 load 20mm SAPHEI rounds into the magazine of an F-14B on the deck of *George Washington* while the carrier sails in the Northern Arabian Sea on August 19, 2002. A gun jam the previous month ruined VF-103's one and only chance to see combat in OEF. (US Navy)

Below: VF-31 liberally adorned its F-14Ds with "Felix and the bomb" markings for every weapon the aircraft dropped OIF I, while five of the squadron's ten jets also featured "Tommy Gun" silhouettes to show that they had fired their cannon in action (the number of rounds expended was noted on the silhouette's "stock"). "Bandwagon 107" (BuNo 163413) was the VF-31 jet that fired the most rounds – 396. (David F Brown)

OIF I–III Strafing

Five Tomcat units were committed to OIF I, and four of them used the 20mm cannon to engage Iraqi forces during March–April 2003. As had been the case in OEF, strafing was a rare event during the conflict in Iraq due to the potent AAA and SAM threat, with VF-2 firing 1,704 rounds, VF-31 1,138 rounds and VF-32 1,128 rounds – the author has been unable to ascertain how many rounds VF-154 fired in its solitary gunnery pass. These numbers confirm that only a handful of strafing runs were indeed made. Six F-14Ds from VF-31 featured "Tommy Gun" silhouettes to show that they had fired their cannon in action (the number of rounds expended was noted on the silhouette's "stock"), while shore-based F-14A "Knight 107" (BuNo 161296) from VF-154 was marked with an M61A1. The three F-14Ds from VF-2 and three F-14Bs from VF-32 that also strafed received no such markings.

Firmly chained down to *Abraham Lincoln*'s fantail as the carrier turns into the wind in the NAG on March 15, 2003, "Bandwagon 111" (BuNo 163895) has a GBU-32 JDAM and a GBU-12 LGB attached to its centerline bomb racks and two AIM-9Ms on its shoulder pylons. This aircraft also used its M61A1 in strafing passes during OIF I, expending 372 rounds. (US Navy)

STRAFING

The most widely reported strafing action of OIF I took place just after dawn on March 30 when a section of VF-2 jets was called upon to make gunnery passes while supporting the stalled US Army V Corps advance in southern Iraq. The lead Tomcat was crewed by Lt Tony Culic and RIO, and mission commander, Capt Craig Geron, DCAG of CVW-2. The latter recalled:

We were flying a night CAS sortie in marginal weather between An Nasiriyah and An Najaf. Shortly after sunrise, an Army FAC on the ground, who was operating with troops heading north-west towards An Najaf, asked my section of two Tomcats [the other jet was flown by Lt Mark Callari and Lt Cdr Michael Bilzor] to take out some Iraqi soldiers who were firing at them from behind a large fence, backed by a treeline. I could tell the guys on the ground really needed our support, so I made the decision to lead my section down through the layers of cloud that blanketed the area in order to employ our GBU-12s. Once overhead, and having been bracketed by AAA, I quickly ascertained the tactical picture below us and directed a series of LGB runs on both the Iraqi troops and a nearby APC. Despite both Tomcats dropping three bombs apiece on the targets, the FAC told us that they were still being fired upon, and he asked us could we strafe?

Carrying plenty of gas, we decided that it was a benign threat environment as there appeared to be no SAMs around. There were no other TACAIR types or attack helicopters available to carry out this request, so after descending still further, we prepared to strafe the enemy positions. Our primary concern during our gunnery passes was the fact that we were now well within the range of MANPADS. Nevertheless, we positioned ourselves for the strafing runs just under the overcast, which started at around 8,000ft, and conducted a modified strafing pattern that bottomed out at 2,500ft. Obviously, we could not achieve pinpoint accuracy when strafing at night, so these passes were more "fire for effect." However, we got the Iraqis to put their heads down, and convinced them that the soldiers they were firing on had robust aerial support.

According to Lt Tony Culic:

We debated as a section whether we should drop below the mandated 20,000ft hard deck [imposed due to the imperative to avoid losses from AAA/MANPADS] when the original troops in contact called through to us. At this time, we did not have authority to go below that altitude as we did in later missions/campaigns. Obviously, we decided to go down, but we really had little choice if we wanted to support the friendlies as there was a solid layer of cloud at 8,000–10,000ft that was an obstacle regardless of whether we were dropping bombs – we couldn't lase for our GBUs through this – or strafing. It was helpful having DCAG in my backseat when it came to making this key decision.

When it came to dropping the GBU-12s, we were so low we did not have time to acquire the target, designate with laser and guide the bombs. So, we dropped them unguided,

Lt Tony Culic and his RIO, DCAG Capt Craig Geron, come to a halt on the flightdeck of *Constellation* in "Bullet 106" (F-14D BuNo 164342) on March 30, 2003 to signal the end of an eventful mission over southern Iraq. Having been forced to drop all three of its aircraft's GBU-12s unguided due to overcast conditions, the crew then resorted to making strafing runs on Iraqi forces near An Najaf after receiving a request to do so by the US Army's V Corps when its advance was halted by stubborn enemy resistance. (US Navy)

World War Two-style, using primarily visual talk on and our CCIP sight, as opposed to a ten-digit grid – thank God for the old school training we had received!

As we dropped our last GBU, DCAG called "Winchester," at which point I said over the internal communication system "but we have 500 rounds of 20mm." DCAG then told the JTAC, "Disregard. We have 500 rounds of 20 'mike mike'." The JTAC replied, "Well bring it on." It was at this point we transitioned to our strafing dive profile, bottoming out at 2,500ft. I think we did four or five passes. We were very concerned about MANPADS, as we were traversing the same piece of sky multiple times in order to visually acquire the target, so we kept the airspeed above 450 KIAS [knots indicated air speed – 833km/h] at high-G and dropped flares pre-emptively as we rolled in and came off target.

After recovering back at the boat, I was immediately called to the Carrier Air Traffic Control Center to explain why I had led my section below the hard deck – I'm pretty sure I had to talk to CAG [Capt Mark I Fox]. When I recited the story about how we felt the need to support our troops

Above: With its gun muzzle and lower nose section smeared with gunpowder residue, "Bullet 106" looks every inch the warfighter. Its pilot for the strafing sortie was Lt Tony Culic, who, by happy coincidence, just happened to have his name applied to the Tomcat's canopy. He posed for this photograph shortly after returning to *Constellation*. Both Culic and RIO Capt Craig Geron were awarded Air Medals following the March 30, 2003 mission, with the citation for Culic's decoration noting "he quickly and skillfully directed multiple visual bomb and gun attacks at very high speed and low altitude, destroying or dispersing all enemy forces and rescuing coalition troops." (US Navy)

Right: Cdr James Flatley, CO of VF-154, was the only pilot in his squadron to strafe with the M61A1 in OIF I. He was flying "Knight 107" (F-14A BuNo 161296) at the time, the aircraft being one of five Tomcats that operated from Al Udeid for much of the campaign. (via Cdr James Flatley)

on the ground (drilled into us from the beginning of the campaign when Vice Admiral [Timothy J] Keating, Commander, Fifth Fleet, gave us the pre-war motivation speech in the fo'c'sle on board *Constellation*), all was understood and forgiven.

The second VF-2 strafing event took place on April 11 when Lt Kurt Bohlken and RIO Lt Cdr Will Burney "strafed some vehicles on the move in and around a bunker complex in northern Baghdad" during the course of a FAC(A) mission.

Although the author has been unable to locate details for the VF-31 strafing attacks, the VF-154 jet ("Black Knight 107") that fired its cannon was being flown by Cdr James Flatley, CO of the squadron. He was leading a four-aircraft detachment based at Al Udeid, in Qatar, at the time. "There wasn't much to the strafing event," Flatley recalled. "We simply took advantage of working well north of the forward edge of the battle area during daylight hours to strafe an abandoned radar site after dropping all of our LGBs. It was a benign environment, and there was no one shooting at us. It was almost as if we were conducting air-to-ground strafing as part of a training evolution."

Two of the three gunnery passes undertaken by VF-32, embarked in *Harry S. Truman* in the eastern Mediterranean, were made by unit CO Cdr Marcus "Hitch" Hitchcock:

The first time I got to use my cannon was against an SA-2 SAM site that we had attacked with bombs from eight aircraft during the course of the day. Despite having expended all of our weapons,

Above: "Knight 107" departs *Kitty Hawk* in Zone Five afterburner, its pilot having deployed the port wing spoilers to gently roll the Tomcat away from aircraft simultaneously launching off the carrier's bow catapults. Aside from being used to strafe an abandoned radar site, the jet would also drop 32 LGBs on enemy targets during OIF I. (US Navy)

Right: VF-154's aircraft had their impressive bomb tallies "tidied up" after the unit returned home to NAF Atsugi in early May 2003. The four surviving jets of the Al Udeid campaign were also adorned with red bombs, these weapons symbolizing a portion of the tally for "Knight 104" (BuNo 158620), which crashed over southern Iraq on April 1, 2003, after suffering a fuel system malfunction. The jet had dropped 18 LGBs prior to its demise. Aside from 38 bomb markings, "Knight 107" also featured an M61A1 silhouette to denote its unique status as VF-154's solitary Tomcat strafer. (Takafumi Hiroe via David F Brown)

there were still parts of the site that appeared to be potentially operational, so I elected to go in and strafe it. I did not descend below 6,000ft during my runs on the target area, despite the aircraft's gun being optimized for attacks at heights of about 1,500ft. Nevertheless, the weapon worked well, and I could see the pattern of shells impacting in the general vicinity of the target. I am not sure how effective my run was, however. Several days later, I also strafed some artillery pieces after I had dropped all of my bombs.

One other crew used the cannon to suppress enemy fire when a SOF team was holed up by Iraqi troops. They found that a strafing run got enemy heads down and allowed our troops to extricate themselves from a potentially dangerous situation.

According to VF-32's Lt Cdr Randy Stearns, "strafing [in OIF I] would have either been a last resort if we were out of LGBs or it was actually requested by the JTAC – which it rarely was in 2003. Also, due to the [AAA and SAM] threat [in OIF I], we rarely went below 10,000ft – we really didn't need to."

Stearns would be involved in the next Tomcat strafing event, which took place on December 11, 2004 during VF-32's OIF III deployment, again embarked in *Harry S. Truman*:

We started flying our first missions on November 19, 2004, after we relieved *JFK* and CVW-17. We got there towards the end of [Operation] *Phantom Fury* in Fallujah, and we remained on station in the NAG until March 21, 2005, when *Carl Vinson* and CVW-9 replaced us. The carrier in the NAG supplied 65 percent of the TACAIR assets to the OIF ATO while we were there. The Air Force had an F-16 squadron at Balad, in Iraq, and some Strike Eagles in Qatar. Otherwise, our four "pointy nose" squadrons [VF-32, VFA-37, VFA-105 and VMFA-115] were it. The Marines had F/A-18 and Harrier II squadrons at Al Asad, in Iraq, but most of the time they were busy supporting their own forces in ongoing operations, while the Air Force and Navy supplied the bulk of the CAS missions around Mosul and central Iraq.

To start off, VFA-37 and VFA-105 each received two [Raytheon AN/ASQ-228] ATFLIR targeting pods and flew in sections of two VFA-37 and two VFA-105 jets – so, essentially, they flew as they had always done, and didn't mix up squadrons unless the airborne spare had to fill in. Their call-sign was "Henchman" the entire time in the NAG. They were supposed to move the then-new ATFLIRs around the air wing so that every squadron got a few weeks with two of them, but after a month of flying in section with each other, everyone wanted to keep it the way it was since we all got used to each other's voices, tactics and routine.

The Marines in VMFA-115 liked flying with us and we liked flying with them. The mixed Marine Hornet and Navy Tomcat sections were christened "Team Vicious," and we always seemed to be in the right place at the right time to support troops in contact. VMFA-115 had one ATFLIR, and we got matched up with them to optimize weapons. They could carry a 500lb [226kg] GBU-38 JDAM and the LMAV, while we could carry GBU-12s, since the only JDAM that the F-14B was cleared for

Left: Cdr Marcus Hitchcock, CO of VF-32, undertakes a visual inspection of the LANTIRN pod attached to "Gypsy 111" (F-14B BuNo 161428) on the ramp at Incirlik on February 21, 2003. He had landed at the Turkish base during an Operation *Northern Watch* mission undertaken from *Harry S. Truman* in the eastern Mediterranean. Hitchcock (who subsequently attained the rank of rear admiral) undertook two of the three gunnery passes made by VF-32 during OIF I. (US Navy)

Opposite: CVW-3's "Team Vicious" operations typically saw a single F-14B paired with an F/A-18A+ from VMFA-115, with the latter aircraft (lacking a state-of-the-art AN/ASQ-228 ATFLIR pod) relying on the Tomcat's vastly superior LANTIRN pod for target marking. The Hornet, in turn, brought the precision-strike capability of LMAV and GBU-38 JDAM to the party – both aircraft are so armed in this photograph. The F-14B's armament usually consisted of a pair of GBU-12s, plus 500+ rounds of 20mm SAPHEI. (Erik Hildebrandt)

STRAFING

was the 2,000lb [907kg] GBU-31. The latter weapon was not the JDAM of choice for urban warfare, as it would take out entire blocks versus just a specific house. For that reason, both VMFA-115 and VF-32 relied on each other to become a potent combat team for the next four months. Our call-sign the entire time over there was "Vicious."

In my opinion, the thing that made "Team Vicious" so special was our can-do attitude and bravado. We happened to be in certain areas where troops were in some deep trouble and really counted on air support to make the difference. On December 11 and 29, we were there, and we didn't hesitate when the call came. After that, I'm sure every ground FAC in Iraq wanted a "Vicious" section overhead when the shit hit the fan.

Stearns provided the author with the following account for the December 11 strafing mission:

Lt Jesse Carpenter and I launched in the lead jet of a two-plane mixed section element consisting of an F-14B [call-sign "Vicious 51"] and an F/A-18A+ [call-sign "Vicious 52," flown by Capt Benjamin Taylor]. Our section was tasked with providing reconnaissance and on-call CAS to elements of the Stryker Brigade in Mosul in support of operations against enemy insurgent forces. *Phantom Fury* in Fallujah had all but ended and surviving insurgents were flowing north, so everyone in CVW-3 was itching to get up to Mosul because that was where the action hot spot was.

On the ingress, we heard over the radio that one of the F-14s from the other section had "fallen out" [due to a technical issue], and his Marine Hornet wingman was a single who didn't have a fourth generation ATFLIR onboard. While on the tanker, we quickly spoke up and told them that we were a FAC(A)-capable F-14 crew with a LANTIRN pod, and that we "should" take over the frag up in Mosul because we carried GBU-12s and the Hornet was armed with JDAM only. It proved to be a good decision because we eventually ended up using our FLIR and GBU-12s that day.

About two minutes after being given the call to head north with the Hornet, we got the word that there were troops in contact. "Vicious" flight was directed to support JTAC "Savage 31" and Stryker Brigade convoy "Deuce 4," consisting of five vehicles and 40 Rangers, patrolling the Mosul area searching for weapons caches. "Deuce 4" had stopped to inspect a container with a large cache of terrorist weapons inside, and they were waiting for the Explosive Ordnance Disposal Team to arrive on station to dispose of it.

While holding overhead the friendly troops, "Vicious 51" observed a large explosion into one of the Stryker vehicles, immediately followed by an ambush from insurgent forces which trapped the friendly unit in a deadly crossfire. The shock of the explosion and ensuing ambush placed the now-disoriented Stryker patrol and JTAC "Savage 31" in a precarious situation – all avenues of approach as well as the unit's position were subject to intense small arms and rocket fire. The Stryker patrol and "Savage 31" were in danger of being overrun, and the JTAC had been unable to call in air support because he was pinned down by enemy fire from all directions.

NFO Lt Cdr Randy Stearns and pilot Lt Jesse Carpenter of VF-32 were the first crew to strafe a target in Iraq post-OIF I when, on December 11, 2004, they provided CAS for US Army Stryker Brigade convoy "Deuce 4" after it was ambushed in Mosul. Both veterans of VF-32's 2003 OIF I deployment, they initially made a strafing pass before hitting a pillbox 80m (262ft) south of a friendly position with a GBU-12. (via Capt Randy Stearns)

we felt that we needed to do something. When the insurgents saw us coming nose-on the firing stopped, and then when we pulled off target, I could see the light show going off again.

After the initial strafing attempt was complete in an effort to keep as much fire off the troops in contact as we could, we became low on gas and told our wingman to stay until we could get back from the tanker – which was in zero/zero visibility at the time – in 15 minutes. He ended up shooting all of his rounds in an effort to suppress fire. He also fired his LMAV into an approaching enemy vehicle to the south of the Stryker's position.

According to the SPINS we were governed by at the time, we weren't supposed to release ordnance unless we were in contact with a JTAC on the ground, so we had already broken rule No 1 because we assessed the situation facing the Stryker patrol to be pretty shitty at the time. In addition, while we were on the tanker, our Hornet wingman (who was a first deployment pilot) was telling us on our tactical frequency that he thought he needed to strafe, and we told him to do whatever was necessary to suppress the fire.

By the time we got back, the shooting was still going on, and other aircraft began showing up with no situational awareness of what was happening on the ground. As we approached the firefight, we heard "Savage 31" screaming "cleared hot" a bunch of times. To an aviator, when a guy on the ground is yelling "cleared hot" in a situation like that, it usually means things are not good. The hair on the back of our necks was standing pretty tall at the time because that meant we needed to be extra careful that we knew precisely where the friendlies were, and not be reckless when expending ordnance.

So, we basically found the heaviest concentration of enemy fire, which was coming from a pillbox approximately 80m [262ft] south of the friendly position, and proceeded out to the east, before heading back in and telling the JTAC "one away," as we dropped a GBU-12. I called "20 seconds," then "ten seconds," followed by "get your head down," then "three, two, one" and "boom." The five-second delay for "Savage 31" to come back and assess our hit was the longest five seconds of my life because it was so close and the chances for fratricide were high – we selected a delayed fuse so that most of the frag blew once the GBU was in the building. Then we heard "good hits" from the JTAC and I asked him if he needed another one or if the firing had stopped. He came back with, "I can't hear a fucking thing you said," so we knew his ears were ringing! He subsequently told us that the pillbox strike had "shut the entire god damn battlefield down."

"Vicious 51" continued to stay on station and support the ex-filtration of the Stryker patrol, maneuvering our aircraft low enough to draw the enemy fire up towards us and away from the friendlies. As the formation lead, our Tomcat continued to be targeted by insurgent forces throughout this period. With clearance from the Stryker patrol commander, we made an additional strafing attack on enemy forces to keep them from pursuing the friendlies back to their staging area. After one more low pass to keep the insurgents' heads down, we then departed the area.

On the way back to the carrier we suffered a massive fuel leak, and our wingman thought that we might have been hit by something because of the abundant small arms and AAA shot at us when

"Deuce 4" soldiers tackle a fuel fire started at the rear of one of their APCs after the Stryker convoy was targeted by intense small arms and RPG fire in Mosul on December 11, 2004. The direct intervention of the "Team Vicious" section from CVW-3 prevented the Stryker convoy from being overrun. (US Army)

Assessing the situation as desperate, and despite the fact that "Savage 31" was unable to communicate for approximately five minutes due to heavy fire, "Vicious 51" acquired the enemy target area using visual cues and the LANTIRN targeting system. The only way that I can describe it from 5,000ft was that it looked like half-time at the Super Bowl, when all the flashes start going off. There must have been 200–300 insurgents lighting their weapons up at the Stryker patrol. Some of the fire was aimed at us too – mostly small arms. In between the tracers were huge orange fireballs – RPGs would be my guess. My first thought upon seeing that was "Holy shit, there's no way a Stryker soldier didn't take a round from somewhere." I had done the initial OIF I push in 2003 and supported the intense Debecka Pass firefight during that campaign, but I had never seen such a concentration of fire in one place as I did that day.

We were a qualified FAC(A) crew, so I broadcast that we were taking dual control with "Savage 31" and rolling in from the east to strafe the 50 or so insurgents in the east–west hardball [TACAIR terminology used to describe a paved road, as opposed to a dirt road, since there were so many of both types in Iraq and Afghanistan]. Whether "Savage 31" heard us say that, I'm not sure, but

Lt Carpenter and Lt Cdr Stearns were flying "Gypsy 110" (BuNo 162703) during the December 11, 2004 mission, this aircraft also being the F-14B marked with Carpenter's name on the canopy rail. The Tomcat was photographed "settling" below the flightdeck after launching off *Harry S. Truman*'s waist catapult four in the NAG on November 21, 2004, 24 hours prior to CVW-3 flying its first OIF III missions. A heavy F-14B/D experiencing "settling" immediately after launching in military power from the waist catapults was a common problem with Tomcats fitted with F110-GE-400 engines. The latter could not be placed in afterburner pre-launch as the heat they generated would melt a raised jet blast deflector. Naval Aviator Lt Dave Dequeljoe of VF-32 read zero feet on his HUD seconds after he launched one night in the NAG in late 2004, although he quickly recovered lost altitude off the catapult with afterburner. His use of the latter resulted in him being yelled at over the radio by the carrier's Air Boss. (US Navy)

we got low on our two strafing runs. He came in close to look at our jet, but couldn't see too well in the dusk light – we had, in fact, suffered a mechanical fuel issue. This meant we had to divert into Al Asad for an overnight stop. As you can imagine, everyone and their brother on the carrier was anxious for the debrief, but we were just happy to not have to go back to *Harry S. Truman* for a night trap after a nine-hour action packed flight.

On this particular mission, I think prior combat experience counted because I believe that if Lt Carpenter and I hadn't supported some fairly rough CAS events in the 2003 OIF I push, we probably wouldn't have done what we did on December 11, 2004. After you achieve any level of combat experience in any capacity in the military, your views on the strict application of RoE tend to be influenced by the thought of "gee, what the hell are the consequences of NOT doing anything, and someone dies because of my inaction and willingness to follow an RoE that I know helps no one in this situation?" That is what was going through our heads that day, even though we didn't realize it until we got back to the ship 24 hours later following repairs to our jet.

Thanks to the efforts of "Vicious 51" and "Vicious 52," only a handful of troops from the Stryker patrol were wounded, and there were no fatalities.

On the afternoon of December 29, "Team Vicious" again found itself supporting a Stryker patrol that had been ambushed by insurgents in Mosul. Two F-14Bs from VF-32 provided CAS for troops in contact that day, with the second Tomcat on the scene being flown by the unit's most junior pilot, Lt(jg) James Consalvi:

My flight lead on the 29th was Capt Shawn Basco from VMFA-115 in F/A-18A+ "Vicious 45," while I was crewed with Maj Dean Castillo [a USAF exchange officer with CVW-3] in F-14B "Vicious 46" of VF-32. The first vul window was overhead surveillance of an oil pipeline in the vicinity of northern Baghdad. Approximately 30 minutes into that mission, the flight was contacted by "Warhawk" [AWACS controller] and told to proceed directly to the tanker, and then overhead Mosul. While we troubleshot some radio issues in our aircraft, "Vicious 45" refueled and headed directly to Mosul. After troubleshooting down to a single bad radio, we too began to receive gas.

While in the basket, a single VMFA-115 Hornet flown by Capt Ben Taylor joined up on the tanker's left wing. He proceeded to give us an excellent situation report and chart talk on to the target area for troops in contact. He explained that a Stryker patrol had been ambushed, with vehicles having been disabled by more than one roadside bomb, after which they began receiving fire from all sides. Capt Taylor had used his 20mm cannon and his single LMAV in support of the troops in contact. A "Gypsy" [VF-32] Tomcat had also been on scene crewed by Lt Jesse Carpenter and Lt Cdr Tom Baker. They too had employed their gun.

Upon reaching Mosul, and monitoring the JTAC frequency, we quickly realized the severity of the situation. During our first look at the target area, "Vicious 45" strafed a bomb-laden pick-up truck and then pulled off the target dispensing flares. Chaotic calls came over the radio. The JTAC, who was taking fire, was yelling for additional support, and more aircraft were trying to both check in and out. Capt Basco then began to talk our eyes on to the target area, indicating where the friendlies were in relation to the insurgents. As I followed "Vicious 45" around the target area, machine gun fire began to erupt from a prominent building on the north-west corner of a four-way intersection. The JTAC, danger-close to the enemy, requested immediate 20mm SAPHEI rounds on that target.

Over a period of 30 minutes, we made six strafing runs against the enemy position [in the building]. On the second run, while pulling off target, Maj Castillo called for me to "break left," and as I turned and dispensed flares, an IR SAM exploded approximately a quarter of a nautical mile [0.5km] aft of our aircraft. On the fifth and sixth runs, as I performed off-target maneuvers, we encountered aimed AAA.

Three of the four F-14Bs seen here parked nose to tail to free up flightdeck space on *Harry S. Truman* for a vertical replenishment evolution in the NAG on December 4, 2004 had been assigned to VF-32 during OIF I. They all have their wings locked in the 75-degree over-sweep position that was the norm for Tomcats when chained down between missions. "Gypsy 107" (BuNo 163224), seen here with its refueling probe partially extended, was the mount of Lt(jg) James Consalvi and Maj Dean Castillo when they strafed insurgents in Mosul on December 29, 2004. (US Navy)

After our gun was emptied out, we commenced an orbit around the target [building], and then used our LANTIRN targeting pod to guide "Vicious 45's" sole LMAV into it. Once the missile hit, no more enemy fire came from that location. The "Vicious" section then climbed out of the fight and headed to the tanker for gas to get back to the boat. While inbound to refuel, we overheard that the Stryker patrol on the ground had mobilized and was heading home.

After returning to the ship, Maj Castillo wrote an e-mail to the troops on the ground, telling them "We hope we helped you out." A reply quickly came from [Stryker officer] Maj Dennis Pearson, who stated "'Helped out' is an UNDERSTATEMENT. You saved a lot of good guys, and put the thump on a lot of those evil bastards! Thank you." Sadly, although it was assessed that more than 25 enemy were killed in action, one Army soldier had lost his life and eight more were wounded.

Flying in a mixed section with a Hornet lead during this mission in particular, and during our time in-theater in general, was a great experience for me. The Marines of VMFA-115 were an extremely talented and professional group of fighter pilots. They taught me a lot, and provided a number of capabilities that we would not have had without a Hornet in the section. Their ATFLIR pods were excellent sensors, their LMAV was a great weapon for areas with high collateral damage concerns and their GPS-guided bombs were ideal for bad weather missions. VF-32 and VMFA-115 worked very well together.

That day, we were doing 30-degree strafing deliveries just 30–50m [98–164ft] from the friendlies. We started our roll-ins at around 6,500–7,000ft, and the majority of the six pulls off-target

"Team Vicious" pose with their CAG jets beneath *Harry S. Truman*'s island during CVW-3's OIF III 2004–05 deployment. The command cadre for VMFA-115 and VF-32 are standing closest to the camera. They are, from left to right, Lt Cols J D Covington (CO) and Peter Ponte (XO) of VMFA-115 and Cdrs Russell Ariza (CO) and Mike Wallace (XO) of VF-32. Both units had also worked closely with each other in OIF I. (VF-32)

bottomed out at approximately 1,000ft – some lower, some higher. Speed in the dive was 500 knots [926km/h] true airspeed. The cannon's targeting system used all of the aircraft's [dynamic] inputs [speed, altitude and angle of attack], as well as radar, to get an accurate solution in real time for the gun symbology that appeared in my HUD. The 20mm cannon was a very good weapon for urban CAS because, if employed correctly, it kept collateral damage to a minimum. It was aimed fire that proved very effective against its target, but it wouldn't blow the shit out of all the buildings nearby.

In respect to the IR SAM fired at us, I think that the flares, combined with a hard break turn maneuver, defeated the missile. Flares were an excellent countermeasure, and could, in some cases, be the only one required to defeat a SAM. However, a break turn maneuver in plane with the missile would cause the greatest line-of-site rate change for the SAM's IR seeker head, and the missile couldn't hack that last-minute correction. Being a heat-seeking weapon, there was no warning sensor for an IR SAM fitted to our jets. Eyeballs were the only sensor available. Thank God I had my RIO. I am convinced that he saved the jet – "Gypsy 107" [F-14B BuNo 163224], which was a truck – that day.

OIF IV Strafing

The final Tomcat strafing events would occur during the aircraft's last operational deployment with the US Navy. VF-31 and VF-213, embarked in *Theodore Roosevelt*, participated in OIF IV with CVW-8 between October 6, 2005 and February 7, 2006 (see Chapter 12 for further details of this historic cruise). Both units would see plenty of action over Iraq, but only aircraft from VF-213 strafed targets on no fewer than five occasions. Naval Aviator Lt Ken Hockycko was involved in two of these missions, and he was interviewed by the author aboard *Theodore Roosevelt* in the NAG in January 2006:

> The rise of the gun as weapon of choice over Iraq for both the Tomcat and the Hornet took me a little by surprise when I first read reports a year ago that spoke about its successful employment in urban CAS by CVW-17 and CVW-3. However, when used properly, the gun can minimize collateral damage issues, and that is the main priority when the decision is made whether to hit a target or not in-theater.
>
> Accurate strafing is a challenge, particularly for junior pilots who generally have only limited experience of firing the gun – I can count the number of times that I have strafed in the jet prior to this deployment literally on one hand. We rarely get the opportunity to fire the gun during training. In an effort to raise our proficiency, the squadron generated a number of sorties on our way out here where we got to strafe the wake of the ship – I flew two such missions en route to the NAG.

Lt Joshua Rose, with Lt Brian Downey as his RIO, was the first to strafe on November 6, 2005 during a mission that also saw the section of VF-213 jets expend two LGBs and two JDAM in support of Operation *Steel Curtain*. Both crewmen were subsequently awarded an Air Medal following this action, and the citation for Rose's award read, in part, as follows:

CVW-8 included two Tomcat units when it embarked in *Theodore Roosevelt* in September 2005 to mark the start of the final operational deployment for the US Navy's F-14s. VF-213 would do all the strafing during the air wing's four months committed to OIF IV in the NAG, its aircraft carrying out five such attacks. The first jet to fire its guns in anger was "Black Lion 213" (BuNo 164602) on November 6, 2005, when Lts Joshua Rose and Brian Downey expended 500 rounds during two strafing passes targeting heavily armed insurgent fighters in the street at Al-Qa'im in north-western Iraq. Both Tomcats involved in this action had already dropped two GBU-12s and two GBU-38s prior to Rose and Downey being cleared to fire "Black Lion 213's" cannon. (Richard Cooper)

Rapidly responding to a US Marine Corps JTAC request for CAS in a small town [Al-Qa'im] in north-western Iraq [near the Syrian border], Lt Rose expertly maneuvered his section of Tomcats to successfully deliver two GBU-12s and two GBU-38s on buildings housing known insurgents. While these weapons were assessed to have achieved several KIAs and to have destroyed enemy strongholds, the Marines continued to encounter heavy resistance from insurgent fighters. In particular, a group of eight US Marine Scout Snipers were pinned down in a house and were receiving hostile AK-47, RPK [light machine gun] and RPG fire from an enclosed enemy position 50m [164ft] away. The JTAC cleared Lt Rose for a strafing run on this enemy position.

As the Tomcat brought its nose to bear and was cleared to fire, a small band of 10–12 desperate insurgent fighters jumped into the street and took aim with an RPG, intent on firing upon the Marine snipers' position. Lt Rose fired a 100-round burst of 20mm Explosive Incendiary ammunition with lethal accuracy, immediately thwarting the insurgents' attack.

Rose and Downey, flying "Black Lion 213" (BuNo 164602), would undertake two strafing runs at Al-Qa'im, expending 500 rounds in total.

On November 17, an unnamed crew in "Black Lion 207" (BuNo 161166) "performed an air strike in the vicinity of Samarra, expending [163] cannon rounds with successful effects against an insurgent's vehicle," according to a press release by CENTAF Public Affairs. Lt Ken Hockycko was involved in VF-213's next two strafing events:

> My first attack [in "Black Lion 201" BuNo 164341] took place near Balad on the night of December 9 when troops in contact with insurgent forces requested aerial support. Our soldiers were on a rooftop, and they had been targeted by RPGs that were being fired at them from a field 100m [328ft] away. There were also two AH-64s orbiting overhead, as well as two UAVs. The Apaches had initially been providing fire support for the troops, but the JTAC decided that he needed more firepower, so he called for fixed-wing support. We answered the call.
>
> I was extremely nervous about the attack as only two other crews, both from VF-213, had strafed up to that point in the cruise, and I personally had never fired the gun at night. I had also never strafed at the angle the JTAC was asking me to approach the target from – we usually practice lower-angle strafing runs in the fleet. However, because of collateral damage issues, I was told that I had to dramatically steepen my approach so as to keep my strafing pattern tightly concentrated on the target area.
>
> I was also very anxious about just how close our troops were to the aim point the JTAC had given me. I really did not want to open fire anywhere near them, as the gun can be the least accurate of our weapons if something goes awry with the targeting

An unnamed crew used "Black Lion 207" (BuNo 161166) to destroy a vehicle being used by insurgents near Samarra on November 17, 2005, the pilot firing 163 rounds. Although the jet fired its cannon in anger on deployment, it never dropped any ordnance while accumulating 460.7 hours of operational flying over Iraq between October 5, 2005 and February 7, 2006. The aircraft was photographed five days prior to departing the NAG, the pilot of the Tomcat, CAG Capt Bill Sizemore, having popped a flare prior to gently banking away from the C-2A Greyhound camera-aircraft. (Richard Cooper)

solution during the course of a strafing run. The troops were literally just 100m (328ft) from the insurgents, separated from them by a road. Fortunately, the soldiers were on the roof of a house that was right on the edge of a built-up area, and their position was clearly visible to me through my NVGs.

We could not get our LANTIRN into the field to lock up the insurgents, so target acquisition came via me scouring the ground with my goggles! The LANTIRN proved ineffective because the JTAC was giving us target co-ordinates that weren't accurate enough for us to acquire the enemy. The pod was rendered totally useless when I was forced to split up the target area for deconfliction purposes so as not to run into the UAVs. They patrolled the eastern half of the target area and we took the western half, which meant that we had to perform left-hand turns – the LANTIRN pod is, of course, mounted on the right side of the jet

Luckily for us, there was a string of street lights marking the position of our troops, then a little bit of field where the insurgents were, and, finally, a second light source to the east of the open ground. My game plan was to roll in and put my gun pipper between the two light sources on the dark patch that marked the field – the latter was about 200–300m [656–984ft] wide.

We ended up being talked on to the target by one of the Apache crews, who were FAC-capable. We performed two recce runs so as to make sure that I was absolutely certain about where I was planning on shooting. While setting up for my third pass, I was cleared "hot" to fire by the ground

VF-213's CO jet, "Black Lion 201" (BuNo 164341) was involved in two strafing actions on December 9, 2005 and January 2, 2006. On the first occasion, Lt Ken Hockycko fired 243 rounds during the course of two runs at night, each time targeting insurgent forces engaging US Army troops near Balad. He was directed on to his target by a FAC-capable AH-64 Apache crew that had previously been providing fire support for the troops. "Black Lion 201's" second strafing action was also the last time a US Navy Tomcat fired its cannon in anger, with Lt(jg) Meagan Varley expending 91 rounds (and a GBU-12) in an air strike near Bayji. (Lt(jg) Scott Timmester)

commander. Although still very nervous about squeezing the trigger, I was happy in the knowledge that I was approaching the target on a parallel run-in heading to the location of the friendly troops. Such an approach drastically reduced the chances of spraying them with any stray cannon shells. I also descended a little lower than I probably should have done in my eagerness to focus all of my rounds on the target area, my radio altimeter going off in the cockpit as the jet bottomed out at below 1,000ft.

Having pulled off the target and safed up the gun, I was anxious to find out how the pass had gone. I immediately got on the radio and asked them "Good hits?" to make sure that we were shooting at the right place. The Apache FACs came straight back with the confirmation "Good hits," and authorization for us to conduct one more run.

Having erred further away from our troops on the first pass, this time I worked the pipper closer to them on the second run. For the third run, the FACs asked me to move the hits further north about

The final VF-213 jet to engage the enemy with cannon fire in 2005–06 was "Black Lion 210" (BuNo 163897), which Lt Ken Hockycko used in a second night strafing attack in an urban area near Balad on December 20, 2005. During the course of two passes he expended 160 rounds. Here, the aircraft is holding station with an S-3B from VS-24 prior to receiving recovery fuel and landing back aboard *Theodore Roosevelt*. Like "Black Lion 207," this aircraft was a strafer but not a bomber during OIF IV. (Lt(jg) Scott Timmester)

50m [164ft]. Such a small change was very difficult for me to gauge at night through NVGs at 5,000ft when looking through the protective glass of the F-14's windscreen while traveling at 400+ knots [740km/h]! Nevertheless, the FACs confirmed that my third pass had put cannon shells exactly where they wanted them, and the RPG and small arms fire had come to a sudden halt in the wake of this run. We were then cleared to leave the area, and I never did see any confirmation from the CAOC stating that the 243 SAPHEI rounds that we fired had indeed inflicted a telling blow on the enemy.

Because I had made my attack at night wearing NVGs, I could actually see each individual round as it was shot out of the cannon muzzles. It looked like a green death ray through the goggles!

On December 20, while flying "Black Lion 210" (BuNo 163897), Hockycko was cleared to use his cannon in support of troops in contact once more:

I was called on to night strafe once again near Balad. This time I did get to use the LANTIRN pod, which made the attack rather different to the first one 11 days earlier. LANTIRN proved crucial on this occasion, being my only form of guidance as I had no idea where the target was from a visual perspective.

The target was in an urban area, which presented significant problems for me on NVGs due to the environmental lighting – it effectively blotted out much of the detail that I would have otherwise broken out through my goggles. Fortunately, I was able to fly right-hand turns over the target area this time, and that meant I could bring the LANTIRN pod into play. The latter quickly locked up a group of insurgents that had been spotted by troops on the ground planting an IED, the pod lacing a target pipper directly over their position.

We subsequently flew two strafing runs [expending a total of 160 rounds] and, upon returning to the boat, received confirmation from our troops after a search of the area that our passes had been effective. I could break out the targets through the LANTIRN as we made our strafing runs, but I could see nothing visually through my goggles.

Despite its age, the F-14 is still the pre-eminent strike fighter in-theater at present. This was proved to me when we were called to Balad to support troops in contact with enemy forces on that second strafing run. We showed up overhead at the same time as sections of F-16s and Harrier IIs, and the JTAC told the Air Force and Marine Corps pilots to stand by, as he would rather work with the Tomcat. I was flying in a mixed section with a Hornet at the time, and the Navy team had a wider variety of sensors and weaponry to offer the guys on the ground, hence their request that we work with them ahead of the other fast jet types.

Cdr Dan Cave, CO of VF-213, gave the following summary of his unit's strafing efforts when interviewed by the author in January 2006:

VF-213 has strafed five targets during the cruise to date, with the gun proving to be an effective weapon when called into play. Our firing runs have been aimed primarily against Anti-Iraqi Forces personnel, either on foot or in vehicles, who have actively engaged coalition troops. We have also strafed buildings in an attempt to drive insurgents out into the open, where they can be engaged either by aircraft or troops in contact. The latter have also called us in to strafe targets simply because that is the effect they want at the time.

Four of our five strafing attacks have taken place at night, which is a unique first for the squadron, as there were no nocturnal gunnery passes made during either OEF or OIF I–III to the best of my knowledge. Surprisingly, crews have not reported too much degradation to their night vision through their NVGs in the wake of these strafing runs simply because the pilot's visibility through the bullet-proof screen is not very good in the first place! We have only been using HEI rounds too, rather than a mix of tracer and HEI. The tracer rounds produce quite a light show, and the pilot really would have been blinded if these had been fired at night.

The LANTIRN pod has proven incredibly useful in these night-strafing runs. The pilot simply locks up the target with the pod, and the latter then projects four dots in the shape of a square in the HUD. He then places the gun pipper in the center of the four dots, and this ensures that the target will be hit with every round that is fired.

The cannon in the F-14 is one of its most reliable systems, as long as you keep it regularly exercised. On our way over to the AOR, I made sure that the squadron was able to conduct plenty of strafing against the ship's wake. This gave both the crews and the guns a good workout. We had also focused on our strafing proficiency during our pre-cruise work-ups following reports from previous air wings that detailed how important the gun had been to their mission effectiveness over Iraq.

The last operational strafing pass made by a US Navy Tomcat occurred on the night of January 2, 2006, when Lt(jg) Meagan Varley and an unknown RIO "performed an air strike in the vicinity of Bayji. One F-14 ['Black Lion 201'] strafed the target with 91 cannon rounds and expended one [GBU-12] precision-guided munition with successful effects against insurgents placing an improvised explosive device."

CHAPTER 11

OPERATION *IRAQI FREEDOM I–III*

In the vanguard of the fight during OIF I, the 52 Tomcats committed to the liberation of Iraq performed air defense, precision bombing, FAC(A), SCAR and photo-reconnaissance missions across the embattled country. The aircraft remained in the forefront of operations once the insurgency began to grip Iraq from the spring of 2004.

The employment of smart weapons such as JDAM and new-generation LGBs, allied with better tactics and more versatile aircraft, allowed the US Navy to make a more meaningful – and visible – contribution to "Gulf War II" than it had done in *Desert Storm*. Driven by the USAF's outstanding use of smart bombs dropped from multi-role aircraft such as the F-15E and F-16C during the 1991 conflict, the US Navy subsequently invested significant sums of money on technologies that ultimately ensured "one weapon, one target destroyed."

As described earlier in this volume, at the forefront of this precision-strike revolution was the venerable F-14 and its highly prized LANTIRN targeting pod. Since the *Desert Fox* four-day war against Iraqi targets in December 1998, Tomcat units had played an integral part in virtually all combat operations involving the US Navy as part of OSW. Indeed,

One of the primary weapons employed by the Tomcat in Iraq in 2003 was the GBU-31(V)2/B JDAM (essentially an unguided 2,000lb (907kg) bomb fitted with a tail section containing an INS and a GPS), which had made its combat debut with the aircraft over Afghanistan during OEF in March of the previous year. Initially cleared for use by the F-14B only, the JDAM was hastily made compatible with the F-14D through the installation of the D04 weapons computer upgrade in the weeks leading up to OIF I. These GBU-31(V)2/Bs are seen on *Harry S. Truman*'s flightdeck on March 21, 2003, waiting to be attached to the under-fuselage racks of the F-14Bs (from VF-32) chained down behind them. (US Navy)

on every carrier deployment to NAG until OIF I commenced, F-14 units on station had ventured into Iraq's southern no-fly zone and attacked targets such as radar, missile, AAA and communications sites operating in contravention of UN security council resolutions. The Tomcat was now seen as the perfect PGM platform thanks to its legendary long range and proven mission effectiveness with the LANTIRN pod.

Operating hand in glove with F/A-18-equipped strike fighter squadrons, F-14 units dropped – or guided – LGBs and LMAVs and provided T3 co-ordinates for JDAM between late 1998 and March 2003. The jet's FAC(A) and SCAR capabilities had also proven to be invaluable, and its upgraded TARPS system (augmented by LANTIRN and FTI) was again the US Navy's primary aerial reconnaissance source.

One of the primary weapons employed by the Tomcat in Iraq in 2003 was the GBU-31 2,000lb (907kg) JDAM, which had made its combat debut with the aircraft over Afghanistan during OEF in March of the previous year. Initially cleared for use by the F-14B only, the JDAM was hastily made compatible with the F-14D through the installation of the D04 mission tape in the weeks leading up to OIF. VF-2, VF-31 and VF-213 would have all of their D-model Tomcats upgraded by the end of February 2003.

"By the time OIF rolled around, the F-14 was the centerpiece of the carrier-borne strike package," confirmed former SFTI instructor Lt Cdr Jim "Puck" Howe (who had left VF-2 just prior to its 2002–03 deployment, only to return to the unit on the eve of OIF commencing as a SFWSL FAC(A) augment pilot). "We had JTIDS, Infrared Search and Track, JDAM, big motors and the best FLIR targeting system in the navy. Add an extra pair of eyes and a Phoenix missile, and the F-14D was as close to unstoppable in the strike role as any jet in history. During that 30-day period in the spring of 2003, the Tomcat's strike capability was never more apparent."

VF-32, based in the Mediterranean aboard *Harry S. Truman*, had had its F-14Bs made JDAM-compatible prior to deployment, but VF-154's F-14As, flying from *Kitty Hawk* in the NAG, remained restricted to LGBs. Unlike the F-14B/D, the A-model had no MIL-STD 1760 databus and associated onboard GPS system to generate computed aim point co-ordinates for JDAM. The F-14A's avionics had no path for the LANTIRN pod to transfer the target data gleaned from a GPS housed in the pod to the JDAM, as the jet lacked an embedded GPS.

Involved in some of the last OSW bombing missions in the early hours of March 21, VF-2 was also in the forefront of OIF from the word go, participating in the "Shock and Awe" strikes on Baghdad the following night that signaled the start of OIF. As well as dropping four GBU-31 JDAM, its Tomcats performed DCA and reconnaissance missions for the remaining aircraft from the air wing. Crews reported seeing continuous AAA following an impressive, non-stop TLAM and Conventional Air-Launched Cruise Missile (CALCM) "airshow."

After the Tomahawks had stopped, the CVW-2 jets, led by VF-2's XO, Cdr Doug Denneny, became the first non-stealth strike package to venture into Baghdad's SuperMEZ (Missile Exclusion

With VF-2's XO Cdr Doug Denneny leading the first "iron bomber" strike on Baghdad in the "Shock and Awe" phase of OIF I, it was natural that the CVW-2 brief would take place in the squadron's ready room onboard *Constellation*. Here, crews from all units assigned to the strike are studying slides of the IADS threat in the SuperMEZ that protected the Iraqi capital. This photograph was taken just hours prior to the March 21, 2003 mission. The green-shaded areas on the maps displayed on the two briefing screens denote the no-fly zones in northern and southern Iraq, while the colored circles illustrate the zonal range for each of the SAM threats in and around Baghdad. (VF-2)

Zone). Here, the aircraft were engaged by up to 12 SAMs – all unguided – and heavy AAA. Nonetheless, the target was destroyed and the jets all returned safely. Cdr Denneny recalled:

I was fortunate enough to participate in the first mission of OIF as a division lead, controlling two Tomcats and two Hornets. My CO [Cdr Andrew Whitson] was designated as the overall strike lead for an operation scheduled to take place later that day, so as per standard procedure in a frontline squadron, the CO and the XO were assigned separate missions.

We had been tasked with dropping JDAM on the Ministry of Information's Salman Pak radio relay transmitter facility at Al Hurriyah, south-west of central Baghdad, and in order to effectively employ our ordnance we had to fly into the infamous SuperMEZ that ringed the city. This had become legendary among coalition aircrew during the years of OSW, as the overlapping rings of Integrated Air Defence Systems [IADS] around Baghdad were so dense that they effectively blotted out the city whenever colour print-outs of the Iraqi capital were produced for threat analysis purposes!

We had tanked in Saudi Arabia en route to Baghdad, and as we flew over southern Iraq, most of the country was blanketed in a thick undercast layer that stretched all the way up to the capital. We were flying the first "iron bombers" to penetrate the SuperMEZ in OIF, and as we headed north, I could see the clouds being illuminated by the explosions of literally hundreds of TLAMs and CALCMs as they struck their targets to signal the start of the "Shock and Awe" phase of OIF. These random bursts of fire looked like lightning strikes in an electrical storm, the cloud cover obscuring my view of the city itself. We were still a long way from Salman Pak, and I nervously watched these explosions taking place for some 30 minutes, fully aware that they were effectively marking our target for us.

At that stage every explosion I saw was a coalition weapon hitting its target, rather than any Iraqi SAMs being fired off in defense of Baghdad. This allayed my fears somewhat, as I found it hard to believe that the IADS threat within the SuperMEZ could survive such a relentless pounding. However, as soon as the explosions stopped – just as had been briefed – prior to our arrival, I was alarmed by the sheer volume of AAA and at least a dozen unguided SAMs criss-crossing the night sky ahead of us. The Iraqi gunners and SAM crews had taken a hell of a beating, yet they still seemed to be capable of returning fire the minute our missile attack had ended.

As we closed on the target, I was kept busy dealing with co-ordination issues for the strike. I had to make go/no go decisions on the attack relative to the weather, which was not good, and our support assets. Our timings began to slip a little too, as our F/A-18 SEAD section had been delayed en route due to the paucity of tankers – the HARM Hornets

VF-2 OSW/OIF I (2002–03)

showed up so late that we ended up passing them in opposite directions as we headed off the target and they shot their HARM into the target! We had prioritized them last on the tankers, which is why they were late. Our Prowler support was not in an optimal position either, although they were still able to jam despite their late arrival on station. Our only dedicated SEAD support was provided by four USAF F-16CJ *Wild Weasel* aircraft that handled themselves very professionally, unloading HARM on our ingress. They then left as planned.

Technically, we were not allowed to enter the SuperMEZ without full SEAD coverage, as the HARM protected us by hitting active SAM sites on the way in. We had lost some DCA assets en route too. Although we felt pretty naked heading into Baghdad with only the quartet of *Wild Weasels* to protect us, I made the decision to press on to the target regardless.

We had to work hard to maintain our SA throughout this phase of the mission, keeping our eyes out of the cockpit in order to locate the SAMs and AAA, as well as the remaining members of the division, and all the while jinking, popping flares and pumping chaff. I had the added responsibility of making sure that we were going to hit our release parameters for our JDAM, as the whole reason we were placing ourselves in harm's way was to drop our bombs accurately on the target.

Flying Cdr Denneny's F-14 on this mission was Lt Cdr Kurt Frankenberger:

As we pressed towards the target in what would be a clockwise flow north, then east, then south, we could see the initial TLAM, CALCM and stealth aircraft weapons impacting in the distance. Our wingman took some great video footage, documenting the target ingress as if he was going to Disneyland – this later made for a nice CNN tape. A medium-altitude cloud layer occasionally obscured the lights of Baghdad, but with or without NVGs, the city could be seen getting hit regularly. This was my first combat experience in 17 years of service with the navy, and my level of anxiety was high.

As the flight lead of a mixed division of two F-14Ds and two F/A-18Cs, we solved some timing issues and flowed from our IP on into the target for weapons release. During the 25-mile run from the IP west

VF-2 NFO Lt Cdr Michael Peterson took this photograph of his section leader's F-14D on the first night of "Shock and Awe," the image being captured by holding a camera up to one of the tubes of the ANVIS-9 NVGs that he was wearing. (via Lt Cdr Michael Peterson)

of Baghdad to the release point, we observed so many SAM launches that we couldn't count them anymore. Effectively, we had to trust our systems and visually confirm that each missile did not appear to be tracking, then disregard it and evaluate the next one. More alarming were the HARM rounds that were launched from the four F-16CJs behind us. These came out of nowhere and into sight very close to our altitude. You needed to recall the location of the shooter and listen up for the launching calls.

JDAM delivery went as advertised from our aircraft, although our wingman's jet had a weapons system failure, which powered down the JDAM just as he attempted to release his bombs. We kept the speed up for egress and continued to monitor the threat, with constant SAM launches and AAA going off below to light up the night sky. The explosions from our weapons were also evident, although the cloud partially obscured them. Then the normal admin issues came back into play – worrying about gas and weather at the tanker, making our recovery slot over the ship and, of course, the gratuitous night trap.

Operating from Fifth Fleet's designated OIF night carrier, VF-2 proceeded to fly the bulk of its missions masked by the cover of darkness. It initially used JDAM to hit fixed targets that had been identified as crucial to the Iraqi war effort such as command and control nodes, SAM and radar sites, airfields and Republican Guard barracks, as well as presidential palaces and Ba'ath party buildings.

Although a veteran Naval Aviator by March 2003, Lt Cdr Howe was one of many Tomcat aircrew dropping JDAM for the first time in OIF while serving with VF-2:

I'll never forget manning my jet late one night in March, with Lt Mark "Fun" Mhley [also an augmentee from SFWSL who had previously served with VF-2] following me out to the aircraft and briefing me on how to use JDAM – he had been sent to the "Bounty Hunters" to help train the unit's Naval Aviators and NFOs in the employment of JDAM. As I finally finished the preflight, I turned to "Fun" and said "Brother, I need the *Reader's Digest* version. Remember, I'm attention deficit, it's dark, and I have DCAG [Capt Craig Geron] in my trunk. Make it fighter pilot-proof!" He did, and three hangars at Baghdad International Airport would later collapse as my JDAM hit dead center. For 28 days, Tomcat bombs rained down on fortified Iraqi positions, and the jet never performed more brilliantly in its distinguished history. We had truly saved the best for last.

During the final week of March, VF-2 switched to flying CAS strikes for troops on the ground as the push north toward Baghdad began to gain momentum. The Tomcat's ability to perform the demanding FAC(A) and SCAR roles was also greatly appreciated as coalition forces engaged the Republican Guard around cities such as Karbala and An Nasiriyah.

Cdr Denneny noted in his command report:

By the end of March, FAC(A) events were going on in earnest. I soon became a little concerned that some of the guys were getting down too low – very dangerous, as a big Tomcat is a heck of a target. I admonished them, and told them only to go low if there was no other option available and it would directly save the lives of our troops on the ground. Early April saw us flying lots of support missions for troops heading for Karbala, where we took out a Tu-124 "Cookpot" airliner [almost certainly a Tu-16 "Badger," for both IrAF Tu-124s were destroyed in *Desert Storm*] on a runway, as well as a building. We were also flying lots of SCAR missions up there, providing the FAC(A)s for B-52s and other coalition assets – Varsity work.

VF-2's Cdr Doug Denneny (left) and Lt Cdr Kurt Frankenberger (right) each received the Distinguished Flying Cross for their mission leadership of the first "iron bomber" strike on Baghdad in OIF I. They are seen here celebrating their 3,000th and 2,000th Tomcat flight hours, respectively, after a mission over Iraq with the traditional cutting of a cake in the ready room. Both men flew as a crew almost exclusively during the 2002–03 OSW/OIF I deployment. (VF-2)

Aside from conducting tactical strike missions, VF-2 completed its share of DCA sorties and TARPS flights for CVW-2 as well. Reverting to flying precision strikes when needed, one of the unit's more interesting targets during OIF was Saddam Hussein's presidential yacht, which

such a mixed load, featuring both bombs and TARPS. This allowed us to act as a stop-gap bomber should anyone need immediate on call support while we were over southern Iraq.

I was in the back seat of the Dash-2 jet, leaving the RIO in the lead aircraft to run the flight in terms of co-ordinating the navigation for all the photo-run targets. My job was to work the radios for my pilot, Lt Sean Mathieson, checking with AWACS controllers and FACs on the ground as to whether anyone needed our bombs. Having bounced around through a series of different nets on various frequencies, I ended up talking to a British Army FAC near Basra. He wanted us to head down the Shatt al-Arab waterway and attack Saddam's presidential yacht, which had been hit by a Maverick fired by a S-3B from VS-38 two days earlier and then missed by two LGB-toting F/A-18Cs. By the time I made contact with the FAC, we had finished our reconnaissance runs and were about to head south over the NAG to hit the tanker and then fly back to the carrier.

The FAC was not actually near the yacht, so he was relaying information to us that he had recently received describing where the vessel was situated. We were at high altitude, scouring the port facility through binoculars looking for the vessel. We soon spotted the burnt-out warehouses that had been hit in error by the F/A-18s the previous day, and these served as a perfect marker for the yacht. It was moored between two freighters, with a third half-sunken vessel nearby. The Maverick missile damage was clearly visible, with smoke rising from the yacht's superstructure.

Lt(jg) Pat Baker and Lt Sean Mathieson pose with their bombed-up TARPS-configured F-14D (BuNo 164350) the day after attacking President Saddam Hussein's presidential yacht in the Port of Basra on March 17, 2003. According to Naval Aviator Lt Mathieson, "As the need for CAS grew in OIF I, it became standard to load Mk 82 500lb (226kg) 'dumb' bombs on all the TARPS jets in VF-2. A mission flown with an aircraft in this configuration was dubbed a 'chum flight' because we flew as low as our SPINS permitted to gain the highest resolution on the imagery we shot. Low, straight and level was not where we wanted to be in an increasingly hostile SAM/AAA-rich environment, however." (VF-2)

it bombed in the Port of Basra on March 27. The RIO in one of the jets that attacked the vessel was Lt(jg) Pat Baker:

We had been conducting a standard TARPS mission along the Euphrates River, looking at two or three air defense sites, as well as a possible command and control facility that our intelligence folks thought was in the area. They needed photographs of the latter in order to confirm its purpose for target assignment. What was different for us that day was the fact that the ATO gave us two jets armed with a pair of 500lb [226kg] Mk 82 bombs apiece. This was the first time VF-2 had carried

Named *Al-Mansur* (*The Victor*), the presidential yacht had been one of the world's largest and most impressive vessels of its kind prior to appearing on the CAOC's ATO. Eight decks high and 350ft long, the Finnish-built ship weighed 7,359 tons and looked more like a cross-Channel ferry than a private yacht. Launched in 1982, it was the largest vessel in the Iraqi Navy, but it had no military use. Permanently staffed by 120 Special Republican Guard troops, the yacht was moved – following a direct order from Saddam himself – from the port of Um Qasr to Basra just days before OIF I commenced in an effort to afford it better protection. *Al-Mansur* was targeted for destruction because the CAOC had received reports that its extensive radio suite was being used for battlefield communications. (US Navy)

The Naval Aviators and NFOs of VF-2 (plus the three augmentees from the SFWSL) pose for a group photograph in early May 2003, having completed their time in the NAG. During the 28 days of OIF I, VF-2 flew 195 combat sorties totaling 887.5 hours. Its ten F-14Ds dropped 221 LGBs (217 GBU-12s and four GBU-16s), 61 JDAM (GBU-31) and four Mk 82 "dumb" bombs. Some 1,704 20mm cannon rounds were also fired in strafing passes. Finally, no fewer than 125 targets were photographed by the unit using its four TARPS-equipped Tomcats. (VF-2)

The lead jet, flown by Lt Mark Callari and Lt(jg) Jeff Sims [RIO], rolled in first, while we provided high cover for it – we were not sure of the AAA or SAM threats in the area. Their first bomb hit the bow, and having been unopposed in the attack, they came in and dropped the second, which struck the vessel just forward of amidships.

The lead jet then swapped places with us, Lt Mathieson following his CCIP crosshairs in the HUD, which were centered on the vessel. We dropped both of our bombs in the same attack, one hitting the hull just above the waterline and the other disappearing among the yacht's superstructure. When we left the target, the ship was on fire, although we knew we had not inflicted sufficient damage to sink it as we were carrying the wrong type of ordnance. Assuming that we were going to be supporting ground troops, we had had our Mk 82s fitted with instantaneous fuses. Therefore, the weapons exploded as soon as they came into contact with the ship, rather than burying themselves into the heart of the vessel before detonating.

I never got to see my bombs hitting home in all the LGB and JDAM missions that we flew in OIF. However, on this occasion, thanks to the diving, rolling and pulling off from the target that we had had to do in order to accurately deliver our Mk 82s, I was able to see the two little grey "blurs" that were our bombs hitting the ship as I peered back over my shoulder at the target.

During the 28 days of OIF, VF-2 successfully completed 195 combat sorties totaling 887.5 hours. Its ten aircraft dropped 221 LGBs (217 GBU-12s and four GBU-16s), 61 JDAM (GBU-31) and at least four Mk 82 "dumb" bombs. In total, 1,704 20mm cannon rounds were also fired in strafing passes. Finally, no fewer than 125 targets were photographed by the unit using its TARPS reconnaissance system.

VF-2's combat experiences were very similar to the remaining Tomcat units in the NAG, namely VF-31 and VF-154, with all three squadrons undertaking a broad mix of missions ranging from precision strikes to CAS and FAC(A). Like VF-2, VF-31, flying from *Abraham Lincoln* (which was the designated day carrier), initially performed JDAM strikes during the "Shock and Awe" phase of the campaign.

Things did not get off to a great start when CVW-14's first strike wave in OIF was stymied by a lack of "big wing" tanker support. The latter was so hard to come by that the entire strike package was forced to turn around and return to *Abraham Lincoln* with bombs still on their racks. Strike lead Cdr Don Braswell, XO of VFA-25, had "told the CAOC prior to launch that there was not going to be enough fuel on-station for everyone to press on into Iraq, and that was exactly what happened."

VF-2's "Bullet 100" (BuNo 163894) was the last of the unit's ten aircraft to receive its scoreboard of 49 LGBs and ten JDAM silhouettes. Indeed, the squadron's maintainers only applied the mission tally 24 hours prior to the unit departing *Constellation* for Oceana on May 31, 2003. Befitting its status as VF-2's CAG jet, it dropped more bombs than any other Tomcat flown by the unit during OIF I. (US Navy)

OPERATION *IRAQI FREEDOM* I–III

Its hull streaked with rust and flightdeck markings heavily weathered, *Constellation* sails at speed in the NAG on April 10, 2003 – seven days prior to departing Fifth Fleet control and heading home. Five of VF-2's ten F-14Ds can be seen, with four jets on an OIF I mission and a single Tomcat in the hangar bay undergoing routine maintenance or rectification work. After returning to North Island, *Constellation* was decommissioned on August 7, 2003 following almost 42 years of operational service. (US Navy)

However, the second and third waves sent to hit targets in and around the Iraqi capital included elements from *Abraham Lincoln*, and on the night of March 22–23 more Tomcats, Hornets and Super Hornets from the carrier ventured into southern Iraq against targets listed on the CAOC's ATO. CVW-14 CAG Capt Kevin Albright (at the controls of an F/A-18E Super Hornet from VFA-115) was on one of those strikes:

The mission called for four F/A-18E Super Hornets and two F-14D Tomcats to strike a missile production facility in the Karbala area, approximately 40 [64km] south-west of Baghdad. I was flying as "Dash 3" in the Super Hornet division. The ATO was well constructed by the CAOC, which had the monumental task of co-ordinating and scheduling thousands of sorties every day. Tanker tracks, offload fuel amounts and timings were all spot on.

Configured as a TARPS jet in the opening stages of OIF I, VF-31's "Bandwagon 101" (BuNo 164600) was restricted to reconnaissance sorties until fitted with a LANTIRN pod as the ground war got into full swing – it would end the war adorned with 11 bomb silhouettes. Photographed while flying a TARPS mission on March 22, 2003, the jet has just topped off its tanks from KC-135R 62-3505 assigned to the 22nd Air Refueling Wing's 344th Aerial Refueling Squadron. The latter unit was controlled by the 379th Air Expeditionary Wing at Al Udeid Air Base during OIF I. Despite the USAF committing 149 KC-135s and 33 KC-10s to the campaign, and the two types flying 6,193 sorties between them, CVW-14's first strike of "Shock and Awe" was stymied by a lack of "big wing" tanker support. (Lt Cdr Jim Muse)

The Super Hornets were loaded with a mix of 2,000lb [907kg] JDAM that included a penetrator variety, as well as the standard Mk 84 bomb body. The F-14Ds were loaded up with two 2,000lb [907kg] JDAM each. All strike fighter aircraft also carried a self-defense air-to-air load-out as well.

Brief, man-up and launch went as planned. We were assigned a USAF KC-10 tanker, along with a US Marine Corps EA-6B, which had sortied from Prince Sultan Air Base in Saudi Arabia.

In addition, we were also slated for a section of USAF F-16CJ HARM-equipped fighters. Everyone was full-up, mission ready, on time and on station. No small feat, and a tribute to the ATO planners in the CAOC and the maintainers in all the services who kept this huge armada airborne "24/7."

Shortly after the strike package established communications with our AWACS controller, he advised us that an Iraqi bomber – a Tu-16 "Badger" – had been located at Al-Taqaddum Air Base,

Although CVW-14 was the dedicated day air wing in the NAG during OIF I, its flying time lasted from noon through to midnight. This, of course, meant that some of the missions undertaken by its units were flown after dark. LGB-equipped F-14D "Bandwagon 110" (BuNo 159618) from VF-31 is heading south in search of a tanker mid-way through a CAS mission. By the time VF-31 flew its final combat sorties from *Abraham Lincoln* on April 14, 2003, this aircraft featured 35 bomb silhouettes beneath the cockpit and a "Tommy Gun" stencil with "193" marked on its stock. (Lt Cdr Jim Muse)

about 30 miles [48km] north-west of our assigned target. The controller passed co-ordinates for the bomber, and we assigned them to the two Tomcats. The mission commander, the F-14D lead and the VF-31 CO [Cdr Paul Haas] quickly worked out a new timing plan en route that allowed the F-16CJs and the EA-6B to provide coverage for both strike packages. The Tomcat crews also reprogrammed the aim points in their JDAM thanks to the weapon's Target Of Opportunity [TOO] mode.

As we flowed north toward the Baghdad area, I was surprised by how eerily quiet the radios were. I was also amazed at how tranquil the Iraqi air defenses seemed to be. We could see downtown Baghdad as we approached from the south-west. When wearing NVGs, any weapons fired could easily be seen – there was no activity. That would soon change, however.

As the Tomcats approached their targets, the F-16CJs prepared to launch several HARM against Iraqi air defense radars. It was easy to spot the missiles coming off the F-16CJs from our position some 20 miles [32km] south-west of their location – I was impressed by the sheer speed of the HARM as they raced towards their targets. After their shots supporting the F-14Ds, the F-16CJs quickly repositioned south-east to protect the F/A-18Es. The Prowler aircrew had previously set up an orbit that allowed them to cover both targets nearly simultaneously.

The Tomcats were on target just as we hit our IP. The four Super Hornets had a great nose-on view of some of the most impressive secondary explosions I have ever witnessed. Clearly the Tomcats "shacked" the bomber, which appeared to be fully loaded with both fuel and ordnance. Those JDAM woke up the entire Al-Taqaddum air defense force, because a huge barrage of AAA and ballistic SAMs began to streak skyward. Fortunately, the Tomcats were able to egress without taking any hits, although they did have at least one SAM come within several hundred yards of their flight as it split their section.

I was impressed with the radio discipline from all the flights. Aircrew were calling SAMs and AAA in calm, clear voices, and keeping leads informed of their intentions. This allowed us to keep the flights together and continue to press home the attacks.

The F/A-18Es hit their assigned target approximately three minutes behind the F-14Ds. Once we completed our off-count and verified eight weapons had been released, the entire package began to egress south-east. The Super Hornet division observed multiple SAM launches on the way home, but none of us were illuminated by target-tracking radars and none of the SAMs appeared to guide on us. We saw significant AAA in the southern Baghdad area, with many secondary explosions from a large coalition strike that was underway.

Return to the post-mission tanking location was uneventful, and we recovered aboard *Abraham Lincoln* four hours after launching following a very satisfying mission.

CVW-14 staff officer Lt Cdr Jim Muse, who had previously served as a RIO with VF-31, also got to employ GPS weapons from an F-14D during "Shock and Awe." He told the author that aircrew would typically conduct much of their mission weaponeering with JDAM prior to walking to the jet:

With JDAM, the real effort came in the pre-flight planning rather than over the target, as was the case with an LGB. There were several different JDAM attack profiles I could choose from once airborne depending on the target being serviced. If I had a vertical target, I wanted the weapon to come in at a low angle, and if I had a horizontal target, then I wanted it to hit at a steep angle. We had a pattern of deliveries for different altitudes, and we could match these up to suit the target being attacked. VF-31 crews usually knew what they were targeting beforehand, and configured the JDAM accordingly prior to launching, but the weapon's TOO mode was flexible enough to allow for airborne co-ordinate changes should the mission requirements alter mid-sortie.

I got to put this flexibility to the test when I dropped my first JDAM of OIF. Although VF-31 had mostly flown fixed strikes for the first few days of the war, by the time I went into action on March 23, we had begun moving into the CAS phase of the campaign. We had also been given a

Lts Paul Brantuas and Rick Burgess smile for Lt Cdr Jim Muse's camera prior to flying an OIF I mission in "Bandwagon 102" (BuNo 163904) in late March 2003. They are both firmly strapped into Martin-Baker SJU-17 (Mk 14) Navy Aircrew Common Ejection Seats (NACES), which, among Tomcat variants, were unique to the F-14D. NACES is also used in several other US Navy fast jet types, including the F/A-18E/F Super Hornet. (Lt Cdr Jim Muse)

"Bandwagon 106" (BuNo 164343), christened *MS HEIDI*, sits on the fantail of *Abraham Lincoln* surrounded by "ordies" and maintainers on March 24, 2003. The GBU-31 and at least one of the GBU-12s seen on the Aero-12C skids in front of the jet would be uploaded to centerline under-fuselage pallets prior to the aircraft flying its next mission. A TARPS pod and LANTIRN store on trolleys are also sat on the flightdeck aft and to the left of the aircraft, respectively. BuNo 164343 ended the campaign with 19 bomb silhouettes beneath the cockpit and a "Tommy Gun" stencil with "179" marked on its stock. (US Navy)

"dump target" list by then, which contained the co-ordinates for things like bunkers, buildings and known SAM sites. We could work down this seemingly endless list if we could not find anyone on the ground who needed our bombs to support their advance to Baghdad.

Upon our arrival over southern Iraq, we checked in with AWACS, who directed us to a location at which tanks had been reported. We scoured the area where they told us they would be, but couldn't find any tanks, so we informed our controller that we had DMPIs [Designated Mean Point of Impact] from the dump list and requested permission to attack those instead. Having received approval in short order, we headed north for the targets.

This was at the beginning of the bad sandstorms that clobbered the area, and visibility was absolutely horrid. We were lucky to find our mission tanker – a VFA-115 Super Hornet from our own boat – on the way in. Flying towards the target, we knew that there were other jets out there, but we sure couldn't see them. We only had datalink with our wingman – no visual reference – and we just stuck to our designated altitude and prayed! It was one of the scariest missions I've ever flown. Having felt the JDAM fall away as we dropped it from 32,000ft in complete brown-out conditions, we turned for home. I assume that we hit the target, but we never did find that out for sure.

On our way back to the carrier, we were contacted by some Marines who needed a jet to fly a low-level road reconnaissance for them. They asked us to look for enemy tanks that they thought were on the north side of a town that they were approaching. We dropped down low and fast, pumping lots of chaff as we descended because we could never tell where MANPADS would show up. We made a high-speed, low-level pass along a river and over a small pontoon bridge, on which I spotted numerous people. Startled by our low pass, they fled in all directions away from the bridge, which they assumed we were going to bomb.

It was very bumpy down low, the visibility was not brilliant, and I have to admit it was a little scary to be flying well within the enemy's AAA and SAM engagement zone – I had never flown so low over Iraq before. If the weather had been better, we could have just scoured the area through our LTS from a nice safe 20,000ft. We couldn't find any tanks there either, so we went home and the Marines eventually pushed through.

Aside from operating with VFA-115 and its brand new F/A-18Es, which were undertaking their debut cruise, VF-31 also flew mixed sections with four forward-deployed Super Hornets from VFA-14 and VFA-41 – answering a CENTCOM request for additional strike and tanker assets, *Nimitz*'s CVW-11 sent four jets to *Abraham Lincoln* on March 30. VFA-41's two F/A-18Fs (with FAC(A)-qualified crews in each jet) were "paired with F-14s [from VF-31] on the *Lincoln* because they were also manned by FAC(A)s," noted Super Hornet pilot Lt Cdr Mark Weisgerber of VFA-41. "We worked as their wingmen initially while they showed us the 'ropes' in-theater. The real reason we did this was because we were taking missions fragged for them on the ATO – essentially filling the spot of a VF-31 jet – for this was the most expeditious way to get the F/A-18Fs into action over Iraq." CVW-11's four Super Hornets returned to *Nimitz* on April 6.

By then VF-31 crews had been predominantly flying FAC(A) and SCAR missions for more than a week, providing targeting for LGBs dropped by F/A-18Cs from CVW-14's VFA-25 and VFA-113 and F/A-18Es from VFA-115. The XO of the latter unit, Cdr Dale Horan (formerly an F-14 pilot), worked with a Tomcat during the ground war phase of OIF:

VF-31 OSW/OEF/OIF I (2002–03)

I got to buddy lase with VF-31 just once. When operating with a Tomcat, the jet's crew would have the target locked up on their FLIR, rather than me having to come in and spend precious minutes trying to find it on my targeting sensor. I would then simply drop my LGB on their instruction and leave them to guide the weapon home.

Having a FAC(A) crew available expanded our capabilities over the battlefield. To have someone on the ground, or airborne, talking you on to targets greatly improved our mission efficiency. Simply flying a single-seat jet over a battlefield was challenging enough in itself, and if you could have a trained individual find the target, identify it and ensure that there were no friendlies or civilians in the immediate area, then this greatly reduced your workload at the most critical phase of the mission. When you had to spend time looking at your sensors in the cockpit or scouring the ground in search of the target, the possibility of you missing a threat such as an IR-guided MANPAD or AAA increased significantly.

As was the case with all the Tomcat units committed to OIF I, VF-31 had a handful of highly valued FAC(A)-qualified Naval Aviators and NFOs within its ranks – some of them were augmentees from NSAWC and SFWSL. Heading up CVW-14's commitment to expeditionary warfare in the campaign was VF-31 RIO Lt Cdr John Patterson, who was chosen pre-war by his CAG, Capt Kevin Albright, to be the air wing's CAS SME. Few boasted better qualifications for the job, with Patterson having performed the role of strike leader in OEF and OSW for both CVW-7 and CVW-14. He had also previously served as an SFTI, a US Marine Corps Aviation Weapons and Tactics Instructor, FAC(A) instructor and Night Vision Device instructor.

Patterson's primary mission in the weeks leading up to OIF was to ensure that CVW-14's TACAIR crews were ready to provide the most effective CAS possible right from the start of the land war. To achieve this, he worked closely with FACs from the US Army's V Corps during pre-OIF exercises staged at bombing ranges in the Kuwaiti desert. Patterson also represented CVW-14 in an emergency planning cell established by 1st MEF that included the 3rd Marine Air Wing (the primary CAS asset supporting the US Marine Corps in OIF) and FAC(A)s and

"Bandwagon 110" is carrying two GBU-12s on its forward tunnel weapons rails and possibly two more in the aft troughs as well during an OIF I CAS mission. The GBU-12 was the F-14's preferred weapon when supporting the ground war, VF-31 expending no fewer than 165 of the 464 dropped by CVW-14's four TACAIR units. Originally constructed as an F-14A and delivered to VF-124 on October 24, 1975, BuNo 159618 was the 17th of 18 A-models rebuilt as F-14Ds in 1990–91. Following a second spell with VF-124 and then VF-101, the jet was assigned to VF-31 in 1995. It remained with the unit until June 2, 2003, when "Bandwagon 110" was stricken and transferred to SARDIP at Oceana. (Lt Cdr Jim Muse)

Airborne Command and Control representatives from CVW-2 and CVW-5. According to VF-31's XO, Cdr Aaron Cudnohufsky:

The cell initiated command and control and CAS allocation procedures for naval TACAIR, which greatly increased the efficient flow of sorties in support of US Marines. As a direct result of these efforts, naval TACAIR was able to compensate for a shortage of Marine direct-support assets, the cell funneling in excess armed reconnaissance and SCAR sorties to ensure the destruction of the Iraqi Army's IV Corps, effectively securing the critical, but lightly defended, supply lines of 1st MEF in their advance on Baghdad. So effective were these efforts that they led to a 100 percent elimination of the combat capability of IV Corps through destruction and desertion, allowing Marine ground forces to later capture IV Corps' headquarters in Al Amarah without having to fire a shot. Moreover, Lt Cdr Patterson also personally participated in this destruction, conceiving, planning and leading a CVW-14 strike package to destroy IV Corps' artillery regimental headquarters.

This sortie was just one of 13 missions that Lt Cdr Patterson flew in OIF as a Strike Leader and FAC(A), during which he personally expended 23 LGB, GPS and general purpose bombs, as well as 500 rounds of 20mm cannon fire against enemy forces, including multiple instances in support of friendly forces in direct contact with the enemy.

In one particularly memorable mission flown on March 30, Patterson acted as the FAC(A) for his section, which destroyed enemy armor and artillery that had damaged three Apache helicopters and halted IV Corps' advance south of Al Hillah. Throughout this mission, Patterson and his pilot remained on-station over the target area despite the presence of enemy AAA and SAM batteries.

NFO Lt Cdr John Patterson, who was chosen pre-war by his CAG, Capt Kevin Albright, to be CVW-14's CAS SME, was one of the most highly qualified RIOs to see action in OIF I. Few boasted better qualifications for the job, with Patterson having performed the role of strike leader in OEF and OSW for both CVW-7 and CVW-14. He also helped VF-31 get to grips with JDAM employment following the installation of the D04 software into the unit's F-14Ds in late February 2003, having had previous experience with the weapon when serving with VF-11 in OEF in 2002. (VF-31)

VF-31 used a generic "Felix and bomb" silhouette to denote each weapon dropped by its aircraft – the marking did not differentiate between LGBs and JDAM, as was the case with several other F-14 units involved in OIF I. Having dropped 42 LGBs and JDAM during the campaign, "Bandwagon 104" (BuNo 163898) also eventually bore a "Tommy Gun" stencil just aft of the cannon-muzzle fairing. This meant it had performed a rare strafing attack, during which 177 20mm SAPHEI rounds were expended. (US Navy)

Due to the complexity of the FAC(A) mission, the US Navy would only allow suitably qualified two-man crews to undertake such sorties. There were only a small number of FAC(A)s in-theater, so in order to keep the battlefield serviced, SCAR and armed reconnaissance missions, rather than dedicated FAC operations, came to prominence early on in the campaign. Both could be performed by virtually any Tomcat crew operating in a two-aircraft section.

By the time *Abraham Lincoln* and CVW-14 were relieved in the NAG by *Nimitz* and CVW-11 on April 14, VF-31 had flown an astounding 585 combat sorties and 1,744 combat hours during its marathon ten-month OEF/OSW/OIF deployment. The unit had dropped 56 JDAM, 165 GBU-12s, five GBU-16s and 13 Mk 82 "dumb" bombs and expended 1,138 20mm cannon rounds.

"Bandwagon 103" (BuNo 164344) provides the backdrop for a group shot of VF-31's officer cadre on April 7, 2003 – two days prior to *Abraham Lincoln* departing the NAG and heading home. Squadron CO, Cdr Paul Haas, is standing in the front row tenth from the right, with his XO, Cdr Aaron Cudnohufsky, facing him. The unit's ten F-14Ds flew 585 sorties totaling 1,744 combat hours during OIF I. BuNo 164344, having survived OIF I, was lost when it crashed into the Pacific Ocean two miles (3.2km) west of Point Loma, California, on March 29, 2004 after suffering fuel transfer problems. The crew successfully ejected and were soon rescued. (VF-31)

"Black Knights" Rule

VF-154's war was undoubtedly the most unusual of any of the Tomcat units involved in OIF I. Deployed on its final cruise with the F-14A as part of CVW-5, the "Black Knights" arrived on station in the NAG aboard *Kitty Hawk* on February 26, 2003. Chosen to be the dedicated CAS air wing by the CAOC, CVW-5 had not ventured into the NAG since July 1999. Despite its unfamiliarity with current operating procedures in OSW, VF-154 completed a handful of successful missions in southern and western Iraq in the three weeks leading up to OIF I.

Prior to CVW-5's combat deployment, VF-154 had tailored its training to optimize the unit's precision strike capabilities, including dedicated training in target acquisition in urban environments. During work-ups, VF-154 was directly responsible for developing standardized precision FAC(A) and CAS tactics for the entire air wing. Once in-theater, the unit's pre-deployment focus on the air-to-ground mission allowed it to work with many different assets as CVW-5 looked to expand its role both in expeditionary warfare and precision CAS in the combat environment. In response to short notice tasking from CENTCOM, VF-154 detached four crews, augmented by FAC(A) instructors recently drafted in from NSAWC and SFWSL, four jets and a small team of maintainers to Al Udeid Air Base, in Qatar.

With combat operations looming, the crews quickly scheduled multiple training events to teach SCAR to coalition and inter-service assets based at Al Udeid. While ashore, VF-154 worked closely with RAF Tornado GR 4s, USAF F-15Es, F-16CGs and F-16CJs and RAAF F/A-18As.

This training evolution subsequently paid huge dividends both in OSW and OIF, as VF-154 crews controlled laser-guided munitions and passed LANTIRN co-ordinates for the successful employment of British Enhanced Paveway II/III LGBs, standard LGBs and JDAM dropped from coalition aircraft. The Al Udeid training detachment also helped maximize the lethality of the support coalition strike aircraft offered to ground forces by reducing the time between PGM impacts to the NSAWC recommendation of one bomb per minute for more than 20 minutes, and less in the case of GPS-guided munitions.

Equipped with the only F-14As to see combat in OIF I, VF-154 was restricted to the use of LGBs or unguided "dumb" bombs during the conflict. As noted earlier in this volume,

Still carrying its GBU-12s, VF-154's "Knight 103" (BuNo 161293) joins the recovery pattern overhead *Kitty Hawk* on the eve of OIF I. This particular aircraft ended the campaign as the unit's high-time ordnance expender, its crews dropping no fewer than 51 LGBs on target. Delivered to the US Navy in late 1981, the jet had served with three fleet units and VF-101 prior to joining VF-154 at NAF Atsugi in early 1998. It remained with the squadron until flown back to Oceana on September 24, 2003. Like the rest of VF-154's F-14As, the aircraft was stricken from the inventory (on December 16, 2003) when the unit became VFA-154, equipped with the F/A-18F. (VF-154)

unlike the F-14B/D, the A-model had no MIL-STD 1760 databus and associated onboard GPS system to generate computed aim point co-ordinates for JDAM. The F-14A's avionics had no path for the LANTIRN pod to transfer its GPS data (gleaned from a GPS housed in the pod itself) to the JDAM, as the jet lacked an embedded GPS. The F-14B/D had such a capability, however, thus allowing the LANTIRN pod to transfer aim point co-ordinates, and much more targeting data besides, to the JDAM attached to its bomb pallets. Although unable to drop GPS-guided PGMs, VF-154 made full use of its jets' LANTIRN pods, and the unit's crews enjoyed a solid reputation as LGB bombers.

The success of the Al Udeid operation pre-war can be gauged by the fact that CENTCOM contacted CVW-5 directly on the eve of OIF I and requested that VF-154 send one-third of its assets to Qatar to support coalition land-based aircraft and SOF teams operating inside Iraq. The unit's XO, Cdr Doug Waters, was ashore at the time when he received word of CENTCOM's unusual request. He recalled:

I was a liaison officer with the CAOC at Prince Sultan Air Base [PSAB], in Saudi Arabia, during the weeks prior to the war. My job was to scrub the OPLAN [operational plan] for OIF from a navy viewpoint, basically making sure that the requested sortie numbers, ordnance loads and tankers would work for carrier-based fighters, and their planned cycle times.

While working the OPLAN scrub, I began to hear about a move to bring some additional FAC(A)-capable Tomcats and crews out to work directly with coalition ground forces. At first it sounded like additional aircraft – rumored to be F-14Bs from VF-11 and VF-143 [both units had undertaken TST mission development with SOF in recent months] – would be added to one of the carriers already in-theater, but then it became apparent that the plan would involve basing aircraft ashore. Some in the Navy didn't like the idea of putting carrier-based aircraft ashore – I guess they felt it was an admission that the Air Force had the better idea. I felt just the opposite, for I thought that it showed the inherent flexibility of carrier-based fighters. We could operate from either venue if required, which is obviously not true for land-based fighters.

NFO Lt Michael Chenoweth of VF-154 checks the guidance vanes on a GBU-24 Paveway III LGB prior to climbing into the cockpit of his jet onboard *Kitty Hawk*. The weapon's large rear-mounted stabilization package allowed it to travel further in flight than any other LGB then in US Navy service. The unit expended 358 LGBs between March 21 and April 14, flying primarily FAC(A) and SCAR missions. (US Navy)

Led by "Knight 100" (BuNo 161866), a division of four VF-154 F-14As form up behind a KC-135R of the Alabama Air National Guard's 106th Air Refueling Squadron on April 15, 2003. The 106th was attached to the 434th Air Refueling Wing at Al Udeid during OIF. (VF-154)

"Knight 103" shares *Kitty Hawk*'s aft port elevator with an EA-6B from VAQ-136 as the jets take a short ride up to the flightdeck on March 24, 2003. CVW-5's five Prowlers would have been put ashore had the US Navy decided to augment its OIF I Tomcat SCAR mission force with F-14Bs from Oceana, as VF-154 was deemed to be the unit most capable of supporting these aircraft from a maintenance perspective. (US Navy)

As the plan took shape, I realized it was my squadron that was going to be sending jets ashore. I would find out later that the reason VF-154 was picked was because of the high serviceability rates that we had achieved with our veteran F-14As – the highest for TACAIR in-theater. Since the squadron CO, Cdr James Flatley, was a FAC(A), and he would be leading the detachment to Qatar, I needed to return to VF-154 onboard *Kitty Hawk*. When I stepped back on the deck three-and-a-half weeks after flying off on the transit over from Japan, I remember thinking, "Thank God I am out of 'PSAB' and back with my unit where I belong."

War Ashore

Five aircraft and five crews were put ashore on the eve of OIF I, VF-154's CO, three department heads, its training officer and several augment instructor pilots and RIOs from NSAWC and SFWSL being charged with the responsibility of waging the impending war from Al Udeid. The unit's maintenance officer came up with a novel plan to keep his elderly charges serviceable while in Qatar, as Cdr Waters explained:

The plan was simple, yet extremely effective – take the minimum number of maintainers possible ashore so that the squadron would not be precluded from maintaining a high ship-based operational tempo in support of normal ATO sorties. This was accomplished by hand-picking the maintainers who would go to Qatar. The shore-based det then created informal relationships with units at Al Udeid to garner the additional external support it needed.

The 30 maintainers sent to Al Udeid worked closely with both the RAAF Hornet detachment and the 157th Fighter Squadron (FS), South Carolina Air National Guard (ANG), which flew F-16CJs. Both units were exceptionally helpful, and without the liaison set up between them and the "Black Knights," it would have been very difficult to achieve the 100 percent combat sortie completion rate that was maintained while operating out of Qatar. The Aussies offered us use of their composite shop to help with any airframe/metalworking issues that came up for our aircraft, and they also fashioned an adapter that allowed us to employ existing USAF servicing equipment to replenish the nitrogen bottles used to cool the seeker heads in our AIM-9Ms.

The South Carolina ANG guys helped the detachment with the support equipment it required, and they went out of their way to make sure our troops were well taken care of while in Qatar. Suffice it to say, without the Aussies and the ANG guys basically "adopting" our maintainers, we would have had to bring more personnel ashore, hurting the squadron's ability to generate ATO sorties from the ship.

Once the war started, the benefits of having VF-154's cadre of FAC(A)s in Qatar soon became apparent. By being able to brief, debrief and operate on a daily basis with the air assets they would control over Iraq, the face-to-face liaison between Naval Aviators and NFOs from the

A division of four VF-154 Qatar detachment F-14As prepare to taxi out for a dusk mission from Al Udeid in late March 2003. Behind them can be seen F-16CJs from the USAF's 389th FS, as well as RAAF F/A-18As from No 75 Sqn and RAF Tornado GR 4s. VF-154 aircrew got to work closely with all three types, as well as the co-located F-15Es of the 4th FW. "Knight 110" (BuNo 161288) expended 35 LGBs, which made it the most prolific of the five shore-based Tomcats in terms of ordnance dropped in OIF I. (VF-154)

"Knight 111" (BuNo 161292) taxies in at Al Udeid at the end of a mission in late March 2003, its bomb racks devoid of ordnance. This aircraft dropped 21 LGBs during the campaign. A veteran of fleet service with VF-1, VF-21, VF-51 and VF-14 (including Operation *Allied Force* with this unit), BuNo 161292 was assigned to VF-154 on December 28, 2001, and remained with the squadron until November 30, 2003. It was stricken the following month. (VF-154)

"Black Knights" and the coalition crews at Al Udeid resulted in a much more effective strike package. The unit's work with F-15E crews from the 366th Fighter Wing's 391st FS (flying aircraft from the 4th FW's 335th and 336th FSs) was particularly successful, with the pairing of the Tomcat and the Strike Eagle proving deadly when conducting secretive TST/Task Force 20 (TF-20) missions in support of SOF.

These operations were arguably the most dynamic TACAIR missions of OIF I. Long, dangerous and incredibly important, the FAC(A) sorties required exceptional SA in a complicated Joint environment. The combination of the F-14's proven endurance and VF-154's NSAWC/SFWSL-boosted FAC(A) crews saw the unit (along with F-14D-equipped VF-2) become the "go-to" FAC(A) asset when it came to TST/TF-20 missions in OIF I. Suitably qualified crews were routinely tasked on the ATO to act as the "quarterbacks" for some of the most dangerous and tactically demanding sorties of the conflict.

The TST/TF-20 missions that VF-154 participated in involved SOF squads taking out targets of opportunity – key individuals in the Iraqi leadership, mobile radar/SAM sites and surface-to-surface missiles. Such elusive targets often presented themselves only fleetingly anywhere across the country, and the SOF teams needed the ability to infiltrate, hit the target with overwhelming force and depart under the cover of blinding firepower and complete darkness.

As the target lists were compiled pre-war, planners realized that many of the prime missions would take place in and around the Baghdad SuperMEZ. SOF MC-130 Hercules and helicopter aircrews tasked with inserting and supporting the TF-20 squads requested dedicated, responsive assistance over and above their organic capabilities to support these dangerous missions. The required ordnance that could reach deep into Iraq and be flexible enough to meet SOF needs could only be delivered from coalition TACAIR assets working the CAS mission.

In fall 2002, SOF air planners joined forces with conventional US Marine Corps, USAF and US Navy aircrews (from VF-11 and VF-143) who were SMEs on Joint CAS (JCAS) and FAC(A) procedures. Their task was to provide a detailed CONOPS on how conventional CAS jets would escort SOF aircraft to and from their targets in a robust high-threat environment. The CONOPS also detailed measures to provide fire support and assistance to SOF ground FACs during missions against their objectives. The CONOPS was then validated by those same FAC(A) aircrew, who put their SOF CAS ideas to the test against USAF and US Navy weapons ranges simulating enemy threat systems.

Face-to-face briefings and, more importantly, debriefings were stressed due to the fact that most SOF-to-conventional air forces interaction ended up being "pick-up" games played out on the battlefield, with both players "arriving at the court" with the JCAS "playbook" already memorized.

After the rehearsals were over, the well-practiced FAC(A) and CAS aircrews were told that they would be assigned to units already in-theater under JTF-SWA control in order to teach what they had learned to the deployed crews. When theater commanders were briefed on how well the rehearsals had been executed, the aircrew were directed to lead the missions as "tactical quarterbacks," integrating other coalition FAC(A) and CAS aircraft into the SOF missions.

When it came to performing TST or TF-20 (also dubbed Task Force *Tawny*) missions, VF-154's FAC(A) crews at Al Udeid were usually paired up with F-15Es from the 4th FW's 335th FS that were being flown by 391st FS personnel – they had been brought in specially from Mountain Home Air Force Base, Idaho, to undertake TF-20 operations. Only ten pilots and weapons system officers from the 391st were "read-in" on these sensitive missions, which were heavily compartmentalized in terms of planning and briefing/debriefing. TF-20 aircrews were allowed to plan their own sorties, as well as secure the assets (VF-154) and weapons (usually GBU-12 LGBs) needed to get the job done.

Most TST/TF-20 missions were carried out at night, and the first such sortie involving VF-154 actually took place prior to OIF I starting. Precise details of what target was hit remain classified, but it is known that a section of two Qatar-based VF-154 F-14s operated alongside aircraft from CVW-2 and CVW-14, as well as USAF and RAF jets, over H-2 and H-3 airfields in western Iraq on March 19, 2003. The jets were providing CAS for SOF elements that had been charged with immobilizing "Pluto" and "Flat Face" early warning radar sites and radio relay antennas sited near the border with Jordan.

These were important targets, as the OIF I battle plan stated that USAF F-16CG/CJs and SOF helicopters would operate over western Iraq from bases in Jordan once hostilities began in earnest. The Intercept Operations Center at H-3 was also destroyed in the same mission by 5,000lb (2,268kg) GBU-28 "bunker buster" bombs dropped by F-15Es operating closely with FAC(A)s from VF-154.

Although the "Black Knights" had sufficient skill in their FAC(A)-qualified crews ashore to carry out these TST missions, their A-model Tomcats were lacking some key systems. VF-2 crews had no such worries with their F-14Ds. Being a part of the designated night carrier air wing, the "Bounty Hunters" played an important role in

VF-154 OSW/OIF I (2003)

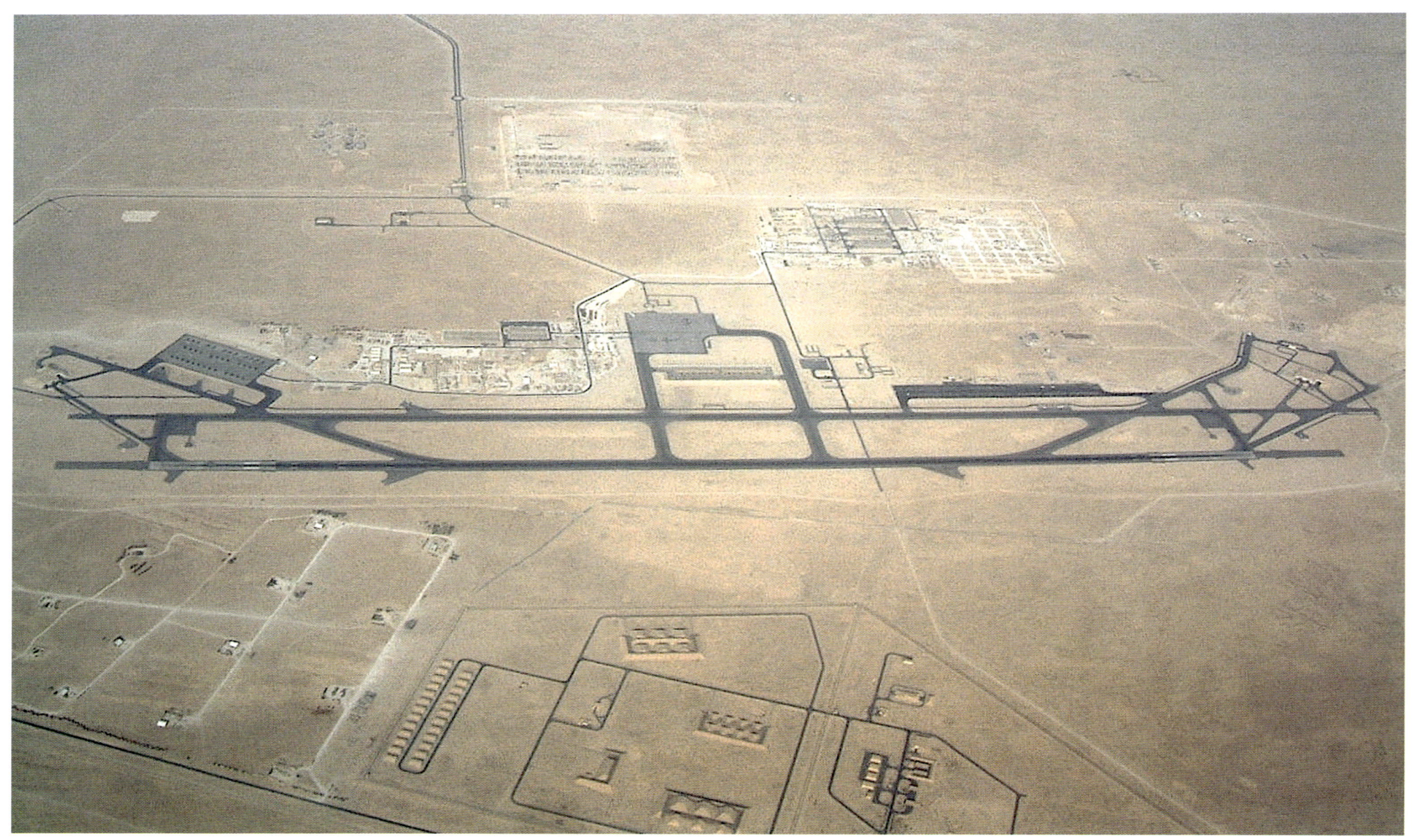

This view of Al Udeid Air Base was taken from a departing F-14 on April 10, 2003, when the VF-154 det returned to *Kitty Hawk* after more than three weeks ashore. Built in the 1990s by French contractors as a base for Qatari Air Force Mirage 2000 fighters, Al Udeid was large enough to house USAF KC-10 and KC-135 tankers, as well as more than 100 fast jets from the USAF, US Navy, Air National Guard, RAF and RAAF during OIF I. (VF-154)

servicing TST targets. VF-2's Lt Cdr Michael Peterson, a veteran of OEF who flew two TF-20 missions, recalled:

> We relieved the VF-154 jets over H-2/H-3 on the night of March 19. I think that our SA in the F-14D was much better than that of the VF-154 crew we replaced. That was a direct reflection of the equipment that we had at our disposal in our upgraded Tomcats, not on the aircrew themselves. On the two TF-20 missions that I flew – a third was cancelled at the last minute – the VF-154 crews did most of the asset set up and covered the rather uneventful ingress into the objective area, while the main effort, target objective and egress were handled by VF-2 aircrew. On one of these missions, I was airborne for 8.6 hours – this was by far the longest sortie of my Tomcat career.
>
> Things were much more dynamic after the kick off of these TST missions, and with our jets having Link-16 JTIDS, we could provide an additional level of deconfliction and SA in respect to where the other assets were at any given time during the mission. VF-154's older F-14As had not received the JTIDS upgrade, leaving crews out of the loop when it came to up-to-the-minute information on how the mission was evolving.

Despite the more austere nature of its jets, VF-154 did, however, fly many CAS, SCAR and FAC(A) missions during the 31 days of OIF I. Aside from the unit's participation in the H-2/H-3 airfield strikes on the eve of OIF I, crews also provided air support for the SOF seizure of an Iraqi presidential palace on the shores of Lake Tharthar, north-west of Baghdad, towards the end of the conflict.

A VF-154 RIO who wished to remain anonymous (he was one of the hand-picked FAC(A)-qualified NFOs chosen to fly from Qatar) provided the author with the following account of a mission he participated in from Al Udeid on April 3:

> It was supposed to be a six-hour "Black Knight" division FAC(A) flight in support of coalition ground forces. Three hours prior to take-off, we received our intelligence update and were then briefed with our strike assets. Tonight, we'd have a division of F-15Es and a section of F-16CJs as our primary strikers. Three weeks of briefing with the same guys made for an expeditious pre-mission routine. After the brief, we geared up and piled into the minivan for the drive to the flightline.
>
> Ten minutes later we arrived at the jets and quickly read the Aircraft Discrepancy Books. We'd been flying the same F-14s for the last three weeks, so we only gave them a cursory glance. We then grabbed a few bottles of water and walked to the aircraft. After start-up, we taxied out to the runway, each jet armed with four 500lb [226kg] GBU-12 LGBs apiece. One of the cool things about taking off from Qatar was the requirement to be above 15,000ft, thus negating the MANPADS threat, prior to actually leaving the airfield boundaries! Two minutes after the first of our Tomcats had rolled, we had all joined up at 23,000ft and proceeded feet wet over the NAG on our way to Iraq.
>
> An hour later, the sun had completely set, we'd finished our first airborne refueling and we were now proceeding west in search of our next tanker. Within 30 minutes all four F-14 crews had topped off for the second time, and we were finally ready to check in with our controlling agency. The latter informed us that we were still on schedule, and that the ground forces were waiting for us in central Iraq. For the purpose of continuity, we split into two sections so that we could constantly have eyes in the target area while the other section refueled.
>
> As the first section worked its way towards the specific element that we were supporting, it became apparent that the friendlies on the ground had taken some casualties from a car bomb that had been set off. Our first mission of the night would be to sanitize the area of any threats so we could medevac out the casualties. After reconnoitering the area, the lead section discovered a hostile vehicle driving towards the helicopter landing zone and eliminated it with a GBU-12. They then proceeded to fly high cover for the medevac helicopter, escorting it back to friendly forces.
>
> The second "Black Knight" section then arrived on the scene and protected the ground forces from any threat proceeding from the lake side of the nearby Haditha Dam, upon which the friendlies had set up their defensive position. We were concerned that the Iraqis would attempt to breach the dam in an attempt to flood the Euphrates River, which ran through the valley below it.
>
> Right about then we were fired upon by several AAA pieces from south-west of the dam. Using NVGs, and slaving the FLIR to the HUD, we were able to roll in on the AAA and destroy one weapon with an LGB. Wanting to save some of our ordnance for follow on "pop up" threats, we called in a section of Strike Eagles and guided their ordnance on to adjacent AAA pieces.
>
> At about this time the lead section arrived back on scene, and after a FAC(A)-to-FAC(A) turnover, they relieved us. We then received word from our buddies on the ground that one of their reconnaissance units had discovered more AAA pieces at an airfield 5km [3 miles] to the south of the dam. They asked us to check it out, and after some searching, we were able to find about ten S-60 AAA weapons dispersed around the airfield. After dropping one LGB to mark the target for the Strike Eagles, we turned the remainder of the AAA sites over to them to destroy. Now reaching "bingo" fuel, the lead section headed for the tanker, and then home, as the second section arrived back on scene.
>
> I don't know whether or not the Iraqis south-west of the dam thought we'd left the area, but they started firing AAA into the night sky soon after the lead section departed. After quickly finding the AAA pieces on NVGs and FLIR, we passed two sets of co-ordinates to a section of F-16CJs and told them to put a JDAM on each of them. We orbited overhead and watched on our FLIR as both JDAM shacked their intended targets. Having destroyed the remaining AAA pieces to the south-west, we passed off the FAC(A) role to a section of A-10s and headed back south to find our second to last tanker of the night.
>
> Unfortunately, the weather was now starting to roll in, and after some attempts to get the tanker to clear air, we finally gave up and rendezvoused in the "goo." Our wingman, being lower on fuel, would tank first. Just as he was plugging in, he experienced an engine compressor stall and disappeared into

VF-154's Al Udeid detachment pose for a group shot on the flightline in late March 2003. The NFO standing at the extreme left in the back row is Lt Cdr Scotty McDonald, who was forced to eject from his Tomcat over Iraq several days after this photograph was taken. Squatting, third from right, is VF-154 CO Cdr James Flatley. Providing the backdrop to this shot is "Knight 112" (BuNo 158624), which dropped 28 LGBs in OIF I. (VF-154)

the darkness below. Not having a lot of fuel, or time, to play with, we decided to tank while he rejoined. After a couple of minutes, and some radio calls, he magically appeared off our right wing. His engine then stalled again, and in the ensuing melee I watched him cross directly in front of the tanker – I instinctively ducked, waiting for the fireball that would kill us all. Fortunately, he just missed the tanker and disappeared down our left side back into the darkness. I can honestly say that was the scariest moment of my life.

Tragedy averted, we managed to finish refueling our section and then headed back east toward our last tanker. After completing our final refueling evolution, we flew back out over the NAG, turned south and tracked down the "aerial highway" to Qatar. Forty minutes later, when we called to check on the status of the airfield, Base Ops replied that they were in the middle of a huge sandstorm, and that we could try to shoot the approach or divert to a different base. We decided that we had enough fuel to try to get down at Al Udeid and divert if we couldn't land.

VF-154 Qatar Det OIF I (2003)

Surprisingly, we were able to see the field at 15 miles [24km] and set up for individual straight-in approaches. Although we could see the runway as clear as day at altitude, we lost sight of everything when we flew into the sandstorm at a height of just 100ft above the ground. Our wingman, one mile in trail, immediately told us that he had lost sight of us. It was as if we had flown into a tunnel. Just as we were about to wave off, we picked up the runway's centerline lights and landed. Relaying this information, and our position on the runway, to our wingman, he was able to land behind us.

The sandstorm was so bad that we were forced to taxi back to our line at a snail's pace, with lighting normally visible at a mile only becoming identifiable when directly abeam us. As it turns out, we were the only aircraft able to land at Al Udeid that night. All of our USAF and coalition buddies were ordered to divert and sit by their aircraft all night, before returning the following morning.

While the specifics relating to exactly what was bombed, and just how the Al Udeid detachment went about servicing its numerous targets, remain largely classified to this day, it has been revealed that these crews were responsible for developing new tactics, techniques and procedures for operating with multi-service SOF teams. The five crews flew daily missions specifically briefed to support individual ground units, and in one 48-hour period the "Black Knights" detachment completed 14 sorties, totaling more than 100 hours of flight time.

According to VF-154's post-cruise summary of its contribution to OIF, FAC(A) crews on the beach amassed more than 300 combat hours and delivered almost 50,000lb [22,679kg] of ordnance [98 GBU-12s] in 21 days of flying with their five crews and five jets. Yet despite its success, this unique operation is unlikely to be repeated according to Cdr Waters: "Since the US Navy does not like to ask permission to use someone else's runway for combat operations, the Al Udeid detachment is not likely a model for the future. However, it did lay the foundation, in both tactics and trust, for future operations between Navy FAC(A)s, JTACs and SOF in support of National Command Authority objectives."

Jet Down

VF-154 did not escape from its shore-based foray unscathed, however. On the night of April 1, NSAWC augment pilot Lt Chad "Vinny" Vincelette and "Black Knights" RIO Lt Cdr Scotty "Gordo" McDonald were forced to eject over southern Iraq when their F-14A (BuNo 158620, call-sign "Junker 14") suffered a port engine and fuel transfer system failure. The latter caused the remaining TF30 to run dry, so the crew, who were two hours into their mission (and having already dropped some of their LGBs), ejected.

Within minutes, a pilot in an orbiting U-2 from the USAF's 9th Reconnaissance Wing picked up a signal from the crew's survival radios and passed the information back to his mission controller at Beale Air Force Base, California. The latter in turn contacted CENTCOM's Joint Personnel Recovery Center, which quickly dispatched a USAF 301st Rescue Squadron combat search-and-rescue HH-60G (supported by TACAIR assets) from Ahmad al-Jaber Air Base in Kuwait. The downed crew was rescued a short while later. This F-14A was the first coalition aircraft to crash in Iraq since the start of OIF.

Lt Vincelette later gave the following account of what went wrong with his jet to a journalist from *Stars and Stripes* magazine upon his return to *Kitty Hawk*:

We were heading out of Iraq after a normal flight, looking for a tanker to refuel from, when the left engine failed. Next came a failure in the fuel transfer system that allows the good engine to use all of the aircraft's fuel. There was nothing we could do but sit and watch the fuel count down. We knew what was coming, and when it dipped below 200lb [91kg], the right engine started to come down, the generator started to hiccup and it was time to go.

"Gordo" yelled "Eject, Eject, Eject!" while I was doing my best to keep the aircraft stable to ensure that we were in a good envelope so that when we punched out the 'chutes would work as advertised. "Gordo" initiated the ejection. It was a fairly surreal experience, as we went from sitting in the warmth and comfort of our own cockpit to a violent windblast and hitting the desert floor pretty hard in our parachutes. Once on the ground, I quickly met up with "Gordo" and asked him whether he could walk as I helped him on to his feet. He replied, "I can *run* – just point the way!"

On the Boat

With the majority of VF-154's senior officers ashore at Al Udeid, it was left to XO Cdr Doug Waters to ensure that the remaining seven jets and ten crews onboard *Kitty Hawk* made a valuable contribution to the CAS mission assigned to CVW-5. Although the "Black Knights" on the boat were primarily junior officers, they proved more than up to the task at hand as they manned the self-proclaimed "JO Flying Club," which expended 246 GBU-12s, ten GBU-16s and four GBU-10s during 27 days of combat.

Cdr Waters detailed a typical carrier-based mission for the unit during OIF:

> We would launch as a section or division from the carrier, transit to a tanker track and top off either in northern Saudi Arabia or southern Iraq, before checking in with an E-3 AWACS or Navy E-2 and receiving tasking. We would then rely on the F-14's LTS to search our assigned area for tanks, artillery and troop emplacements. When we located the enemy, we engaged them with our standard load of four GBU-12s, and if there were enough targets, we would call in other jets and provide them with precise co-ordinates in order to let them finish off the job. We would also buddy-lase for aircraft such as CVW-5's F/A-18s so as to ensure the quick destruction of smaller, hard-to-find targets that our LTS pods were able to see but other platforms had difficulty acquiring.
>
> Normally, having attacked a target, we would then head south to the tanker to replenish our fuel, before pressing back into Iraq in order to have one more go round in the same – or different – kill box, prior to hitting the tanker again on our way back to the ship. Our missions typically lasted between three and three-and-a-half hours.

The shore detachment returned to *Kitty Hawk* on April 10, and by the end of the aerial campaign on the 14th, VF-154 had dropped no fewer than 358 LGBs, buddy-lased 65 more and passed target co-ordinates for 32 JDAM during the course of 286 combat sorties. These totals meant that the squadron had dropped more ordnance than any other unit in CVW-5.

Having selected full military power, which has ignited the afterburners, the crew of "Knight 105" (BuNo 161271) prepare for a night launch from *Kitty Hawk* in early April 2003. Operating as the dedicated CAS air wing, CVW-5 performed an equal number of day and night sorties in OIF I. This particular F-14 dropped 24 LGBs during the campaign. (US Navy)

"Knight 105" heads north into central Iraq on April 14, 2003, the aircraft being configured as a "quad bomber" with a typical OIF I war load for a VF-154 jet – four GBU-12 LGBs. No fewer than 7,114 GBU-12s were expended during the campaign by coalition aircraft. Note the solitary AIM-9M on the Tomcat's port shoulder pylon and the LANTIRN pod on the starboard pylon. This aircraft was one of seven F-14As from VF-154 that remained aboard *Kitty Hawk* throughout the campaign as part of the self-styled "JO Flying Club." The latter expended 246 GBU-12s, ten GBU-16s and four GBU-10s during 28 days of combat. (VF-154)

Post-war, VF-154's successes in OIF I both ashore and at sea garnered individual and unit awards from both the US Navy and the Association of Naval Aviation (ANA). The latter organization presented its prestigious annual Outstanding Achievement in Tactical Aviation award to VF-154 in November 2003, the ANA stating that the "squadron's actions during OIF dramatically demonstrated the resourcefulness and perseverance of Naval Aviation. Its unequalled ability, intense drive and determination to be the best in the Navy made the unit most deserving of recognition by this award."

No fewer than 27 Air Medals and a single Bronze Star (awarded to VF-154's 25-mission OIF CO, Cdr Flatley) were presented by the US Navy to aircrew of VF-154, with a further four going to the augment crews who saw action with the "Black Knights." One of the latter was Lt Cdr Rik Crecelius, who received an Air Medal for his efforts during the March 29, 2003 mission that he led from Al Udeid. The citation that accompanied the award read as follows:

Crecelius launched as the lead aircraft of a division of four strikers tasked with performing night CAS for the 1st Marine Expeditionary Force. Working with a FAC, he located and positively identified an enemy tank brigade poised to engage the Marines along their planned line of advance. While receiving enemy AAA and rocket fire from the towns of Turabah and Duwa in southern Iraq, he and his wingman precisely guided eight 500lb [226kg] LGBs, destroying six enemy tanks and two ammunition storage bunkers. At great personal risk, he remained in the target area to co-ordinate the delivery of six more 500lb [226kg] LGBs from the rest of his division, ultimately neutralizing the enemy tank brigade.

A veteran of 12 missions from Al Udeid, VF-154 NFO Lt Clay Williams received his Air Medal for FAC(A) work that he and his pilot performed for the US Army's V Corps near Baghdad on 2 April. His citation noted:

As with all TACAIR coalition units participating in OIF I, VF-154 relied heavily on "big wing" tanker support throughout the campaign. One such asset was No 216 Sqn's TriStar K 1 ZD951, which was photographed over the NAG about to refuel "Knight 102" (BuNo 161280) towards the end of the campaign. (VF-154)

Williams located, identified and prosecuted portions of an armored column before being re-tasked on a time-sensitive strike to destroy a chemical decontamination vehicle. He encountered accurate AAA throughout this mission. After refueling, he was sent to a different target area, where he controlled attacks by a section of F/A-18C Hornets. The latter destroyed two pieces of self-propelled artillery and multiple support vehicles, and both the F/A-18Cs and the F-14s again met resistance from AAA fire.

As this citation points out, VF-154's crews diced with enemy air defenses virtually until war's end. Twelve-mission veteran Cdr Waters recalled:

The Iraqis were very defensive-minded in their use of SAMs – I suppose ten years of getting pounded any time they brought one up during OSW had left a pretty good impression on them. Our main concern was AAA and rocket fire, and when we were operating in the lower altitude environment in support of troops on the ground, MANPADS came into play too. We would also face the occasional SAM, although the majority of these were fired ballistically. Ballistic or not, they could get your attention – my wingman captured some interesting video footage of a probable SA-6 in his LTS field-of-view as it passed directly in front of our section while we transited south of Baghdad. More often than not though, most close calls were due to AAA.

VF-154's ordnance team also received the Precision Guided Bombing Award post-OIF I, as the unit's Gunner, CWO4 Kim Williams, explained to the author:

VF-154 delivered a record number of GBU-12 PGMs in OIF I. However, what is more impressive is that we did not have one single hung bomb or a weapon that failed to detonate "High Order." To the best of my knowledge, this significant achievement was unmatched by any other US or coalition command that participated in OIF I. As VF-154's ordnance officer at Al Udeid, I operated with only four junior ordnance personnel ashore, which resulted in my selection as the Commander, Fighter Wing Ordnance Officer of the Year for 2003. My award citation noted that the training I had provided allowed my ordnance personnel to operate in dual locations supporting combat operations.

The various awards presented to VF-154 post-war simply served to confirm the unit's remarkable achievements in OIF I. However, the "Black Knights'" success did not come as a surprise to Cdr Waters, who was fulsome in his praise of the squadron in the statement that was released by the ANA following VF-154's receipt of its award: "This is the finest group of naval officers, Naval Aviators and NFOs I have ever served with. They possess a rare blend of poise and skill, an unsurpassable knowledge of strike fighter tactics and unparalleled leadership in every environment. They are backed by what can only be described as the finest maintenance department in Naval Fighter aviation."

VF-154's XO, Cdr Doug Waters (far right), enters details of his flight time and any technical discrepancies with his jet into the unit's Naval Aviation Logistics Command Management Information Systems maintenance program via a laptop computer post-sortie. The aircraft would be listed as either UP or DOWN for the next flight by the aircrew as part of the post-flight report. Should there have been a significant problem with the jet, a workshop was assigned – meaning the Integrated Weapons Team if it was a radar gripe, for example – and the failure details listed (fault codes, description of the failure, aircrew corrective procedures taken), along with the name of the aircrew that generated the issue in case there were follow-up questions. The uncorrected and corrected problems were then printed out and placed in the Aircraft Discrepancy Book, which was required reading by the aircrew prior to the next flight of that F-14. Despite working with the oldest Tomcats involved in OIF I, VF-154's maintenance personnel lived up to their billing as COMFITWINGLANT Golden Wrench award winners for 2002 by keeping enough jets "up on the roof" of *Kitty Hawk* to allow the unit to almost triple its monthly flying hours for the duration of the campaign. (VF-154)

For many years VF-154's late lamented F-14s bore the acronym "BKR" on their twin fins, these letters standing for "Black Knights Rule." This was indisputably the case during OIF I.

CO's Perspective

Cdr James Flatley, VF-154's CO during OIF I, provided the author with the following insight into his unit's operations during the campaign:

Our January 23, 2003 deployment for OIF would prove to be the last time the unit embarked in *Kitty Hawk*, its aircraft, aircrew, maintainers and families returning home to the US eight months later to commence the squadron's transition from the venerable F-14A Tomcat to the F/A-18F Super Hornet. Little did we know when we departed Japan for the NAG that the squadron would play a unique role in OIF that effectively put an exclamation mark on the squadron's last F-14 Tomcat deployment. Our achievements on cruise would be a fitting end to VF-154's 20-year relationship with the jet, the unit's combat record in OIF providing a remarkable demonstration of the Tomcat's capability both as a precision bomber and as a FAC(A) platform.

Following our arrival in the NAG in the latter part of February 2003, we were kept busy conducting flight operations in support of OSW. We had been in-theater less than three weeks when VF-154 was tasked to deploy a small detachment of aircraft, aircrew and maintainers to Al Udeid. In the lead up to us sending jets ashore there had been discussions about one of the three carriers in the NAG supporting a detachment of F-14Bs manned by aircrew from NSAWC and the SFWSL. The augment pilots and RIOs had participated in the development of tactics, techniques and procedures for a specific set of missions involving SOF. In the end, it was decided that VF-154 would deploy to Al Udeid instead.

The VF-154 aircrew involved would all be FAC(A) qualified given the nature of the operations they would have to undertake. We would be augmented by the NSAWC and SFWSL aircrew, the latter from NAS Oceana. I would be included among the VF-154 aircrew deploying to Al Udeid, but I needed to get my XO, Cdr Waters, back aboard *Kitty Hawk*. Since the carrier's arrival in the NAG he had been CVW-5's representative at the CAOC at "PSAB."

Meanwhile, back in Washington, D.C., the Navy's senior leadership was not overly keen on the idea of land-basing Tomcats to support ground operations, believing it would be counter-productive at a time when the Navy was placing greater emphasis on sea-basing assets. The position of the Navy bureaucracy was understandable though, given that there was very little experience of thoroughly integrating Naval Aviation with SOF to the extent that we were about to experience. Just showing up and checking in wasn't going to hack it. The need for aircrew and ground operators to plan and brief together was critical to the success of each mission.

With the XO back aboard the boat, we had less than 24 hours to get him up to speed before my departure – he had been ashore for several weeks "flying a desk." I was taking three department heads and the squadron's training officer, leaving the XO with two department heads and the junior officers to conduct combat operations from the boat. None of us had green ink [combat missions] in our logbooks prior to OIF. If you had done a risk assessment at the time on how successful VF-154 would be at waging war from two locations, I think the analysis would have shown a high risk of failure!

VF-154 OIF I (2003)

Cdr James Flatley runs through his preflight cockpit checks prior to taxiing out in "Knight 103," which boasts the most impressive mission tally markings seen on any VF-154 Tomcat following OIF I. (Jinno Yukihisa)

As a forward deployed squadron [being part of CVW-5, VF-154 then called NAF Atsugi, 20 miles (32km) south-west of central Tokyo, home], one of the advantages we enjoyed over the Oceana-based units was permanent custody of our LTS pods. We capitalized on this luxury, always flying with a pod regardless of the training mission. When flying from Atsugi, it was a requirement on every sortie heading to and from the warning areas south-east of Japan that VF-154 crews prosecute three target sets. The Izu Islands, located off the south coast of Japan and in line with our flightpath to the warning areas, afforded us with ample ground targets to prosecute. We also took advantage of the dense urban areas around Atsugi to train our FAC(A)s and aircrew in urban CAS.

The time dedicated to perfecting all aspects of air-to-ground tactics, continually grooming the weapon system and learning how to operate the LTS pod in degraded modes, would manifest itself in the successful operations conducted by the squadron during OIF.

For the short time you're in command of a fleet squadron you relish the opportunity to lead and to capitalize on every opportunity to excel as a team. With VF-154 establishing operations ashore, we were splitting up the team. It certainly created some challenges and unknowns, but I was confident that the squadron would perform well, entrusting the XO to run the show on the boat. My Department Heads, "Gordo," "Chili," "Kombat," "Hurl," "Sato," and the training officer, "Tuffy", had done an outstanding job leading and training the squadron over the previous 12 months. The junior officers were exceptional in every way, being skilled tacticians and solid aviators that were confident in their abilities.

Master Chief Keith Hulbert, back on active duty after two years in retirement, had masterfully guided the maintenance department to new heights of production, aircraft availability and weapons system reliability. The chief's mess was exceptional in leading and guiding the young men and women in the work centers. The VF-154 team was the finest group of Naval Aviators, NFOs, chiefs and sailors I had had the pleasure of working with. Receipt of the Fighter Wing Golden Wrench [awarded to the F-14 unit with the best serviceability rates for its jets], the 2003 Battle "E" [for efficiency in the unit's designated role] and the Chief of Naval Operations' RADM Joseph C. Clifton award [for meritorious achievement by a fighter squadron while deployed aboard a carrier] would confirm as much.

During 21 days of dedicated combat operations that the VF-154 detachment executed from Al Udeid, each pilot and RIO flew an average of 80 hours, with a typical sortie lasting eight hours. We would usually fly for three consecutive nights, with a day or two off, before getting back into the fight. Planning for missions typically began around noon, followed by launch after sunset. The standard load-out was four GBU-12s and two AIM-9M Sidewinders, two turkey wraps and a couple of bottles of water.

During the week prior to the start of combat operations the det stayed busy getting accustomed to its new surroundings, setting up logistics support from the boat and establishing working relationships with the various US and coalition organizations we were flying alongside. Soon after our arrival, the first order of business was to get the augmented aircrew F-14A NATOPS qualified, since all of them were from the F-14B. A quick sortie overhead the field to get re-acquainted with the aircraft, along with a signature on the relevant paperwork, completed the qualification process. With quals out of the way, we spent the next couple of nights flying rehearsals to learn the tactics, techniques and procedures for the missions we had been assigned to support. We also made a trip back out to the boat to keep our night quals current.

One of my most memorable flights during our operations ashore occurred on the last night [April 7–8] we flew combat missions from Al Udeid in OIF. "Tuffy" and I had logged more than 15 hours over the previous two nights, and on this occasion we would log a further 11.5 hours. The flight would be uneventful for the most part, but in the final hours that would all change.

With the briefing complete, we launched after sunset and flew north to rendezvous with a tanker over central Saudi Arabia. It would be the first of six tankers we would hit. Once we had topped-off we pushed into Iraq to check in with our customers operating just south of Haditha Dam. When we arrived, there was a casualty evacuation in progress. We provided the necessary support until the helicopter cleared the airspace. For the next seven to eight hours, we would remain overhead, providing protective cover for the guys on the ground as they moved through the desert. Utilizing the LTS pod to reconnoitre their route of travel, we provided recommended deviations so that our troops avoided detection.

In order to maintain constant coverage and comms with the guys, I decided that my wingman and I would cycle on and off the tanker as singles. "Tuffy" did a great job of getting additional tanker assets to support our efforts. I think we hit every kind of US and coalition tanker that night that was configured with a basket. With the guys on the ground stopping for the night, we wished them well and pushed south to rendezvous with our last tanker.

With our wingman already on the tanker, he let us know that the weather was an issue – we'd be tanking in the goo. "Tuffy" was able to get a lock on the tanker to set us up for a rendezvous. Our wingman was on the right wing of the tanker as we came up on the left side. As we were about to plug into the basket the right engine stalled. I immediately dropped the nose to get the necessary airspeed to execute an air start. Once we had the engine back on-line we climbed back up through the goo to find the tanker on radar. As we executed another approach to the basket, the right engine stalled again.

Four sides of the box were now in place. We were still over Iraq, in bad weather, with an engine problem, and now below bingo [fuel]. We had one last chance to get back on the tanker. While remaining focused on the immediate task, I was reliving the events of the night when my wingman went down in southern Iraq. Was the squadron about to suffer another loss? After several failed attempts to get the engine on-line, we were finally back in

Tomcat Precision Baby!!! OIF I (2003)

business, but low on fuel. Once rendezvoused with the tanker, I left the right throttle at a constant power setting, flying into the basket with the left engine. We were in and receiving, the fuel totalizer showing 2,500lb [1,134lb]. It wasn't the prettiest plug, but it got the job done. Topped off, we headed south-east. Once feet were wet, we turned south and watched the sun rise over the NAG.

Although much has been written about the exploits of the F-14 Tomcat and its aircrew over Iraq during OIF, there has been little coverage given to the men and women who worked tirelessly to ensure the veteran aircraft were maintained, turned around, armed and readied in time for the next launch. From the work centers to the flightdeck, they made things happen day in, day out. The dedication, pride and teamwork on display was nothing short of spectacular.

Behind the scenes, squadron leadership trusted and relied on the chief's mess to provide the critical leadership necessary to keep the sailors focused, safe and alert. As the commanding officer of VF-154, I looked to one of the most experienced maintainers in the Tomcat community to get the job done, Master Chief Keith Hulbert. Indeed, the success of the "Black Knights" during OIF can be attributed to him. If you had to select one individual who had the greatest impact on VF-154's ability to simultaneously conduct combat operations both at sea and ashore, it would be Master Chief Hulbert. He was the key to our success.

Master Chief Hulbert arrived at NAF Atsugi in the early fall of 2001, two years after his retirement from active duty. He had successfully petitioned the FITWING commander to come back on active duty, although the latter had one condition – Master Chief Hulbert was to take on the job as Maintenance Master Chief Petty Officer [MMCPO] of VF-154. Prior to his arrival, the "Black Knights" had been without an MMCPO for nearly two years. As a result, the squadron had been experiencing more than its share of challenges trying to maintain some of the oldest F-14As in the fleet while permanently forward deployed to Atsugi. Master Chief Hulbert willingly took on the job.

Soon after his arrival in Japan, the MMCPO quickly went about whipping the maintenance department into shape. His knowledge of the Tomcat, decades of operational experience and no-nonsense style of leadership quickly set the battle rhythm for the maintenance department. By the spring of 2002, VF-154 was back on step, negotiating with CAG [Capt Pat Driscoll] to deploy with 12 fully mission-capable aircraft aboard *Kitty Hawk*.

While the MMCPO set about fixing aircraft, he was also busy at work developing future leaders. Anyone who has had the pleasure of serving with Master Chief Hulbert will attest that his old school style of leadership was inspiring and motivating in many ways. The results speak for themselves, as more than a decade later there are nearly a dozen former VF-154 maintainers serving as MMCPOs in various maintenance departments today. Most, if not all, will attribute their success to the lessons learned under Master Chief Hulbert during the 22 months he served as the squadron's MMCPO.

As *Kitty Hawk* steamed into the NAG in February 2003, I was called to CAG's office. The Navy was operating with three carriers in the Gulf. Each CAG was asked to provide their input about their air wing's ability to support a small detachment of F-14B Tomcats. *Constellation* and *Abraham Lincoln*

Master Chief Keith Hulbert, VF-154's legendary MMCPO during OIF I, is seen "in action" dealing with maintenance gripes during the campaign. (VF-154)

both had F-14Ds embarked and *Kitty Hawk* had F-14As. Although the airframes were basically the same, there were some distinct differences that needed to be considered if the F-14B was going to be integrated into air wing operations in the NAG, since none of the three carriers was properly configured to support the B-model. And the aircraft weren't going to be showing up with a big box of spare parts.

Some of the key differences were as follows. The F-14D had Martin-Baker Mk 14 NACES ejection seats, while the A- and B-model jets were fitted with Martin-Baker GRU-7 seats. The F-14D canopy was different from that fitted to the A- and B-models. The B- and D-models had GE 110 engines, while the F-14A had the original TF30s. The F-14D boasted the AN/APG-71 radar, along with a glass cockpit and JTIDS. A- and B-model jets were still operating with the venerable AN/AWG-9.

Master Chief Hulbert and I discussed what options we could provide CAG. From the MMCPO's perspective it was obvious that *Kitty Hawk* could best support the F-14B. It came down to the commonality of the ejection seats and avionics. Our answer to CAG was that the "Black Knights" were the best-suited squadron to take on the F-14Bs if they were to be deployed onboard a carrier

VF-154's full complement of aircrew, administration and maintenance personnel come together in front of "Knight 101" chained down over waist catapult three in early May 2003, immediately prior to the Tomcats departing *Kitty Hawk* for the last time. The F-14 was a labor-intensive jet, hence the requirement for the large number of green-shirted maintainers visible in this photograph. (VF-154)

in the NAG. The trade-off would likely be that CVW-5's Prowlers might have to operate from the beach in order to make room for additional Tomcats on the flightdeck. It was eventually decided that the detachment of F-14Bs would not be based aboard one of the carriers in the NAG. Instead, VF-154 would establish a det of four aircraft ashore at Al Udeid.

On March 12, VF-154 sent four jets to Al Udeid from *Kitty Hawk* for a brief training detachment, giving the det's aircrew and 30 maintainers less than a week to establish operations ashore, train to the tactics, techniques and procedures for the coming operations and NATOPS-qualify the guest aircrew in the F-14A that would augment squadron pilots and RIOs. Meanwhile, VF-154, *Kitty Hawk* and CVW-5 now had to sort out the challenge of supporting the det ashore, while simultaneously conducting combat operations from the boat.

Prior to our second departure for Al Udeid [on March 18, when five jets went ashore for OIF I], Vice Admiral Timothy Keating, Commander, Fifth Fleet, paid a visit to *Kitty Hawk*. After he finished speaking with the air wing he tracked me down, as he wanted to meet Master Chief Hulbert. We made our way from the second deck wardroom up to the 0–3 level of the ship, where Vice Admiral Keating engaged the MMCPO in conversation, remarking that he had heard great things about him and that he appreciated all the hard work he had put into making VF-154's maintenance department so successful.

Weeks later, when combat operations were complete, the admiral flew back out to *Kitty Hawk* to congratulate the ship and air wing. Once again, he went out of his way to pay a visit to the "Black Knights'" Maintenance Control Desk to personally thank Master Chief Hulbert for his remarkable performance.

Sister-ships *Kitty Hawk* and *Constellation* hold formation at slow speed in the NAG on April 13, 2003, by which point OIF I operations had begun to wind down. Seven of VF-2's ten F-14Ds can be seen on *Constellation*'s flightdeck, although only five of VF-154's surviving 11 Tomcats are visible on *Kitty Hawk*. This would suggest that a number of the unit's aircraft were undertaking a mission into Iraq at the time. Both carriers had left the NAG by April 17. (US Navy)

VF-154, flying some of the oldest aircraft in the Tomcat inventory, had managed to execute flight operations from the beach and the boat. Furthermore, the unit had delivered more ordnance than any squadron deployed aboard the three carriers in the NAG and the two carriers in the Mediterranean. Such achievements stand as a testament to the incredible leadership, hard work and relentless pursuit of perfection of one incredible MMCPO, and the team he was able to inspire in the short time he spent while back on active duty.

Soon after VF-154's return to NAF Atsugi, Master Chief Hulbert went to work prepping the Tomcats for a transpacific flight to NAS Oceana. With the departure of the squadron, the MMCPO faded into retirement, his encore performance complete. For the 15 months I was in command of the "Black Knights" I was blessed to have worked with one of the legends in the Tomcat community. I am forever grateful to Master Chief Hulbert, and consider our time together in 2002–03 as the highlight of my career as a Naval Aviator.

Northern War

OIF was fought on two fronts by US Naval Aviation, with a pair of eastern Mediterranean-based carriers attacking targets in northern Iraq. The campaign waged from the flightdecks of *Theodore Roosevelt* and *Harry S. Truman* contrasted markedly with that fought by the trio of vessels sailing in the NAG, as CVW-3's Public Affairs Officer, Lt(jg) Jason Rojas, explained in the air wing's cruise summary:

The war over northern Iraq was quite different from the one in the south. With Turkey denying the US Army's 4th Infantry Division use of its territory as a jumping-off point, northern front activities centered around SOF activity, with some teams as small as three individuals. The teams relied heavily on CAS from CVW-3 and CVW-8, the latter embarked in *Theodore Roosevelt*, which was also positioned in the eastern Mediterranean. Aircraft from both air wings flew CAS missions in support of SOF units, often putting ordnance dangerously close to friendly forces. The support these aircraft provided undoubtedly saved the lives of coalition forces on the ground, and eventually led to the capitulation of nearly 100,000 Iraqi soldiers.

Prior to immersing itself in CAS missions with SOF, both VF-32 and VF-213 completed a number of conventional strike missions with JDAM and LGBs against fixed targets in Iraq. These sorties, flown at the start of the conflict, were some of the longest of the war, covering distances up to 800 miles (1,287km) one way. As the Tomcat had proven in OEF in 2001–02, it was more than capable of handling such sorties, and the mission lead for these more conventional strikes was often an F-14 crew. Indeed, the first CVW-3 operation of the war was led by *Desert Storm* veteran and VF-32 CO, Cdr Marcus Hitchcock. He explained the complex routing problems that the carrier air wings in the eastern Mediterranean had to grapple with for the first 72 hours of the war:

In the lead up to our first mission, the political situation in our area was a little topsy-turvy to say the least. We didn't know whether we would be heading in via Turkey or not. This meant that we had to plan a series of different routes into Iraq – northerly, central and southerly. This uncertainty prevailed until the diplomats figured out which way we could go. Launching from just off the Nile Delta, we were allowed to proceed down the Sinai Peninsula, around the southern tips of Israel and Jordan, across the Saudi Arabian desert and then finally into Iraq. Approval for this route was given just 24 hours before the start of OIF, and a lot of the supporting tanker assets were not told that they needed to change locations in order to facilitate our first strikes.

Nineteen aircraft – six of which were supporting E-2s and S-3s – launched on the first mission on March 22, and 13 pressed south to Iraq on a 1,400-mile [2,253km] transit to the target. Our tanking with USAF assets en route was interesting to say the least, as the KC-135s showed up at the designated rendezvous point so late that we were on the verge of deciding whether to divert jets because they were running out of gas. Two of the Hornets were in fact "timed out" and could not press into Iraq.

Pre-OIF, we had trained with our Tomcats loaded up with three JDAM. No other F-14 unit had sortied with more than two weapons up to that point, as the jet was very heavy on the controls at

Capt Mark Vance (in the tan flightsuit), CAG of CVW-3, and Cdr Marcus Hitchcock, CO of VF-32, brief the embedded press corps onboard *Harry S. Truman* following their return from the March 22, 2003 mission to Al-Taqaddum Air Base. Thirteen of the 19 aircraft that launched that night attacked the target, these jets each covering a distance of 2,800 miles (4,506km) in total. NFO Vance was strapped into the rear cockpit of an F-14B during the mission. (US Navy)

F-14 "BOMBCAT": THE US NAVY'S ULTIMATE PRECISION BOMBER

cruising altitude when fueled for combat, carrying defensive missiles and three 2,000lb [907kg] bombs. We trained hard in this configuration once in the Med, and this paid off in OIF.

When attacking our targets at Al-Taqaddum Air Base on that first night, my crews did spectacular work with their JDAM. We knew about the weapon's capability in theory, but it was not until we had each delivered our three bombs smack on to our targets, spread across the airfield, within a matter of seconds that it became readily apparent that this was a new kind of weapon the likes of which we had never seen before.

We were opposed by localized AAA as we approached the target, but we easily avoided this at our run-in height of between 28,000–33,000ft. As we flew across another airfield on our departure from the target, we saw three SA-2 SAMs launch. We had already dropped our JDAM by then, and as

Left: VF-32 OIF I (2003)

Below: A "Covey" flight of F-14Bs from VF-32 and F/A-18A+s from VMFA-115 (one of which is being flown by the photographer, Capt Eric Jakubowski) fly over the snow-covered Zagros Mountains that form the border of south-eastern Turkey and northern Iraq on March 31, 2003. All the aircraft lack ordnance, bar a single AIM-9M each, indicating that they have already attacked their targets and are heading back to *Harry S. Truman* in the eastern Mediterranean. The weather in the area was rarely this clear during OIF I. (Capt Eric Jakubowski)

the bombs did not need targeting support, we were able to go to 100 percent power and defeat the SAMs through tactical maneuvering. We had a very aggressive Prowler squadron [VAQ-130] within CVW-3, and it made sure that we were not threatened too much by SAMs through the employment of both HARM and EW jamming.

Within 48 hours of flying their long-range "Shock and Awe" missions, both *Theodore Roosevelt* and *Harry S. Truman* were sailing north-east for the Turkish coast following the freeing up of airspace restrictions. The role played by the TACAIR crews of CVW-3 and CVW-8 would be very different from now on, as Cdr Hitchcock explained to the author:

> After we had flown two missions routing in from the south, Turkish airspace was finally freed up for us, so we took a day off and repositioned the carrier further north-east in the Mediterranean. Once CVW-3 started the second phase of its campaign by launching attacks on northern Iraq, we found ourselves consistently flying 12 sorties per day for the rest of the war.
>
> We were pleased with the way the jets and the crews had performed in the first two missions of OIF, but we were now unsure of how we were to fight the war once we got word that we were heading north to start flying through Turkey. The missions flown on the first day of the war had been pre-planned fixed strikes. Each crew knew exactly what it was going to attack well before it launched, having trained to hit these targets for several weeks in advance. As we headed for Turkey, our mission changed, and we were no longer sure of what we were going to attack. On most missions during the war, when we strapped into our jets on the flightdeck we were not entirely sure of what we would encounter over northern Iraq, where the target was and what would be defending it.

VF-32's "Gypsy 111" (BuNo 161428), "Gypsy 107" (BuNo 163224) and "Gypsy 112" (BuNo 161608) sit with their engines idling on the flightdeck of *Harry S. Truman*, waiting for their deck chains to be removed. All three jets appear to be armed with a single GBU-31 JDAM and GBU-12 LGB. "Gypsy 107" features a Space Shuttle silhouette and STS titling in honor of NASA's Mission STS-107 crew, who perished when their spacecraft *Columbia* broke up upon re-entry into Earth's atmosphere on February 1, 2003. All three aircraft have had their refueling probe covers removed, thus preventing them from being ripped off when tanking from KC-135s. (US Navy)

For the first few days in the north, we were scratching around for fixed targets to hit so that we could at least conduct some rudimentary pre-mission planning, thus allowing us to familiarize ourselves with the procedural aspects of working in this theater. We soon realized that such mission preparation was a waste of time, however. We would brief to attack a certain target, but by the time we had launched, the situation on the ground had invariably changed. Effectively, this meant that the only pre-flight planning we could perform on the ship centered on where the tanker support was located, what frequency it would be using and what weather we could generally expect to encounter over the target. If it was cloudy we would take JDAM and if the weather was good we would take LGBs, as they were more mission flexible. We also flew with a mix of both on occasions in order to hedge our bets.

VF-32 conducted limited brief missions over northern Iraq with SOF teams on the ground from March 24 to April 18. Unlike in the south, where there was a clear line between the good guys and the bad guys, in the north, we never had a clearly developed battlefield that boasted a discernable front line. We therefore tailored our operations towards servicing one or more of the 40 to 50 kill boxes that we had divided northern Iraq up into. Crews rarely knew which one they would be assigned to prior to launching from the ship. We would be allocated a kill box to service en route to Iraq, the decision on where we would be heading being made by our AWACS controllers after they had checked "on trade" with SOF guys on the ground.

Occasionally, we would break out of the kill box mentality and hit targets within a small geographic region that was perhaps being worked over as part of a ground offensive by Peshmerga militiamen, or we would alternate between targets in several areas during the course of one sortie.

The Iraqis had three corps of ground troops in the north. Our mission was to keep them occupied so that they could not head south to help defend Baghdad from V Corps and 1st MEF. This task was given to CVW-3 and CVW-8, controlling 72 Hornets and 20 Tomcats, and 1,000 SOF guys on the ground, supported by Kurdish freedom fighters. The SOF squads operated throughout northern Iraq, and they were responsible for finding us targets.

"Gypsy 114" (BuNo 161424) comes under tension on bow catapult two in early April 2003, the aircraft carrying a single GBU-31 JDAM and a GBU-12. Flying 90 combat hours during the course of 21 sorties, the jet dropped four GBU-31s in OIF I, as well as ten GBU-16s and 12 GBU-12s. The aircraft also participated in VF-32's OIF III deployment in 2004–05, after which it was retired to AMARC. The Tomcat was sold to HVF West for scrapping in May 2012. (Erik Lenten)

OPERATION *IRAQI FREEDOM* I–III

In general, we would launch from the carrier and head to the border between Turkey and Iraq, where we tanked for the first time – we were at half gas when we meet the tanker inbound to the target area. We then headed into Iraq for about 45 minutes and worked the targets. If we expended all of our weapons, we would fly straight home. If not, we found the tanker once again and then headed back into Iraq for another 45 minutes before returning home, again via the tanker.

We effectively covered a two-hour block on a typical sortie, whether we deployed our weapons or not, and were then replaced on station by a second wave from the ship. Each wave usually consisted of a section of two Tomcats and up to four sections of F/A-18s [eight Hornets in total]. These aircraft would be supported by a single EA-6B, with an E-2C providing AWACS control and several S-3B tankers airborne with tactical gas should we need it over the carrier.

CVW-3 was designated as the day carrier, while CVW-8 handled much of the night work. VF-213 soon dubbed its nocturnal missions "Vampire" sorties, the constant night operations eventually inspiring the Tomcat aircrew to coin the phrase "living after midnight, bombing 'til

A solitary F-14B is shot off waist catapult four while *Harry S. Truman* undertakes a replenishment at sea evolution with the Military Sealift Command (MSC) fast combat logistics ship USNS *Arctic* (T-AOE-8) in the eastern Mediterranean on 10 April 2003. The MSC ammunition ship USNS *Mount Baker* (T-AE-34) is holding station off the carrier's port side. A second Tomcat is immediately behind the raised jet blast deflector, awaiting its turn to launch, and six more are chained down aft of the island – traditionally "fighter country" on a carrier flightdeck. (US Navy)

the dawn"! More often than not, CVW-8's strike aircraft would launch from *Theodore Roosevelt* into poor weather conditions. One of those grappling with solid cloud and extreme turbulence on a near daily basis was VF-213's XO, Cdr John Hefti:

> We routinely had to tank in the clouds, being buffeted by thunderstorms and turbulence in weather fronts that stretched from ground level up to 40,000ft. Once in Iraq, the bad weather made it far more difficult for us to acquire targets with our FLIR. As it transpired, flying to and from the targets at night in these poor conditions posed a far greater danger to us than the Iraqi military.

Although I had seen combat in the Tomcat in both *Desert Storm* and OEF, those night missions in OIF were some of the most challenging flights I have ever had to perform as a Naval Aviator. I had experienced bad weather in *Desert Storm* on only two or three occasions, when we had to tank in clouds, and in Afghanistan the skies were generally clear. In OIF, by contrast, at least half of our sorties were flown in poor to bad weather, where we stayed IFR virtually from the minute we got over Turkey until we came back out over the Mediterranean and headed for the boat four or five hours later. Thankfully, for some reason the weather never actually seemed to reach the carrier. We were still faced with a night trap though, which is always far more difficult than a daylight recovery.

"Black Lion 106" (BuNo 163893) participated in CVW-8's first OIF I strike, VF-213's aircrew summing up its wartime experiences with the phrase "living after midnight, bombing 'til the dawn" following its myriad nocturnal "Vampire" missions. The unit completed 198 sorties totaling 907 flight hours during the conflict. Achieving a 100 percent sortie completion rate, VF-213 delivered 102 LGBs and 94 JDAM during OIF I. (US Navy)

Despite the constant threat posed by the weather, both VF-32 and VF-213 did their best to maintain 24-hour TACAIR support for the SOF teams taking the fight to Iraqi forces. One of the more unusual missions flown by VF-213 saw the unit providing CAP for the airborne landings made by 1,000 paratroopers of the US Army's 173rd Airborne Brigade on March 26. Conducting the largest parachute drop since World War Two, the soldiers jumped from a fleet of USAF C-17s on to Harir airfield in Kurdish-controlled northern Iraq. The Globemaster IIs were escorted by three waves of strike aircraft from *Theodore Roosevelt*, with one of the F-14Ds being flown by Lt Cdr Larry Sidbury:

> I shot some nice FLIR video of the paratroopers as they jumped out of the back of the C-17s that night. It was sobering to think that these guys would be relying almost exclusively on us to support their efforts on the ground in the coming days as they pushed out to attack targets around Tikrit, Kirkuk and Mosul. We had ordnance on our jets ready to support the landings should they be resisted either from the air or the ground, but there were no enemy forces to be seen. We eventually left the area once the transports had headed back into Turkey and dropped our bombs on some pre-planned targets around Kirkuk.

As the war progressed, CAS for SOF squads became the staple mission for VF-32 and VF-213. "Dropping precision-guided ordnance for a SOF team was a mission that gave immediate gratification," recalled CVW-8's CAG, Capt David Newland. "Our aircrew were told where to drop the munitions, and they got direct feedback from the troops after they had deployed their ordnance." The SOF teams on the ground also expressed their appreciation in no uncertain terms immediately after OIF I had concluded, Col Charles Cleveland, Commander Joint Special Operations Task Force-North, e-mailing the following message to CVW-3 CAG Capt Mark Vance (who flew as an F-14 RIO with VF-32 throughout the war) in late April 2003:

> On behalf of Special Forces A teams and the rest of us here at Task Force Viking, I want to say thanks for being there when we needed you. You were instrumental in our dismantling of three IZ Corps and the ultimate capture of the third and fourth largest cities in Iraq. This fact says a lot, considering the coalition ground component largely consisted of the 10th Special Forces Group (Airborne) and our Kurdish allies. We took some big risks knowing that when we needed you, you'd be there. You never failed us, and as a direct result we never lost a position and had only four casualties during the entire operation.

VF-213's Lt Cdr Larry Sidbury was one of those Naval Aviators who worked closely with SOF FACs at night during OIF I. Here, he explains how a typical CAS sortie in northern Iraq was performed:

Cdr Marcus Hitchcock is given the signal to start the port engine of "Gypsy 101" (BuNo 161860) by his plane captain while the jet is chained down on the fantail of *Harry S. Truman* on April 11, 2003. By the end of the campaign, this aircraft had dropped more bombs (26 GBU-12s, six GBU-16s and 18 GBU-31s), flown more sorties (37) and completed more combat hours (178.9) than any other jet in VF-32. (US Navy)

> We would launch from the boat and head south-east over Turkey, hitting the tanker over the border with Iraq. With our tanks topped off, we would call up our AWACS controller and tell him we were ready for mission tasking. We would let him know what type of aircraft we were flying, what ordnance we were carrying, our fuel state and whether we were FAC(A) qualified or not. We would then be given a tasking, which usually consisted of location co-ordinates and a frequency for our ground FAC, or the co-ordinates for a kill box.
>
> Heading into northern Iraq as a section of two jets or occasionally as a division of four, the strike lead would head toward the ground FAC's location and then attempt to raise him on the radio. In most cases, the SOF FACs were using hand-held radios, and you almost needed to be overhead their position to talk to them, particularly when they were working in the mountains east of Mosul and Kirkuk. They usually had tasking for you right away, so they were ready to accept your ordnance as soon as you checked in. The nine-line brief that the FAC would give you would be short, clipped and concise – these guys were big on comms brevity, and you certainly got the impression that they did not want to talk on the radio for any extended period of time.

A nine-line attack brief was the key element in effective CAS, as it had to have all of the information needed to ingress, conduct the attack and egress the target area. The components of the brief were as follows:

1) Initial Point [IP] – location where the CAS aircraft began its attack run from.
2) Bearing/offset – bearing from IP to target and offset direction (left or right).
3) Distance – distance to target from IP in nautical miles for fixed-wing aircraft.
4) Elevation – elevation of target in feet or meters AGL.
5) Target Description.
6) Target Location – co-ordinates for the target.
7) Mark – type of visual or laser/IR mark that the FAC put on/near the target.
8) Friendlies – bearing and distance from target to nearest friendly unit (not co-ordinates so as to minimize confusion with target co-ordinates).
9) Egress – direction and IP for the CAS aircraft to fly after attacking the target in order to clear the target area.

Additional information was provided after the nine-line brief to include ToT (time when CAS aircraft's bombs must hit the target) and Final Attack Heading/Cone (either a heading or an arc of headings that the CAS aircraft had to fly down when delivering weapons on to the target).

This information would be relayed to you via radio, and we literally had a phonebook of frequencies on which the FACs would be operating, as there were hundreds of small two- or three-man teams servicing targets across northern Iraq. Each team had at least two radios, and these all needed to have unique frequencies. Nine times out of ten, the frequency you would end up working on wouldn't be in the book! One of the nice things about flying the F-14 was that you had a RIO who could run the radios while you concentrated on flying the jet. This proved vital in OIF when dealing with AWACS controllers and ground FACs.

The latter would not talk to the AWACS controller at all, relying instead on another agency with more powerful communications equipment that would in turn contact the AWACS to inform them that they had some tasking in-theater. We would be passed this information, along with the FAC's frequency, and we would talk directly with the controller on the ground. Both the RIO and I would write down the FAC's nine-line brief on to our kneeboards. We then read the co-ordinates back to him to check that we had copied them down correctly, and he would confirm with a simple "Yes", or a mic click. We then started to punch the target co-ordinates into the bomb guidance system, after which I verified that I was working with the same target information up front as my RIO was in the back.

With all the targeting parameters confirmed, we then started working up a game plan on how best to attack the target. A lot of the time we worked the target in a textbook way, having briefed for just this type of mission back on the ship. We would tailor our general tactics to suit the situation at hand, confident in the knowledge that we had our training to fall back on should the target prove difficult to hit.

This amazing photograph of a GBU-12 heading for its target was taken by a TARPS-equipped VF-32 jet. The aircraft's RIO was Lt Cdr David Dorn, who recalled, "We were flying the TARPS jet when we were fragged to hit an SA-2 site with an LGB. Skipper Hitchcock was flying the aircraft that lased the target for our weapon. We got some great TARPS footage of the LGB coming off a jet and heading down to impact the missile site. I got to replicate this mission several days later, when we again hit a target with an LGB while flying the TARPS jet. The F-14 could only carry LGBs when equipped with a TARPS store as the JDAM was just too big. And when carrying TARPS, we could not employ the LANTIRN, so we had to rely on our wingman to buddy lase for us." VF-32 flew very few TARPS missions in OIF, and because the unit was so pressed with servicing ground targets, Cdr Marcus Hitchcock decided to load TARPS jets with bombs too so that crews could fly dual missions. (VF-32)

VF-213's "Black Lion 111" (BuNo 159629) is directed back to the fantail soon after recovering aboard *Theodore Roosevelt* following CVW-8's first OIF mission against targets near Fallujah on March 22–23, 2003. The aircraft launched with three JDAM, two AIM-9s and a single AIM-54. Only the missiles returned with the Tomcat to the carrier. VF-213's XO, Cdr John Hefti, participated in this mission, subsequently recalling, "I flew a DCA sortie for the main strike package, which was actually launched on the second night of the war – the carrier did not participate in the opening night of the campaign. VF-213 did not make a huge contribution to the 'Shock and Awe' phase of the war simply because there were insufficient tankers available to support aircraft from the Mediterranean-based carriers. Once we moved north, the tanker assets from Akrotiri came into play, allowing us to up the level of our participation." (Troy Quigley)

Both VF-32 and VF-213 regularly carried mixed bomb loads in OIF I, this "Black Lions" jet boasting a GBU-12 and a GBU-31(V)2/B. VF-213 pilot Lt Cdr Marc Hudson explained the thinking behind such a loadout. "With the Tomcat capable of employing both JDAM and LGBs, we could service two different targets during the course of one CAS mission – we often sortied with a mixed load towards the end of the war. This proved to be a lethal combination for the enemy because we had instances where the ground FAC was giving us GPS co-ordinates for the JDAM and we were able to lock the target up with our LANTIRN pod. We then chose to go with the LGB instead. If the target was not destroyed first time around by the 500lb [226kg] weapon, we could make a correction with the JDAM off the LGB drop and come back and hit it again with clinical precision using the appreciably larger GBU-31." (VF-213)

With the jet set up correctly and the target locked up by the LTS, the bomb symbology in the HUD would tell us exactly when to drop either the JDAM or the LGB – we still had to physically hit the bomb release switch, as there was no automatic release system in the F-14. With all the release parameters met, we would tell the FAC that we were set up for our run in to the target, giving him the "wings level" call, which would prompt his reply of "cleared hot." We now had his authority to release our ordnance – we could not do this without his say so. Once the FAC had declared our targets "hostile," I was happy to drop my bombs. His view of the enemy forces was far better than mine, so the final call on whether to release the ordnance or not was ultimately his to make. I never questioned his decision at all.

I often dealt with FACs who were under fire. In those cases, you might see enemy vehicles on the ground, and you would be fully aware of his situation. I would be in constant communication with him, telling him what I could see from my jet, and he would use this information to confirm my targets. I would be cleared by the FAC to engage those targets, after which I required no further information from him in order to do my job. Usually, when you hit vehicles or troops on the move that were in close contact with a SOF team, things would settle down for a short while as the enemy took cover. This allowed the FAC to tell you where the "next alligator closest to the canoe" was, and you could start working those targets.

Things could rapidly get dynamic in close contact situations, and you had to be very sure that you were targeting bad guys and not friendlies before you dropped your bombs. I would remain in continual comms with the FAC during such engagements, constantly questioning him on the radio as to the orientation and direction of the vehicles that we could see, or where his placement was in relation to a nearby river or road.

We helped a FAC out one night who had positioned himself on a small hill in order to improve his view of a main road that ran between Mosul and Kirkuk. Somehow, his position was compromised to the extent that a convoy of Iraqi Army vehicles detoured off the road that he had under observation and surrounded him. They jumped out of their trucks and started engaging him with small arms and mortar fire. The FAC was screaming on the radio for some assistance, and he gave us a hasty plot of his location. As it turned out, we found it easy to target the enemy troops thanks to the way their vehicles ended up oriented around the FAC off the road. He was able to give us a very quick visual talk-on, and we soon had bombs on the enemy positions that allowed the FAC to make good his escape.

On a typical OIF CAS mission, a two-ship Tomcat formation would operate with one to two miles [1.6–3.2km] of separation between jets, with my wingman keeping a visual lock on us at all times, and vice versa. Such separation ensured that you had the room to maneuver in combat without the fear of running into your wingman. Despite operating as a formation at all times over Iraq, you usually worked a separate target to your wingman, although you would both be flying in the same general area.

We would also be employing the services of the same FAC, who normally had multiple targets in the same area. They might be a mile or so apart, or sometimes they were closer to each other than that. The FAC would have a list of targets that needed servicing, and looking out on the battlefield from his position, they would all be in the same general area from his perspective.

My wingman and I were sent to work with a FAC near Kirkuk towards the end of the war. On our last run in on the target, three SAMs came up in front of my section, launched with plenty of lead. We started to defend against the missiles, and there was some excitement on the radio between the two jets as we performed tight maneuvers to defeat the apparently ballistic SAMs. I was trying to talk the other crew's eyes on to the missiles, and although they never saw them, they reacted positively to our calls and broke off in the right direction.

We then came over the top of their jet in an effort to reorient our section and continue our flow to the north. I was in afterburner at the time, and the junior RIO in the second jet shouted over the radio that he could see the missiles right above his cockpit! The pilot looked up in response to the call and quickly informed his RIO that his "missiles" were their strike lead! The RIO caught a lot of heat about that call for the rest of the cruise.

Parked on the fantail of *Theodore Roosevelt*, both "Black Lion 107" (BuNo 161166) and "Black Lion 100" (BuNo 164602) behind it have the mixed LGB/JDAM load-out so favored by VF-213 during the 2003 campaign. The sixth F-14A to be converted into a D-model Tomcat following more than a decade of fleet service, BuNo 161166 was subsequently assigned to VF-2 (twice), VF-11 and VF-101 after being upgraded in 1991–93. It was eventually passed on to VF-213 in April 2002, and the aircraft remained with the unit until stricken in April 2006. (US Navy)

Above: Flying over mountains on the Turkey–Iraq border on April 10, 2003, "Black Lion 105" tops off its tanks from an Akrotiri-based KC-135R of the 100th Air Refueling Wing. Such clear conditions were exceptional on the northern frontline of OIF I, as Lt Cdr Larry Sidbury recalled. "Tanking was a miserable experience pretty well throughout the campaign, as you would invariably have to join on a tanker that was socked in the clouds. Some nights you would get to within a mile of the tanker before you could break out its lights and control your closure speed. You had to get your approach right the first time in such poor weather, for disorientation could have easily proven fatal in the solid cloud we often had to rendezvous in." (VF-213)

Left: A diminutive Fly 2 plane director signals to the pilot of "Black Lion 107" to raise all eight of the jet's overwing spoilers for visual inspection by the deck checkers. They will remain open until Fly 2 lowers her palms from the wrists, signaling to the pilot to close the spoilers. Meanwhile, a member of the catapult crew is checking that the nose gear launch bar is correctly placed to receive the catapult shuttle (which is still out of shot). The aircraft, carrying a mixed LGB/JDAM loadout, was photographed on April 1, 2003.

Having refueled over the snow-covered mountains of northern Iraq, VF-32's CAG jet (BuNo 162916) drops away from a 163rd ARW KC-135R on April 11, 2003. This aircraft dropped 30 LGBs and five JDAM during 35 OIF missions. Retired after VF-32's final Tomcat deployment in 2004–05, BuNo 162916 has been on display at VFW Post 8896 in East Berlin, Pennsylvania, since 2008. (Paul Farley)

By the time VF-213 ceased combat operations in OIF I on April 15, its crews had flown 198 sorties totaling 907.6 combat flight hours, with a 100 percent sortie completion rate. The unit had dropped 196 PGMs weighing 250,000lb (113,398kg) – 102 LGBs and 94 JDAM.

VF-32 completed an impressive 268 sorties totaling 1,135.2 combat flight hours, dropping 247 LGBs and 118 JDAM (402,600lb/182,616kg). Its crews also expended 1,128 rounds of 20mm SAPHEI in strafing passes.

"Black Lion 111" is towed by an A/S-32A-31A flightdeck tractor towards *Theodore Roosevelt*'s stern after a recent recovery cycle that saw the aircraft return from a mission to northern Iraq on the night of April 13–14, 2003. To the jet's right, "Black Lion 110" (BuNo 163903) sits fully armed and fueled, awaiting the arrival of its crew for the next OIF I push during the early hours of the morning. One of the S-3Bs from VS-24 assigned to provide frontside fuel for that mission has already been manned and its engines started – Vikings were usually the first aircraft launched when conducting such operations. (US Navy)

OIF II–III

Although OIF I officially ended on May 1, 2003, combat operations steadily increased in their intensity as al-Qaeda stepped up its insurgency across Iraq. The US Navy, in turn, relied on carrier strike groups to take the fight to the enemy, which meant that the Tomcat would remain a key player in the Global War on Terror in Iraq until its withdrawal from fleet service in 2006.

On May 1, 2003, President George W Bush stood on the flightdeck of *Abraham Lincoln* as the carrier sailed off the coast of southern California and declared that major hostilities in Iraq were over. Just four weeks earlier, 52 Tomcats had been in the vanguard of OIF I, flying into the heart of Iraq from carriers in the NAG and the eastern Mediterranean. By the time the President made his now-famous proclamation, the only carrier on station in the region was *Nimitz*, and the vessel was bereft of Tomcats, as its embarked air wing – CVW-11 – was making history by becoming the first to venture into the NAG without an F-14 component.

Indeed, it was not until October 23, 2003 that the unmistakable shape of the Tomcat was seen in the skies over Iraq once again following the arrival of VF-211 in-theater onboard *Enterprise*. Assigned to CVW-1, this unit was also making history, as it was conducting the last operational cruise of the F-14A. Unable to employ JDAM, which had again become the weapon of choice in Iraq post-OIF I, VF-211 saw very little action while in the NAG, or during the unit's brief spell in support of OEF in Afghanistan.

As with all Tomcat units that participated in OIF II–III in 2003–04, VF-211 spent much of its time flying TARPS missions or performing show of force and convoy patrols over main/alternate supply routes in Iraq. Very occasionally, the squadron also used its LANTIRN pod to provide target co-ordinates for CVW-1's trio of Hornet squadrons, which dropped a handful of JDAM. By the end of the deployment in February 2004, VF-211 had flown 220 combat sorties. Despite the unit not expending any ordnance, its CO, Cdr Mike Whetstone, proudly noted VF-211's performance in his post-deployment report:

> Most of our accolades have come due to our TARPS missions for the strike group commander and the task force commander on the ground. Each day we'd send two or three aircraft over Iraq or Afghanistan, while the rest conducted training missions. Our F-14s were the oldest in the navy's inventory, so it took a lot to get them into the air. But the maintainers treated them like they would

VF-211's "Nickel 115" (BuNo 161297) is launched from *Enterprise*'s waist catapult three in December 2003. The unit spent much of its time on station in OIF II performing ISR patrols of southern Iraq, often in partnership with F/A-18A+s of VMFA-312. A participant in VF-211's OEF deployment of 2001–02, this aircraft had been delivered new to VF-2 in March 1982. Later serving with VF-194, VF-114 and VF-213, it was stored for a short while at NAS Jacksonville, Florida, before joining VF-211 in early 2001. The fighter was retired by the unit to AMARC in September 2004 and eventually sold to HVF West for scrapping in July 2009. (US Navy)

a classic car. It was a challenge for them, but they handled it like pros. Each time we were assigned a mission, we were able to fulfill it thanks to the efforts of our maintainers.

CVW-1's CAG, Capt Mark Mills, elaborated on VF-211's reconnaissance missions during his Tailhook 2004 convention address as part of the symposium's OIF panel. He recalled: "VF-211 flew 22 TARPS missions and produced 325 target images as requested by the CAOC. The unit also performed TST thanks to the Tomcat's unrivalled ability to relay digital imagery in flight to CENTCOM's dissemination module in-country. VF-211 supplied some 115 digital images while patrolling over Iraq."

On March 1, 2004, *Enterprise* was replaced on station in the NAG by *George Washington*, with CVW-7 embarked. The latter boasted two F-14B units within its ranks, VF-11 and VF-143 making their last deployments with the jet prior to transitioning to Super Hornets.

Replacing *Enterprise* on station in the NAG on March 1, 2004 was *George Washington*, with CVW-7 embarked. The latter included two F-14B units within its ranks, VF-11 (whose "Ripper 201" and "202" are seen here formating with a USAF tanker) and VF-143 making their last deployment with the jet prior to transitioning to Super Hornets. Unlike VF-211, which did not drop a single bomb in anger while in-theater, both Tomcat squadrons would experience considerable action during their time in the NAG following a dramatic increase in insurgent activity in Iraq. (USAF)

Unlike VF-211, both squadrons would get to deliver ordnance in combat during their time in the NAG following a dramatic increase in insurgent activity across the country.

The first bomb drops by F-14s post-OIF I took place on April 28–29 when CVW-7 was called on to provide direct fire support for 1st MEF, which had troops in contact with enemy forces in Fallujah – a stronghold of the insurgency in Al Anbar province, in central Iraq. A total of 17 GBU-12 LGBs were dropped by F-14s and F/A-18s during the 40 sorties flown in this 48-hour period.

For the two months prior to CVW-7's TACAIR assets engaging insurgency targets, the air wing had seen little action, as DCAG Capt Rob Field explained at Tailhook 2004:

When we first arrived in the NAG, things were pretty quiet on the ground in Iraq. Indeed, Tomcat and Hornet crews initially spent most of their time in-country running security patrols over oil pipelines, railroad tracks, high-tension power lines and highways. The F-14 units also flew near-daily TARPS missions, collecting strategic data for target sets required by the coalition troops on the ground.

On a typical mission during this early phase in our deployment, we would launch a mixed package of F-14Bs, F/A-18Cs, EA-6Bs and E-2Cs. The latter would support the Tomcats and Hornets, with the Hawkeyes effectively controlling the airspace in the bottom third of the country due to southern Iraq then lacking any mature surveillance or command and control structure permanently sited in country. E-2C crews would "triple cycle" during patrols deep into Iraq, running the radars and controlling the mission tasking for our jets in response to requests emanating from troops on the ground.

Each TACAIR section typically consisted of a single Tomcat and Hornet paired up for maximum operational flexibility. After launching from the carrier, the section would fly north for about an hour into Iraq and immediately hit one of several tankers operating on pre-briefed tracks in-country. Once topped off, the section would be assigned to run a security patrol over one of the many transport routes heading north towards Baghdad. You would drop down to 5,000ft and keep your speed up as you overflew the route, keeping an eye out for AAA or hand-held SAMs. Using binoculars during daytime and NVGs at night, our mission was simply to flush out the routes in order to make sure that there was no one on the ground attempting to sabotage them.

After completing an hour on-station patrolling these routes, you would hand over to another two-jet section and head to the nearest tanker in order to replenish your fuel. With tanks topped off, you would return to route patrolling, often heading as far north as Mosul or Kirkuk. Eventually heading

"Ripper 201" (BuNo 162912) and "Ripper 202" (BuNo 161418) fly in a stepped-up formation for the benefit of the USAF photographer occupying the boom operator's seat in this April 22, 2004 shot taken over Iraq. Both aircraft are armed with two GBU-12s. For the first two months of the deployment, CVW-7 lived up to its nickname of the "peace wing." However, on April 28–29, CVW-7 was called on to provide direct fire support for 1st MEF, which had troops in contact with enemy forces in Fallujah – a stronghold of the insurgency. F-14Bs and F/A-18Cs from CVW-7 duly dropped 17 GBU-12 LGBs during the 40 sorties flown in this 48-hour period. (USAF)

Carrying two GBU-12s, VF-11's "Ripper 201" rides *George Washington*'s bow catapult one in the NAG in June 2004. Only the third F-14B built, BuNo 162912 was delivered to VF-24 in April 1989. It subsequently served with VF-101, VF-143 and VF-102, prior to joining VF-11 in late 1997 when the latter unit swapped its F-14Ds for B-model Tomcats. The jet served as the squadron's CO jet from 1999 until it was put on display at the Grissom Air Museum in Bunker Hill, Indiana, in June 2005. (Erik Sleutelberg)

back south after a further 60 minutes on-station, you would hit the tanker once more and then return to the carrier. A typical mission during this period lasted around 4.7 hours for a "pointy-nosed" jet.

When the insurgency started to escalate from late April onwards, we became more involved in providing direct support for troops on the ground that were in contact with the enemy. This initially started with Tomcats and Hornets providing show-of-force overflights and the occasional bombing strike for coalition patrols that were being attacked in the so-called "Sunni Triangle." By May–June, CVW-7 was also being called on to support vehicle convoys that appeared to be falling victim to co-ordinated attacks by insurgents using RPGs, mortars and improvised explosive devices [IEDs].

OPERATION *IRAQI FREEDOM* I–III

We quickly developed some training techniques and procedures that allowed us to effectively escort convoys throughout Iraq. While on patrol, if we received a report that "friendlies" were in contact with the enemy, we were authorized by the CAOC to drop down below 500ft in order to make plenty of noise in an effort to neutralize the threat. If this did not work, we were cleared to prosecute positively identified targets in order to get the friendly troops or vehicles out of harm's way.

The Tomcat units in particular were kept very busy during this phase of the deployment, with most pilots and RIOs getting more than 65 hours of flying time a month during *George Washington*'s spell in the NAG.

From mid-April onwards, CVW-7 became immersed in high-density urban CAS as a result of increased insurgent activity in and around Fallujah. These operations proved very challenging

During OIF II–III most Tomcats returned from ISR patrols with their bombs still aboard. "Ripper 210" (BuNo 162911), seen in the final stages of its recovery in June 2004, was no exception. Delivered new to VF-24 in April 1989, the aircraft spent time with the NFWS and VX-9 before being transferred to VF-11 in 1997. Also a veteran of OEF, the jet was retired to the Estrella Warbirds Museum in Paso Robles, California, in December 2004. (Erik Sleutelberg)

for our Hornet and Tomcat crews, and the learning curve was incredibly steep. We had done some training at home for such ops with Carrier Groups 4 and 1 prior to deploying, conducting "haystack" sorties where crews would be tasked with identifying specific buildings in a high-density urban environment. Such training was of limited value, however, as it is difficult to find urban areas in the US that replicate the towns and cities in Iraq. Typically, they will put around three times as many buildings in the same amount of space as we do!

The pipeline patrols were effectively replaced by CAS stack missions from mid-April through to mid-May. The sorties were run much the same way, however, with mixed sections of jets observing the same kind of tanker drill once in Iraq. We would then check in with the AWACS controller and be held in a CAS stack overhead in an area known as "Eight-Mile" between Fallujah and Baghdad. All you had to do then was sit and wait for your turn to be called down to get involved helping troops in contact on the ground.

Again with its LGBs still firmly attached to their shackles, VF-143's "Dog 102" (BuNo 162921) hits the deck of *George Washington* and smokes the tyres as it snags an arrestor wire at the end of another ISR patrol over Iraq. Delivered new to VF-103 in December 1989, this aircraft served as the unit's CAG jet when with CVW-17 during *Desert Storm*. It remained with VF-103 in this capacity until the squadron assumed the identity of VF-84 in late 1995, after which it was transferred to VF-143. Following nine months with VF-101 in 2002, BuNo 162921 returned to VF-143 and served with the unit until its retirement to AMARC in December 2004. The Tomcat was eventually sold to HVF West for scrapping in June 2008. (Erik Sleutelberg)

These CAS missions were run in a similar way to those flown in OIF I, with the Tomcats and Hornets being controlled by the JTAC on the ground. He would talk your eyes on to the target that he wanted you to see, and you would be cleared to run in "Hot" to drop your ordnance – should it still be required – once the JTAC was satisfied that you were looking at the same target as he was.

CVW-7 soon found that urban CAS was somewhat different to operations we had carried out in the past both in OSW and OEF. A lot of the targets the JTACs were asking us to find were so well hidden that it was taking crews multiple passes at low altitudes to locate them. Orbiting down low and using binoculars or simply our "Mk I eyeballs" to ID the targets, we would then have to climb back up to 10,000ft in order to drop our GBU-12s. Our F/A-18s also got to strafe on several occasions too.

CVW-7 found that the key to providing effective urban CAS was that the designated target had to be serviced perfectly the first time. There was no margin for error, for a lot of the targets we attacked were wedged in between schools, mosques or residential areas – the insurgents chose such sites in order to further complicate our mission tasking in response to their attacks on our troops.

With CVW-7 fielding near equal numbers of Tomcats (22) and Hornets (24), the two types operated very closely with each other over Iraq, as VF-11 CO Cdr Scott Moyer explained:

In order to enhance our mission capabilities, we decided to fly mixed sections over Iraq. We also mixed our weapon load-out too, with the Hornet carrying two GBU-32(V) 1,000lb [454kg] JDAM and the Tomcat two or four GBU-12 500lb [226kg] LGBs. More often than not, we returned to CVN-73 with our bombs still aboard due to the fact that we were called on to fly less traditional missions over Iraq. Instead of dropping bombs, we would fly "show of force" sorties for troops that had either come under attack or were faced with a gathering mob situation. We would be asked by the JTAC to fly low passes in afterburner so as to make plenty of noise. This usually got the crowd running for cover.

Even on the odd occasion when we were cleared to deliver bombs, these missions could also prove to be non-traditional. More than once my crews were instructed by the JTAC to drop LGBs or JDAM a short distance away from insurgent positions in urban areas so as to minimize collateral damage, but still register a presence with nearby enemy forces.

On July 10, 2004, *John F. Kennedy*, with CVW-17 embarked, arrived on station in the NAG to relieve *George Washington*. As with the three previous Tomcat units to serve in-theater, the air wing's VF-103 was conducting its last deployment with the F-14B. Just ten days after flying its first mission over Iraq, the squadron dropped a single GBU-12 on an insurgent position near Baghdad. This set the tone for VF-103's four months in the NAG, with the unit seeing far more action than any other Tomcat squadron since the end of major hostilities in May 2003.

Mirroring previous air wing operations in the region post-OIF I, CVW-17 routinely paired up Hornets and Tomcats over Iraq. This was primarily because two of the three F/A-18C units (VFA-34 and VFA-83) onboard *John F. Kennedy* were equipped with the first production examples of the US Navy's newest targeting pod, the Raytheon AN/ASQ-228 ATFLIR.

For almost a decade, the LANTIRN pod had been viewed as the premier targeting system within a carrier air wing, but according to VFA-83's Lt Cdr Matt Pothier, who used the

VF-143 pilot Lt Javier Lee inspects the laser codes on a GBU-12 attached to the port forward BRU-32 bomb rack during his aircraft walkaround check on April 28, 2004. This weapon was dropped on an insurgent position in Fallujah later that same day. (US Navy)

Above: CVW-7's TACAIR units flew round-the-clock missions in support of operations in Fallujah, the Tomcat's LANTIRN capability and NVG-compatible cockpit proving critically important during nocturnal sorties. Its anti-collision lights blazing, "Dog 111" (BuNo 162701) is just moments away from being launched off into the inky-black night sky, the crew bound for western Iraq. Another OEF veteran, this aircraft was originally delivered new to VF-14 in April 1987. Rebuilt as an F-14B, it then served with VF-103 and was eventually passed on to VF-143 in 1999. The unit retired the fighter to AMARC in March 2005, and it was subsequently sold to HVF West for scrapping in September 2008. (US Navy)

Right: A pilot and RIO from VF-103 walk past "Victory 100" (BuNo 162918), complete with anniversary/final cruise titling on the nose of the aircraft, and head for their assigned jet on *John F. Kennedy*'s flightdeck on July 9, 2004. The veteran carrier, with CVW-17 embarked, relieved *George Washington* in the NAG 24 hours after this photograph was taken. (US Navy)

AN/ASQ-228 in action over Fallujah, "the ATFLIR made the LTS look cheap! We would patrol predefined positions, tanking three to four times during a five- to seven-hour sortie. We usually patrolled hot spots and protected high-interest targets such as coalition convoys, talking to our ground-based JTACs if they were embedded in the area where we were working. VFA-81, operating the older non-ATFLIR Lot X F/A-18Cs, always flew in a mixed section with VF-103, as the unit relied on the Tomcat's LANTIRN pod to provide primary targeting for its LGBs."

Fallujah continued to prove a hot bed of unrest throughout CVW-17's time on station, and on October 8, 2004, the air wing helped provide aerial support over the city for Operation *Phantom Fury/Al-Fajr*. Some 10,000 Marines from 1st MEF, supported by 5,000 Iraqi Army soldiers, were tasked with flushing out an estimated 3,000 insurgents in a bitterly contested house-to-house campaign. One of the NFOs involved in this operation was VF-103's Lt(jg) Matt Koop:

> When *Phantom Fury* kicked off, CENTCOM was concerned that large numbers of foreign fighters would come streaming in from Syria and Iran to aid the insurgency in Fallujah. Coalition forces had set up outposts along the borders with both countries in order to prevent this from happening, and an increase in the number of skirmishes in nearby towns was anticipated. Sections of fighters were pre-positioned to provide CAS to our troops in these areas, if needed. And it was on one of these missions that I saw my first real action of the deployment.
>
> Ironically, when my pilot and I were told that we would be conducting a Syrian border patrol, we were more than a little disappointed not to be working with the Marines in Fallujah, since that was where all the action seemed to be taking place. Our Tomcat was the lead aircraft that afternoon, flying in a mixed section with a Hornet wingman from VFA-81. I checked in with the [US Marine Corps'] Direct Air Support Center [DASC] controller to tell him what our mission was, and where we had been told to patrol – we were hoping that he would give us a last-minute tasking to Fallujah, but that was not to be the case. We continued westward and contacted the JTAC that we had been assigned to work with. He described the area that we would be patrolling and pointed out a few outposts that had intermittently received fire in the previous 48 hours.
>
> Once we were on-station, the troops on the ground requested a show of force over their positions to either ward off or stir up any insurgent activity that might be brewing nearby. We bumped up the speed and dropped down to overfly their outposts low, fast and loud. Once we had completed our pass, we climbed back up and waited for their next request. Everything remained quiet, and after a while it was time to go hit the tanker and top off the tanks.
>
> When we checked back in, we were disappointed to hear that we had just missed some action. Apparently, two mortar rounds had been launched from a nearby field, and a pair of Marine [AH-1W Cobra] helicopters had been tasked with finding the culprits. We could see a Cobra and a Huey conducting their search below us, but we were flying too fast and too high to offer much assistance. It was at this time that the JTAC told us, "Sorry boys. It just doesn't look like we have much for you fixed-wing guys to do." He told us that if we had any alternate missions, we were cleared to proceed. We didn't have any other assignments, so we told him we'd stick around in case anything else popped up.
>
> Literally minutes later the JTAC received a report that some of our troops patrolling a nearby town had been attacked by terrorists in a blue van. Having exchanged shots, the insurgents had sped away in their vehicle. The Marines called for immediate air support, and both our section and the helicopters were tasked with locating the blue van. We were told to contact a different JTAC who was actually in the town, and he gave us targeting information relating to where the van had last been seen. While I was working the FLIR in our LANTIRN pod, my pilot was scouring the streets with his binoculars. We soon spotted an abandoned blue van, and the helicopters came in to confirm that this was indeed the insurgents' vehicle. With this confirmation, the Cobra was cleared to destroy the van with rockets.
>
> While the AH-1W was firing at the vehicle, new reports were coming in and being passed to us on the radio that additional Marines had been engaged by insurgents who were holed up in a "café." The troops had been subjected to both machine gun fire and RPG rounds, and they were in need

VF-103's Lt(jg) Matt Koop sits on the cockpit sill of his F-14B, "Victory 112" (BuNo 161422), after making an emergency arrested landing in Kuwait in September 2004. Shortly after the event he recalled, "We were flying over Iraq, north of Baghdad, near Balad, on a routine XCAS mission when we lost our left engine. The oil system had a catastrophic failure and the engine seized up. We had the option to put the jet down right there in Balad, but the other engine and the hydraulics looked like they were holding up fine, so we decided to try to make it back to Ali Al Salem, in Kuwait. It would have been impossible to replace the engine in Balad, so it was a good thing that we made the decision that we did – a spare engine and a team of squadron maintainers was quickly flown out to us and we were soon back on board *Kennedy*." (Lt(jg) Matt Koop)

Above: The pilot of "Victory 100" has extended the fighter's tailhook in preparation for landing back aboard *John F. Kennedy* on July 13, 2004 – by which point the anniversary titling previously seen on its nose had been removed. Exactly one week after this photograph was taken, VF-103 dropped its first GBU-12 of the deployment on an insurgent position near Baghdad. A further 20 would be expended in anger by the unit over the next four-and-a-half months. On January 5, 2005, Lt(jg) Matt Koop (who took this shot) was the RIO in BuNo 162918 when the fighter made its final flight (with Lt David Reade at the controls) from Oceana to AMARC, where it was placed in storage. The aircraft was sold to HVF West for scrapping in March 2008. (Lt(jg) Matt Koop)

Left: VF-103's Lt Anthony Walley inspects a moveable fin for the MAU-169/B Paveway II Computer Control Group for one of two GBU-12s attached to his F-14B onboard *John F. Kennedy* in August 2004. Once dropped from the aircraft, the MAU-169/B provided terminal guidance control for the LGB. (US Navy)

Opposite: Its upper surfaces streaked with hydraulic fluid and grime from the boots worn by the maintainers charged with keeping it airworthy, "Victory 101" (BuNo 162705) drops away from a USAF tanker after receiving mid-mission fuel on August 14, 2004. The crew has already expended one of its GBU-12s, and is now heading back to western Iraq in search of another target. A veteran of *Desert Storm* with VF-33, this aircraft was also retired to AMARC in January 2005. It was eventually scrapped by HVF West in June 2008. (USAF)

of immediate air support. This "café" was less than one kilometer from where we had found the van, and after a quick talk-on by the JTAC, we confirmed that we were "tally the target." He then requested that we provide laser designation for a Hellfire missile that was to be fired by the gunner in the Cobra, since his line-of-site for missile guidance was poor.

Neither my pilot or I had ever done anything like this before, having never been briefed on how to lase for a Hellfire missile! But we had been briefed on buddy lasing for our Hornet wingman's LMAV, and we figured that the two laser-guided weapons were similar enough to expect success if we employed the same tactics. We were right. The Hellfire guided to the dead center of our crosshairs and

VF-103 worked closely with the three F/A-18C-equipped units assigned to CVW-17 while in the NAG, flying mixed section patrols in-country that provided a variety of weapon options for troops in contact on the ground in Iraq. Holding formation off the port side of a USAF tanker, two Hornets from VFA-34 lead "Victory 101" during a mission on August 14, 2004. The jet in the foreground is armed with an LMAV and the Hornet behind it is carrying a GBU-32 JDAM. The Tomcat is devoid of any bombs, having already dropped its two GBU-12s earlier in the mission. (USAF)

blew right through the front door of the building. That hit stopped the fire our troops were receiving, and the weapon's small warhead caused minimal damage to the structure of the building itself.

With the possibility of more insurgents hiding deeper in the "café," the order was given for us to destroy the building with our two GBU-12s. We were told to target each end of the building with one bomb, so we would have to make two passes with as little time in between as possible.

As soon as we had received permission to drop our LGBs, we raced out to an appropriate run-in position that minimized the danger posed to our troops nearby. Fortunately, our LANTIRN pod was producing a crisp image, and the target was easily identifiable from more than five miles away. We stepped through the checklist we had memorized and made sure that all our parameters were correct before dropping the first bomb. The weapon guided with perfect precision to the center of my laser spot, destroying the east wing of the building. We immediately turned outbound and set up for our second run-in. This bomb came off just as well as the first, and it guided precisely to the target, levelling the structure.

As in OIF I, JDAM quickly became the weapon of choice in the fight against the insurgency as it grew in size. "Our F-14Bs were JDAM capable," VF-103's Lt(jg) Koop confirmed, "and we had undertaken a lot of training with the weapon in our pre-cruise work-ups, but the decision was made once in-theater that for maximum flexibility we would have our Hornet wingman carry a JDAM and one other bomb [usually another GBU-12, but later an LMAV]. This left us carrying GBU-12s exclusively throughout our time on station. This was viewed as the 'Maxflex' load-out for a mixed Tomcat/Hornet section, as we could now deal with the typical "pop up" targets that proliferated in Iraq without having mensurated co-ordinates. And we only carried 500lb [226kg] LGBs in order to minimize collateral damage."

Former CVW-3 CAG (and ex-F-14 RIO) Capt Mark Vance spoke about this shift in weapon emphasis in OIF II/III at Tailhook 2004: "The increase in urban CAS, and the associated risk of collateral damage, has forced us to take a serious look at the size of the warheads we are employing against the insurgency. Creative fusing by air wing armorers has seen the frag blast area drastically reduced, and large weapons such as 2,000lb [907kg] JDAM are not being employed at all in built-up areas."

Such weaponeering changes had a direct impact on the bombs cleared for use by the F-14, which at that time could not carry the new, urban CAS optimized, 500lb (226kg)

VF-103's "Victory 112" (BuNo 161422) leads "Victory 102" (BuNo 161419) into the break overhead *John F. Kennedy* on a typically hazy day in the NAG in August 2004, both jets returning home with their mission load-out of two GBU-12s apiece still intact. (Lt(jg) Matt Koop)

GBU-38 JDAM debuted in OIF III by CVW-17's trio of Hornet squadrons in August 2004. The 1,000lb (454kg) GBU-32 JDAM was also incompatible with the Tomcat for the same reasons that the GBU-38 had not been cleared for use with the aircraft – the umbilical cord that transferred the data to/from the weapon when mated with the bomb rack did not interface with the F-14's bomb computer. NAVAIR had originally decided in 2002 that it would be too expensive to pay for this integration in the jet's "twilight years" with the fleet. However, as detailed in the next chapter, the US Navy's last two Tomcat units, VF-31 and VF-213, would have their F-14Ds "wired up" to take the GBU-38 prior to the jets' final operational deployment with CVW-8 in 2005–06.

VF-103's support of Operation *Phantom Fury* lasted well into November 2004, as 1st MEF continued with its bloody campaign to rid the so-called "Sunni Triangle" of insurgents. The unit persisted with flying mixed formations throughout this period, with the following account being related by an unnamed FAC(A)-qualified Tomcat pilot who was involved in just such a mission:

> While leading a Hornet from VFA-83 on a routine *Phantom Fury* standby CAS mission in early November, with a second mixed section in-country with us, we were told to look at a building on the outskirts of Fallujah. It was one of many targets we had received imagery and information for prior to launching, the second CVW-17 section receiving instructions to investigate another dwelling nearby. Once both buildings were confirmed as being safe houses for insurgents, their destruction was approved. After locating our targets, we were told to deliver a single 1,000lb [454kg] JDAM on each building. We carried LGBs on our Tomcats, so the JDAM-equipped Hornets would be the primary strikers.
>
> We joined both sections together into a four-jet division so as to facilitate precise timings for the attack. Essentially, the F/A-18s joined as a lead section, with the F-14s in trail capturing BHA. This proved to be an ideal arrangement for us, as it made the best use of our LANTIRN sensor, which turned out to be key to our successful attack. The Hornets' runs on the targets went well, with single bombs impacting each of the buildings virtually instantaneously. However, the JDAM from my Hornet wingman did not explode, or was a dud. Luckily, we could confirm this using LANTIRN imagery in the cockpit, as both my RIO and I spotted the small puff of smoke on the roof of the building as the bomb hit home.
>
> We passed this information on to the decision makers on the ground, along with the fact that the building was still standing. I recommended a re-attack, and within minutes we were authorized to expend our two 500lb GBU-12s on the target. This we did with the minimum of fuss, scoring a direct hit. The size of the resulting explosion verified that not only had we hit the building, but also the JDAM inside. Surprisingly, given the fact that three bombs had now struck the dwelling, most of the resulting damage was restricted to within the walls of the target itself – a testament to the accuracy of both the JDAM and the LGB.

While this was not the most intense mission flown over Iraq during the course of our final Tomcat cruise, it provides a good illustration of the mixed section concept, and its strengths. Flying with the F/A-18s allowed CVW-17 to bring a more diverse range of weaponry to the fight. The Tomcat's sensors, employed by a dedicated operator in the form of the RIO, proved invaluable in target acquisition, allowing precision and flexibility in targeting, and keeping everyone honest as a BHA platform. Combine this with the jet's longer loitering time and the SA of a second aircrew, and the F-14 became arguably the CAS platform of choice for operations in Iraq.

By the time *Harry S. Truman* relieved *John F. Kennedy* in the NAG on November 19, 2004, VF-103 had flown 384 OIF sorties totaling 1,913.4 hours. The unit had also dropped 21 GBU-12s in anger while on-station.

The Tomcat presence in-theater was then assumed by VF-32, embarked with CVW-3 aboard *Harry S. Truman*. As with VF-103 before it, the "Swordsmen" flew mixed formations with the air wing's trio of Hornet units, and VMFA-115 in particular. The latter unit's veteran F/A-18A+s were not equipped with ATFLIR pods, so in order to comply with a CENTAF requirement that all bomb droppers in-theater have access to advanced targeting FLIR imagery prior to attacking a target, VMFA-115 relied on VF-32's LANTIRN designation. This partnership worked well, as according to Maj Guy Ravey from VMFA-323 (embarked in *Carl Vinson* as part of CVW-9, which replaced CVW-3 on station in the NAG in mid-March 2005):

> VMFA-115 did a whole lot of dropping, especially early on with the mop up of Fallujah. To my understanding the guys operated in mixed sections [Tomcat/Hornet] to take advantage of the LANTIRN pod's higher resolution, and to accommodate a requirement from CENTAF that only ATFLIR- and LANTIRN-equipped aircraft could drop in-country. VMFA-115 had the upgraded F/A-18A+s that had AN/APG-73 radar and a lot of other mods to keep them on a par with the rest of the fleet's more modern Hornets, but they were not capable of displaying the ATFLIR images on their Data Display Indicators – hence their reliance on VF-32's LTS-equipped F-14Bs.

With the assistance of the Tomcats, VMFA-115 flew a handful of pre-planned strikes near the Iraqi border town of Al Qa'im (just five miles [8km] east of Syria) against a known foreign fighter and weapons smuggling ring. Continuing to operate closely with VF-32, the US Marine Corps unit was also heavily involved in providing on-call CAS for Marines and soldiers in central and northern Iraq.

Like VF-103, VF-32 was making its final Tomcat deployment. Unlike CVW-17, CVW-3 was only occasionally called on to directly engage insurgents, although it continued to fly "show of force" operations – including overseeing the Iraqi elections on January 30, 2005 – and convoy

John F. Kennedy sits at anchor off Manama, the capital of Bahrain, at dusk on October 5, 2004 during a brief R&R visit. Seven of VF-103's ten Tomcats can be seen on the flightdeck. During the carrier's final operational deployment, CVW-17 undertook 8,296 sorties for a total flight time of 21,824 hours. While the carrier was assigned to Fifth Fleet in the NAG (from July 7 to November 19), the air wing had generated 4,396 sorties totaling 11,607 flight hours. (US Navy)

From November 19, 2004, the Tomcat presence in OIF III was assumed by VF-32, which was part of CVW-3 embarked in *Harry S. Truman*. As with VF-103 before it, the "Swordsmen" flew mixed formations – "Covey" flights – with CVW-3's trio of Hornet units, and, as had been the case in OIF I some 20 months earlier, F/A-18A+-equipped VMFA-115 in particular due to a lack of ATFLIR pods for its ageing jets. "Gypsy 103" (BuNo 162915) is armed with a pair of GBU-12s and the Hornet on its wing is carrying two GBU-38 JDAM. (Erik Hildebrandt)

protection missions on a near daily basis. These sorties primarily took the form of Intelligence, Surveillance, Reconnaissance (ISR) and XCAS missions across the country from *Harry S. Truman*. Tomcat crews tasked with flying XCAS provided on-call (alert) CAS, which meant that they were a CAS asset for the period they were assigned to patrol over Iraq, but they were not allocated to support a particular ground unit at the time the ATO was written. According to VF-32 NFO Lt Cdr Randy Stearns:

Each TACAIR squadron would fly six OIF sorties a day on average, so that translated to about six sections over the beach each day for VMFA-115 and VF-32. We would go up to Mosul or patrol around central Iraq in the vicinity of Fallujah and Baghdad. Occasionally, when there was a gap in Marine Air coverage in Fallujah, they would give the CVW-3 jets a call to come help out.

Both VFA-37 and VMFA-115 dropped a few GBU-38s through bad weather in the Fallujah area and up in north-western Iraq in the first week of December 2004. On the 10th, VF-32 expended

its first ordnance of the deployment when a single jet hit a target in Fallujah with two LGBs – the VMFA-115 aircraft in the same "Team Vicious" section dropped two GBU-38s and fired its LMAV. They were supporting the Marines who had some insurgents held up in a house, and they decided the most efficient way to deal with them was to use TACAIR.

As described in the previous chapter, a "Team Vicious" VF-32/VMFA-115 section supported an ambushed Stryker patrol in Mosul the following day. Both jets dropped all their ordnance and also strafed. Eleven days later, VF-32 was again in the thick of the action, as Lt Cdr Stearns explained:

The next time that CVW-3 dropped was on December 22. That's when two F-14s got the call to target positions in north-west Iraq with four 2,000lb [907kg] GBU-31 JDAM. The Tomcat could bring back both JDAM if they didn't drop, but the Hornet would have to jettison one of its bombs if it didn't drop. So, it was a no brainer for VF-32 to get the call because the unit had expended 118 JDAM in OIF I, and it therefore had experience with the weapon. The mission went off without a hitch.

I found it interesting that no other F-14B squadron – VF-11, VF-143 or VF-103 – had dropped JDAM in-theater after we left OIF I in 2003. Being the only B-model Tomcat unit to have dropped a significant number of JDAM helped us get our foot in the door when it came time to deliver 2,000lb [907kg] GBU-31s almost two years later.

Mosul was again the location for more bombing and strafing by VF-32 and VMFA-115 on December 29, and this event was also covered in the previous chapter.

According to Lt Cdr Stearns:

The last time that ordnance was expended by CVW-3 on the 2004–05 deployment was in the first week of January when a "Team Vicious" section of jets dropped two JDAM [VMFA-115] and two GBU-12s [VF-32] on some suspected mortar positions just south of Mosul. Overall, VFA-37 dropped two JDAM, VFA-105 dropped nine weapons, VMFA-115 dropped 20 bombs and fired a handful of LMAVs and VF-32 expended five GBU-12s and four JDAM – not a lot of ordnance, but we did support a lot of troops.

Once the bomb-dropping stopped in the run up to the Iraqi parliamentary elections on January 30, we pretty much got into "groundhog day" until we left the NAG. Although we got a lot of calls to fly in low and use our loud noise to scatter suspicious activity, there were always a

Armorers from VF-32 check the MAU-169/B Paveway II Computer Control Group fins fitted to two GBU-12s shackled to the BRU-32 bomb racks of a VF-32 Tomcat onboard *Harry S. Truman* in December 2004. The unit dropped five GBU-12 LGBs and four GBU-31 JDAM during its four months committed to OIF III with CVW-3. (US Navy)

Above: A section of F-14Bs from VF-32 fly their final orbit over *Harry S. Truman* from marshall before pitching up into the break and then entering the landing pattern. The wingman would aim to land just 45 seconds after his section leader in "Gypsy 107" – another VF-32 jet that had flown with the squadron since fall 2000. (Erik Hildebrandt)

Opposite: VF-32's "Gypsy 101," flown by staff officers from CVW-3, circles over the Swords of Qādisīyah and the Monument to the Unknown Soldier in central Baghdad during the Iraqi elections on January 30, 2005. This aircraft (BuNo 161860) was VF-32's CO jet from September 2000 until it became the final Tomcat on strength with the squadron on October 5, 2005. (Erik Hildebrandt)

ton of armed helicopters flying around the country whose weapons [specifically AGM-114 Hellfire air-to-surface missiles] were better suited for use in the urban environment than our PGMs. If they were calling us in to drop ordnance in a city, then things were usually pretty bad on the ground.

Our OIF III missions in-country lasted about five to six hours depending on what we were doing. Sometimes guys got extended beyond that. We typically would take off from the ship in the NAG, rendezvous and press out to our first tanker for frontside gas. We'd go off and do our mission, and then hit the tanker for mid-cycle gas, go back and check-in with the ground FAC again, before hitting the backside tanker and then heading out of country.

VMFA-115 and VF-32 would switch leads every other event, and when we got into the target area, the F-14 crew normally worked the LANTIRN and talked to the FAC since they had the good FLIR for target acquisition. We did a lot of car chasing, building watching during insurgent take-downs, convoy support and routine pipeline patrols. The 2004–05 cruise was a long haul, with a full four months in the NAG and very few breaks. There were hours of boredom punctuated by minutes of sheer terror.

As noted earlier in this chapter, *Harry S. Truman* was relieved in the NAG by *Carl Vinson* on March 19, 2005. Reinforcing the fact that the F-14's days in the fleet were now well and truly numbered, CVN-70's flightdeck was devoid of any Tomcats – the carrier's CVW-9 included three F/A-18C squadrons and a single unit equipped with the F/A-18F. Fifth Fleet had not seen the last of the F-14 just yet, however.

The final act in the Tomcat's long career with the US Navy was played out in the NAG in fall 2005 when CVW-8, embarked in *Theodore Roosevelt*, arrived on station to support OIF IV. Assigned to the air wing for this historic cruise were VF-31 and VF-213, both conducting their final deployments with the F-14D prior to transitioning to the Super Hornet in 2006.

VF-32's "Gypsy 100" (BuNo 162916) accelerates down *Harry S. Truman*'s waist catapult three to signal the start of a dusk patrol over Iraq on February 9, 2005. As was standard for Tomcats undertaking such a mission at this stage in the OIF campaign, the aircraft is armed with two GBU-12s and 600 rounds of 20mm SAPHEI ammunition. "Gypsy 100" served as VF-32's CAG jet from November 1998 until it was stricken in September 2005. (US Navy)

OPERATION *IRAQI FREEDOM* I–III

Right: VF-32 OIF III (2004–05)

Below: "Gypsy 101" snags the three-wire with its tailhook after landing on *Harry S. Truman*'s well-weathered flightdeck on February 14, 2005. By the time VF-32's OIF III commitment had ended, the unit's mission tally stood at 413. This aircraft's departure from Oceana to an aviation museum in Kentucky brought VF-32's association with the Grumman fighter to an end after 31 years, three months and 21 days – a Tomcat squadron record. (US Navy)

CHAPTER 12

LAST CATFIGHT

The Tomcat's final operational deployment saw the jet at the "tip of the spear" during CVW-8's commitment to OIF IV in 2005–06, flying from the familiar flightdeck of the "Big Stick", Theodore Roosevelt.

On March 25, 1986, an F-4S Phantom II from VF-151 launched from USS *Midway* (CV-41) as it steamed in the East China Sea, thus bringing to an end the frontline fleet service of a US Navy stalwart. A decade later, on December 19, 1996, VA-75 made the last ever cruise fly-off by an A-6 Intruder squadron when it departed *Enterprise* as the vessel returned to NAVSTA Norfolk at the end of a six-month-long deployment to the Mediterranean and the NAG. Repeating the cycle that has seen a major US Navy type withdrawn from service every ten years (admittedly, the A-7 Corsair II was retired from the fleet in June 1991!), on March 10, 2006, F-14D-equipped VF-31 and VF-213 returned to NAS Oceana at the completion of the Tomcat's last operational cruise.

Neither the Phantom II nor the Intruder saw combat during their final deployments. However, VA-75 (and fellow A-6 unit VA-196, which was also in the NAG at the time aboard *Carl Vinson*) came mighty close as part of Operation *Desert Strike* on September 3, 1996. At the last minute, President Bill Clinton chose to attack air-defense targets in southern Iraq with TLAMs, rather than with manned strike aircraft. This was certainly not the case for the Tomcat, which was in the thick of the action from literally the day the aircraft arrived on station in the NAG on October 5, 2005 aboard the 97,000-ton *Theodore Roosevelt*.

The 22 F-14Ds in-theater with CVW-8 were part of the air wing's 64-strong force of combat aircraft that had relieved CVW-11 aboard *Nimitz* – the latter air wing had the distinction of being the first not to have dropped any ordnance in Iraq since December 1998. It soon became apparent that this comparative lull in insurgency activity would not last, however. Within 48 hours of *Theodore Roosevelt* being declared mission ready to the CAOC at Al Udeid, which controlled all coalition aircraft operating over Iraq, a section of F-14s had provided CAS for US and Iraqi troops in the vicinities of Al-Hawija, Al Hillah and Al Muqdadiyah. This pattern of operations was set to continue unabated for the next four months.

VF-213 RIO Lt Cdr Robb Soderholm recalled the type of missions undertaken by both F-14 units during *Theodore Roosevelt*'s time in the NAG:

We flew between 14 and 18 sorties per day, with only three five-day port visits to Jebel Ali, in the United Arab Emirates, to break up this routine during "TR's" commitment to OIF. At least a third of these flights were six-hour marathons up into Iraq. Generally, the two Tomcat units each put six to eight jets over the beach every day. The remaining ten sorties that VF-213 generated on a typical day would see crews remain "around the boat," with some of these jets being air spares for OIF pushes that then flip-flopped into maritime surveillance patrols as part of Operation *Sea Dragon*, had they not been required for the main event over Iraq.

Crews also conducted proficiency training for the myriad missions that VF-213 performed in-theater, including practicing our air-to-air work with the E-2s in an effort to keep both us and them at the top of our games. The squadron also carried out task group protection flights, as our admiral was big on making sure that the vessels under his control were not taken by surprise by a fishing dhow packed with explosives. Finally, we manned CAPs and provided Alert 30 jets on a regular basis.

Squadronmate, and fellow RIO, Lt(jg) Scott Timmester also shared his impressions of the cruise from a junior officer's viewpoint:

OIF sorties during the deployment typically meant a six-hour hop across the beach about every third day for each crew. Missions were generally flown in support of friendly ground forces that needed an airborne presence to deter or disrupt insurgent activities throughout the country. Most hops were long and uneventful, which was good for the guys on the ground, but less than thrilling for us. We typically ended up getting gas three times a flight from a variety of big-wing tankers, including international players. When not flying over Iraq, we were conducting hour-long flights around the NAG, "pressurizing the maritime environment" – official terminology for making our presence known and preventing illegal activities in international waters.

VF-213

VF-31 pilot Lt Justin Halligan explained how a typical flying day in the NAG was planned out by CVW-8:

> The 1120hrs launch, which was the first mission sent over the beach by CVW-8 every day, returned from Iraq at 1730hrs, which in the fall/winter-time meant a fully blown night recovery. During the course of this mission, all aircraft that pushed on into Iraq would have refueled three times. We received fuel soon after we had made landfall, then headed off-station to the tanker about 90 minutes later. Our last aerial refueling took place 90 minutes after the second top up, and we then headed home. The longest mission that we would fly on a daily basis was event three, which was scheduled to last six-and-half hours from 1345hrs to 2015hrs.
>
> There was a significant overlap between packages in Iraqi airspace, with jets from the previous event still heading south for the NAG as you were getting yourself established in-country.
>
> Typically, the air wing launched four events during a 12-hour period in order to fulfill its commitments to the CAOC. Each event was usually made up of 16 tactical jets [F-14Ds and F/A-18Cs], supported by EA-6Bs, S-3B tankers and an E-2C Airborne Early Warning [AEW] aircraft. Half of the 16 tactical jets were launched as air spares for the primary aircraft, so only eight pressed on into Iraq for the full six-hour mission. The rest would come back for the next recovery cycle after first performing a *Sea Dragon* patrol, which usually only lasted between one to two hours. We could sortie one or two air spares depending on what the air wing wanted us to do, and a single jet was cleared to conduct a *Sea Dragon* patrol should the squadron not have enough aircraft available to launch two air spares to cover a standard OIF section.

"Felix 107" (BuNo 163902) of VF-31 comes under tension on *Theodore Roosevelt*'s waist catapult three prior to flying an OIF IV mission on October 17, 2005. The jet is armed with a GBU-38 JDAM and a GBU-12 LGB, which, combined with the 500 rounds of 20mm HEI ammunition for the M61A1 Vulcan cannon, was the standard armament for CVW-8's F-14Ds during the Tomcat's final combat cruise. (Gert Kromhout)

VF-31's "Felix 105" (BuNo 159619) is only milliseconds away from trapping back onboard *Theodore Roosevelt* at the end of a six-hour patrol over Iraq. Both VF-31 and VF-213 maintained a high sortie tempo during the 2005–06 deployment, flying between 14 and 18 sorties per day. This particular aircraft – which was VF-31's leading bomb dropper on cruise, with three GBU-38s expended – was one of three Tomcats assigned to CVW-8 that had been built in 1975, the jet serving with VF-124, VF-24, VF-1 and VF-111 prior to its remanufacture as a D-model. BuNo 159619 was then assigned to VX-4, after which it flew with VF-2 and VF-213, before being allocated to VF-31 for its final fleet service. The fighter has been on display at the Florida Air Museum in Lakeland since its retirement in September 2006. (Gert Kromhout)

The focus was very much on VF-31 and VF-213 during CVW-8's 2005–06 deployment to the detriment of S-3B-equipped VS-24. Like the two Tomcat units, the "Scouts" were making their final cruise with the Viking. Having operated the humble ASW "Hoover" since July 1976, VS-24 embarked eight jets onboard *Theodore Roosevelt* for the OIF IV deployment. Aside from being the sole provider of recovery fuel for CVW-8, the unit's S-3Bs flew more than 220 overland non-traditional intelligence, surveillance and reconnaissance missions in support of British-led coalition ground forces in southern Iraq. Having extended its hose and drogue, "Scout 705" (BuNo 160149) prepares to pass recovery fuel to "Felix 101" following a training mission on October 1, 2005. (Lt(jg) Scott Timmester)

Proving that the F-14 was still at the "tip of the spear" during its final deployment, VF-213 had the distinction of dropping the first ordnance to be expended in anger by CVW-8 on October 11, 2005. One of its crews used a single GBU-12 to destroy rocket and mortar positions used by insurgents to shell coalition forward-operating bases (FOBs) in the vicinity of Ar Ramadi. Lt Cdr Soderholm explained how the Tomcat units had gone about detecting such positions during patrols over Iraq:

> We were routinely called in to overfly urban areas such as Baghdad and the surrounding "Sunni triangle" in the wake of random mortar attacks on our FOBs. The insurgents liked to lob rounds into these secure areas usually at night, when most of the occupants of the FOBs were asleep. Troops on the ground would work out roughly where the attack had come from and then ask us to go and check these areas for "hot spots" – we would be looking for the hot mortar tube(s) and/or people running away from this general location.

Above: "Black Lion 201" (BuNo 164341) returns to *Theodore Roosevelt* after an OIF IV patrol on October 17, 2005, its ordnance still secured to the under-fuselage racks. In the foreground is "Black Lion 206" (BuNo 163893), which did not drop any bombs or strafe during the deployment. VF-213 "ordies" have chalked "1x500 HEI" on the fairing covering the latter jet's M61A1 cannon. (Gert Kromhout)

Left: CVW-8

On several occasions our squadron was called on to hit such a target after it had been pinpointed, the crew on station quickly locating the "hot spot" in the target area assigned to them by the troops within the FOB that had been attacked. Although the insurgents that had fired the weapon had usually fled the scene by then, the Tomcat crew was occasionally given authorization to drop a 500lb [226kg] GBU-38 JDAM on the mortar equipment that they had discovered.

JDAM to the Fore

The GBU-38 had been specifically cleared for use by the Tomcat on the eve of its final deployment, the availability of the new weapon combining with the state-of-the-art ROVER system and the tried-and tested LANTIRN pod to make the F-14 CVW-8's primary surveillance and force protection platform while on station in the NAG. Indeed, the 22 Tomcats became the ground forces' combat aircraft of choice during *Theodore Roosevelt*'s time in-theater.

The GBU-38 had made its service debut with the US Navy in the NAG in October 2004 when it was used in combat over Fallujah by the trio of F/A-18C Hornet units assigned to CVW-17 and embarked in *John F. Kennedy*. The air wing's Tomcat squadron at that time was F-14B-equipped VF-103, and the unit was replaced in-theater by VF-32 as part of CVW-3 aboard *Harry S. Truman*.

Both squadrons had discovered during their time in the NAG that the 2,000lb (907kg) GBU-31 JDAM was simply too large in terms of blast footprint for safe use in an urban environment when supporting coalition troops fighting the insurgency. The laser-guided 500lb (226kg) GBU-12 thus became the only "small" weapon left open to the Tomcat force, and with question marks hanging over the performance of the LGB (a number had dropped inexplicably short in Iraq during OIF I and in the fight against the insurgency that followed), the F-14 squadrons appeared to be out of the bomb-dropping business.

Having seen how restricted both VF-103 and VF-32 had been when it came time to putting smaller "warheads on foreheads" in Iraq in 2004–05, VF-213's then CO, Cdr Brian Kocher, was determined not to allow his unit to become a bit-part player during the aircraft's final combat cruise. In early February 2005, he contacted the US Navy's F-14 Tomcat Program Manager and expressed his interest in having the D-model jet cleared to employ the GBU-38 JDAM.

The pilot of "Felix 101" (BuNo 164603) prepares to step into the front cockpit of the jet while his RIO goes over some last-minute paperwork with one of VF-31's maintenance chiefs. This aircraft was one of four squadron jets to exceed 500 flying hours on deployment (September 1, 2005 to March 11, 2006). The penultimate F-14 built, BuNo 164603 was delivered new to VF-124 in May 1992, and it subsequently served with VF-2 and VF-213, seeing combat in OEF with the latter unit in 2001. Transferred to VF-101 the following year, the aircraft became VF-31's last "Bandwagon (later 'Felix') 101" and participated in two more cruises prior to retirement. BuNo 164603 made the last flight by a US Navy Tomcat on October 4, 2006, when it was flown from Oceana to Republic Airport, in Farmingdale, New York, for eventual display as a memorial to all Northrop Grumman workers at nearby Bethpage. (Gert Kromhout)

This request saw a team of F-14 specialists from NAVAIR, assisted by VF-213, quickly analyze the level of testing required for the integration effort with bomb manufacturer Boeing. According to an article by Chuck Wagner that appeared in the NAS Patuxent River base newspaper *Tester* on October 6, 2005:

> There had not been a Tomcat test pilot team for four years when this evaluation began, nor was there a Tomcat at NAS Patuxent River, where the aircraft's weapons testing takes place. Plus, the program had only enough funding to cover the scheduled missions and final decommissioning. The way Lt Mike Doxey of the F-14 Class Desk Team explains it, "the program pulled together disappearing resources to find a solution, and quick." Pilots with Patuxent River-based test unit Air Test and Evaluation Squadron 23 (VX-23) who had Tomcat experience leapt at the opportunity to be involved in renewed weapons testing.

A GBU-12 LGB and GBU-38 JDAM are dwarfed by the sheer size of the VF-213 F-14D to which they have been attached. Receiving clearance to drop the small blast footprint JDAM on the eve of the deployment was critically important to both VF-31 and VF-213, for troops on the ground had become increasingly wary of asking for LGBs after a handful of GBU-12s had fallen short of their targets in Iraq leading up to *Theodore Roosevelt* arriving in the NAG. By the end of the cruise only VF-213 had dropped LGBs in anger, with six being expended. Both units delivered GBU-38s, however, with VF-31 dropping nine and VF-213 five. (Danny Coremans)

With VF-213 busy conducting its pre-cruise work-ups at NAS Fallon, it fell to VF-101 to supply aircraft – and aircrew refresher training – to VX-23 so that the unit could conduct two weeks of flight-testing with the GBU-38 in May. The first two sorties flown from Patuxent River checked that the JDAM remained functional on the aircraft as it performed a series of extreme maneuvers, including high-speed runs and steep dives. The second flight tested the weapon's clean separation with the successful dropping of two inert weapons, while the third sortie put the armed aircraft through a series of high rate of decent carrier landings.

In June, a mixed VF-101/213 crew hit targets with two live GBU-38s on the range at NAWS China Lake. The completion of this final trial ultimately saw the F-14D receive clearance to employ the 500lb (226kg) JDAM on August 30 – just 48 hours prior to *Theodore Roosevelt* leaving Norfolk for the NAG.

Developed by precision weapons pioneer Boeing in the mid to late 1990s, JDAM differs from other GPS weapons in that it guides completely autonomously after being released. In its original form, the weapon could not be steered or fed updated targeting data once dropped, but development of the Laser JDAM in the years after the F-14's retirement means the bomb can now also target moving objects. Unlike laser-guided or electro-optical munitions, its accuracy remains unaffected by bad weather or poor targeting solutions.

A clinically accurate weapon originally developed to strike fixed targets (which proliferated in Iraq post OIF I), JDAM is effectively a standard Mk 82 (500lb/226kg), Mk 83 (1,000lb/454kg), Mk 84 (2,000lb/907kg) or BLU-109 (penetrator) unguided bomb fitted with a GPS guidance control unit (GCU), ventral strakes (nose-mounted on the 500lb/226kg weapon and mid-body on the remaining bombs) and a tail unit that has steerable control fins.

The "baseline" JDAM is considered to be a "near precision" weapon, the bomb's GCU relying on a three-axis INS and a GPS receiver to provide its pre-planned or in-flight targeting capability. The INS is a back-up system should the GPS lose satellite reception or be jammed. With GPS guidance at its heart, JDAM can only be employed by an aircraft fitted with an onboard GPS system so that GPS-computed co-ordinates can be downloaded to the weapon for both the target itself and the weapon release point. That way the jet's onboard INS remains as accurate as possible while the weapon is acquiring a GPS signal after being released over the target. This effectively means that the aircraft has to have a MIL-STD 1760 databus and compatible pylon wiring in order to program the bomb's aim point, intended trajectory shape and impact geometry.

Achieving initial operational capability in 1997, JDAM made its frontline debut during the NATO-led bombing campaign in Serbia and Kosovo during Operation *Allied Force* in 1999. It was then progressively employed during OSW, primarily by the US Navy, until the weapon really began to capture headlines during OEF thanks to the exploits of US Navy Hornet units

Making the most of a rare break in flight operations while *Theodore Roosevelt* transits south through the Suez Canal on September 27, 2005, the personnel of VF-213 come together for a squadron photograph in front of a then spotless "Black Lion 213." (VF-213)

operating from carriers assigned to the conflict. JDAM finally made its combat debut with the Tomcat (F-14B only) in March 2002, again in OEF, and on the eve of OIF I with the F-14D.

Aside from its stunning accuracy in OEF, JDAM also proved popular with crews because it could be released in level flight from high altitude, thus allowing aircraft to stay well above any SAM or AAA threats. Depending on the height and speed of the delivery platform, JDAM can be released up to 15 miles (24km) away from its target in ideal conditions.

Following several mishaps with LGBs in Iraq post-OIF I, JDAM once again assumed the position of dominance that it had previously enjoyed during OSW as the preferred weapon for precision strikes on targets in Iraq. This was quickly proven to CVW-8 on October 18 and 19, 2005, when three GBU-38s were dropped in just 24 hours on targets near Karabilah and Al Muqdadiyah. Validating the effort put in by the Tomcat community to get clearance to drop this weapon operationally, two of the JDAM were released by aircraft from VF-31.

Above: "Felix 101" accelerates along bow catapult one during a rare unit level training mission. Such sorties were fitted into a busy flight schedule dominated by OIF IV missions. The F-14D was restricted to non-afterburner take-offs, as the thrust created by its twin F110-GE-400 turbofan engines in reheat would warp a raised jet blast deflector. (US Navy)

Left: VF-213 F-14D

Over the next four months, the GBU-38 would be ever-present on the F-14's bomb pallets. One of those to drop a JDAM in combat was VF-213 CO, Cdr Dan Cave:

The availability of the GBU-38 was a blessing for us during the cruise, as this was our principal weapon of choice while in the NAG. If the GBU-12 had been the only 500lb [226kg] bomb available to us, I don't think we would have dropped more than 2,000lb [907kg] of ordnance for the whole deployment – VF-213 ultimately expended more than 5,000lb [2,268kg] in total.

I got to drop a single GBU-38 in support of troops in contact during Operation *Steel Curtain* in early November [VF-213 expended two GBU-38s and two GBU-12s on the 6th and two GBU-38s

VF-213's "Black Lion 201" (BuNo 164341) takes on mid-mission fuel from a KC-135R over central Iraq on October 7, 2005. This aircraft dropped two GBU-12 LGBs and conducted two strafing runs (expending 334 20mm cannon rounds) during the cruise. In one of the more bizarre missions flown by an F-14 in US Navy service, the crew of the aircraft was told to bomb a cow on an island in the seasonal lake of Bahayrat-ar-Razazah with a GBU-12 exactly one week after this photograph was taken – they achieved a direct hit. The jet would expend a second GBU-12 against an insurgency target on January 2, 2006, and also conduct two strafing attacks (on December 9 and January 2 – it was the only Tomcat to strafe twice on cruise). A veteran of OIF I with VF-213, this aircraft was retired to AMARC by VF-31 in September 2006. (Lt(jg) Scott Timmester)

on the 7th]. Unusually, my pilot and I were called on to strike a TST during daylight hours – most of our bomb drops occurred at night on this cruise. The weather was pretty awful that day, and we were in the process of conducting a routine security patrol over Baghdad when we got the call to head north-west at speed to help Marines in contact with the enemy.

Once over Husaybah, in the Anbar province, we checked in with the Marine JTAC on-scene and he gave us a standard CAS nine-line brief that included the target co-ordinates. I punched these into the single GBU-38 slung beneath the jet, and with JDAM being pretty much a "no brain" weapon, we simply flew to our release point and "pickled" it off. We were bombing through solid cloud, which was more than a little unnerving, as the JTAC had given us no details about the weapon's intended target. I zeroed the clock in the jet as the bomb was released and counted down until it was about the time that the JDAM was due to be hitting the target. To our collective relief, seconds later the JTAC came on the radio and told us "Good effects. Target destroyed." We then cleared the area and continued with our patrol over Baghdad.

Carrying a GBU-38 and a GBU-12, an F-14D from VF-213 heads north into Iraq at sunset on October 23, 2006. *Theodore Roosevelt*'s senior Naval Aviator, Rear Admiral James A "Jaws" Winnefeld, Commander, Carrier Strike Group Two/*Theodore Roosevelt* Carrier Strike Group, described a typical OIF IV sortie as follows. "These were tiring missions that involved a long flight from the NAG into the heart of Iraq and two periods working with a JTAC. In addition, the flight required three trips to a USAF tanker – one on the way in, one with a tanker orbiting high over the battlefield, and one during the long flight home. These seven-hour-plus flights culminated with a landing back on the ship." (Lt(jg) Scott Timmester)

Aside from five GBU-38s, VF-213 also expended six GBU-12s. There is a little bit of talent and skill involved in accurately dropping an LGB, as opposed to a GBU-38. With the latter weapon, if you can type using a computer keyboard you can drop a JDAM – it really is that simple.

VF-31's Lt Dan Komar was also fulsome in his praise for the 500lb (226kg) JDAM:

> The GBU-38 proved to be a crucial weapon in our arsenal while in the NAG due to the focus on operations in the urban environment in Iraq. Collateral damage estimates usually dictated whether you got to employ ordnance or not, and having a GPS-guided munition that had a small blast footprint allowed the Tomcat community to keep its foot in the door when it came to neutralizing the insurgency threat. The GBU-38 could hit a target in an urban environment with clinical precision, and its small warhead meant that both casualties and the loss of nearby infrastructure was kept to a bare minimum.
>
> Being only a 500lb [226kg] weapon, it also had very little impact on our maximum trap weight when we came back to the boat with unexpended ordnance. We almost always returned with our bombs – typically a single GBU-38 and a GBU-12, weighing 1,000lb [454kg] in total – still aboard, and the weight of this ordnance had to be offset by a reduced fuel load in order to allow the jet to reach a safe landing weight.

Perhaps the ultimate endorsement for the GBU-38 came from *Theodore Roosevelt*'s senior Naval Aviator, Rear Admiral James A Winnefeld Jr, Commander, Carrier Strike Group Two/*Theodore Roosevelt* Carrier Strike Group. A career fighter pilot with fleet time in the F-14 stretching back to 1981, he noted, "Clearing the Tomcat to use the GBU-38 on its final deployment was a very wise and courageous decision on Vice Admiral James Zortman's part, the Commander, Naval Air Forces realizing that it was worth spending precious funds to give an outgoing platform an added operational capability that has proven its worth in combat on this cruise. Having the flexibility to use either the GBU-38 JDAM or an GBU-12 LGB proved crucial on several occasions following our arrival in-theater."

Unlike many officers of flag rank who led carrier groups in the NAG, Rear Admiral Winnefeld had made a concerted effort to undertake regular combat missions over Iraq during his time in-theater, as he explained:

> I usually tried to fly once or twice a week if I could, conducting both training missions over the NAG and patrols in Iraq. I was the senior Naval Aviator flying in-theater, and my USAF equivalent was Maj Gen

VF-31

Its tanks topped off after receiving mid-mission fuel, VF-31's "Felix 107" drops away from a USAF KC-135R over central Iraq and heads back out on patrol in early October 2005. This aircraft flew 160 sorties during VF-31's 2005–06 deployment, after which it was retired to the Sabre Society's Hickory Aviation Museum in North Carolina. (USAF)

> Allen Peck, who flew the F-15E as part of his job running the CAOC. Being current on the Tomcat, and having 22 examples embarked in "TR", I felt that it was very important for me to fly as often as my schedule allowed me to.
>
> Such missions really helped me to get a full understanding of what my crews were doing both in Iraq and over the NAG. Having flown these sorties for real, I had a lot more credibility when I talked to my counterparts, such as Maj Gen Peck, in-theater. Having seen Iraq from the pilot's perspective, I knew what I was talking about when I put forward ideas to the CAOC – who we worked for when over the beach – about things that we wanted to try in order to better protect our troops on the ground. I could also explain to them difficulties that we might have been encountering when it came to performing a certain mission tasking or communicating with other friendly forces in-theater.

JDAM in action

The GBU-38 initially proved its worth during CVW-8's only pre-planned strike during *Theodore Roosevelt*'s time in the NAG – an attack on an IED factory in Al Muqdadiyah, 50 miles (80km) north-east of Baghdad, on the night of October 19, 2005. According to a report that appeared in the US armed forces newspaper *Stars and Stripes* the following day:

> The abandoned slaughterhouse alongside the main road into the city center had been a problem for months, a hideaway and staging ground for insurgents planting roadside bombs, according to soldiers here. During the past year, soldiers found more than 30 bombs on the road in front of the large brick structure – and shortly before last week's elections, five bombs were set during a two-day period.
>
> "That was kind of the last straw," said Lt Col Roger Cloutier, commander of [US Army unit] Task Force 1-30 [formed around 1st Battalion, 30th Infantry Regiment] at FOB *Normandy*. After warning local officials here, the defunct slaughterhouse was reduced to a pile of rubble in a large, muddy crater after two navy F-14s flew in from the Persian Gulf and dropped two 500lb [226kg] bombs through its roof.
>
> The bombing also had a public relations element, being designed to show local residents that US and Iraqi armies are determined to disrupt insurgents. "The bottom line is we are trying to create a safe environment," said Maj Marc "Dewey" Boberg, who led the bombing operation.
>
> The mayor of Al Muqdadiyah joined the team of US soldiers on a rooftop about 600m [a third-of-one-mile] away from the slaughterhouse to watch the precision-guided bombs explode. Iraqi army trucks with megaphones drove through the city streets blaring a message that the building would be demolished because insurgents were using it. The recorded voice of the local Iraqi army colonel urged residents to stay indoors.
>
> The bombs left a crater the size of a large swimming pool – a portion of wall that remained standing was later demolished to remove all cover for insurgent activity. Soldiers had cordoned off the area surrounding the building several hours before the explosion to ensure no residents were injured.

One of the pilots involved in this mission was VF-31's Lt Justin Halligan:

> I was pretty excited to be asked to fly on that mission as a wingman, for it was only the second time that VF-31 had dropped ordnance on cruise. The strike had been planned for some time by the CAOC, but we only found out that the mission was to be given to VF-31 the night before. The whole event had been scripted long before it was given to us to perform, right down to how we were to make our attack runs

Rear Admiral James A Winnefeld (seen here in 2011 as a four-star admiral while in charge of NORAD) was also a Tomcat pilot, with fleet service stretching back to 1981. His familiarity with the F-14 meant that he routinely flew operational missions over Iraq in the jet during *Theodore Roosevelt*'s commitment to OIF IV. (USAF)

VF-213/VF-31 Last Tomcat Cruise (2005–06)

The crew of "Felix 102" (BuNo 163904) wait patiently while VFA-87's F/A-18C "War Party 407" (BuNo 164628) tops off its tanks during a mid-cycle rendezvous with an Al Udeid-based KC-10A on November 8, 2005. BuNo 163904 dropped a GBU-38 on a building used by insurgents in the vicinity of Husaybah during the course of this mission, VF-213 having also expended ordnance in the same town in the previous 48 hours as CVW-8 was committed to Operation *Steel Curtain* – an offensive that ran from November 5–22, 2005 aimed at restoring security along the Iraq–Syria border. (VF-31)

on the target and the co-ordinates we were to dial into our JDAM in order to achieve destruction of the IED factory.

We talked to guys on the ground as we approached Al Muqdadiyah, and they had been notified of our intentions well before we launched. With so much advanced warning, the soldiers had been able to clear residents from the surrounding area. This meant that there was no chance the local population would suffer casualties due to collateral damage inflicted on buildings adjacent to the target.

The original plan had called for us to take out two sides of the building, "pickling" all four GBU-38s at the same time. One bomb from each jet would hit each side of the structure, but the ground commander changed the plan en route to the target. He told us that they would take one JDAM and see what happened, followed by a second bomb.

Upon reaching the target area, we discovered that the JTAC observing our attack was so close to the IED factory that I felt a little uncomfortable about dropping our bombs. We feared inflicting injuries, or worse, on friendly troops. Each of the four corners of the factory was being watched by separate observation posts set up on top of nearby buildings. They each provided a radio check, and then "roped" our sensors, and our eyes, onto the factory with individual infrared targeting beams visible to us through our LANTIRN and our NVGs. It was only then that I realized just how close these guys were to our target! They were outside the stipulated collateral damage areas for the 500lb [226kg] JDAM, but they still looked close to me. I would say that they were no further than half-a-mile from the factory, and this ensured that they got a "good show" when the target was hit.

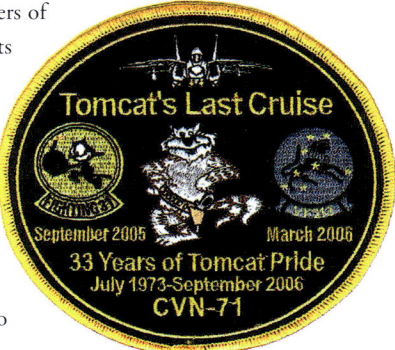

VF-31/VF-213 Tomcat's Last Cruise (2005–06)

We dropped our bombs from as high an altitude as possible in order to get some good kinetic energy behind the weapons. Having climbed up through cloud cover to reach our optimum delivery altitude, we could not see the target when we "pickled" our GBU-38s. However, by the time my bomb had almost reached its aim point, the clouds had opened up sufficiently for us to see the target though the LANTIRN pod.

My section leader had descended below the cloud after expending his ordnance, and he was circling the target area when I commenced my bombing run. I continued straight ahead after weapon separation from the jet, and seconds later the LANTIRN suddenly broke out of the cloud and highlighted the target. I immediately banked the jet up on its left wing and started circling around in order to maintain the LANTIRN picture until impact. We got to see the bomb detonate.

We delivered our weapon about six or seven minutes after our section lead had dropped his GBU-38. I had followed him in and stayed in formation as he completed his bombing run, before breaking off and circling back around so as to approach the target from the same direction. The JTAC had told us between drops that he was happy to take another bomb – meaning ours – and then he would let us know if he needed any more to complete the job.

My RIO and I used exactly the same target co-ordinates as the lead Tomcat crew, and both bombs hit pretty much the same spot in the small building. The LANTIRN footage of the first GBU-38 to hit the factory showed the bomb going into the building and the top blowing out, immediately after which the whole structure collapsed. Having seen the level of destruction wrought by this single 500lb [226kg] bomb, I thought to myself that they would never need my GBU-38, thus robbing me of possibly my only chance to drop ordnance in anger from an F-14. However, the JTAC called us in just minutes later. It took about 30 seconds for the bomb to reach the aim point, and throughout this time I was quietly saying to myself "please hit the target, please hit the target," over and over again! Then I saw the explosion in the LANTIRN and I was happy.

When it came to the delivery of JDAM as a weapon, I didn't have much to do with it. It was the RIO in the back who was pushing all the buttons to ensure that the bomb hit its chosen target.

Three well-weathered F-14Ds from VF-31 head north over the NAG towards Iraq in early January 2006. Only "Felix 102" would expend any ordnance during this deployment, dropping a single GBU-38. Each of the aircraft are equipped with LANTIRN pods (there were enough available for all 22 Tomcats embarked in *Theodore Roosevelt*), and the closest two jets are armed with GBU-38s – "Felix 112" (BuNo 163417) is carrying at least one LGB. (Erik Hildebrandt)

He had to make sure that the target co-ordinates dialed into the bomb matched those given to us pre-launch or on the target run-in by the JTAC.

Lt Halligan was one of VF-31's seven first-cruise aircrew, and he was paired up with a more seasoned RIO for this particular mission:

My squadron had only four "nugget" [first cruise] pilots and three "nugget" RIOs. Most of our aircrew had completed multiple cruises thanks to VF-31 having been on deployment a lot between 2002 and 2005. As a junior officer myself, I was the direct beneficiary of this because I was able to learn plenty about blue water ops from these multi-cruise veterans. We had so many "senior" junior guys in the squadron who had completed one or more cruises in the previous three years that I never got to fly over the beach with another "nugget" RIO in the back seat. Exposure to such experience on a regular basis gave me an accelerated learning curve when it came to mastering the F-14 in a blue water ops environment. It was, of course, fun to crew up with your flight school buddy, but it was usually during such sorties that mistakes happened and accidents occurred.

The swift progress made by VF-31's "nuggets" saw them thrown into action virtually as soon as *Theodore Roosevelt* commenced NAG operations. This was a deliberate policy devised by the senior personnel within the unit, as squadron CO (and former Intruder pilot) Cdr Rick LaBranche explained:

I was keen to schedule the more junior guys to fly the Iraq missions as often as possible in order to build up their operational experience. To this end, our squadron was the first in CVW-8 to have all of its aircrew theater qualified. We planned the flight schedule in such a way that all of our lieutenants had completed at least one mission into Iraq within a week of arriving in the NAG. I was not sure how many opportunities there would be for these guys to see action on this cruise, so I wanted those Naval Aviators and NFOs who had not previously experienced it to be the first over the beach.

VF-213/VF-31 DD's Last Time Baby...! (2005–06)

Operation *Steel Curtain*
VF-31's "nuggets" remained in the vanguard of the action when CVW-8 was committed to Operation *Steel Curtain* in early November 2005. According to press releases issued by CENTCOM at the time:

The aim of this offensive is to restore security along the Iraqi–Syrian border and destroy the al-Qaeda elements in Iraq's terror network that have been operating throughout the town of Husaybah. Approximately 2,500 Marines, sailors and soldiers with Regimental Combat Team 2 and 1,000 Iraqi Army soldiers began Operation *Al Hajip Elfulathi* (*Steel Curtain*) in western Al Anbar on November 5. Terrorists have used the region's porous borders to smuggle foreign fighters, money and equipment into Iraq to be used in their ongoing attacks against the Iraqi people and coalition forces. The offensive is part of Operation *Sayaid* (*Hunter*), designed to deny al-Qaeda in Iraq the ability to operate in the Euphrates River Valley and to establish a joint permanent security presence along the Syrian border.

The combined Iraqi Army and coalition force have been clearing Husaybah and nearby Al-Qa'im house by house, discovering weapons caches, terrorist propaganda and improvised bombs. The arms, munitions, bomb-making material, artillery and mortar shells converted to homemade bombs found in these cache sites continue to validate suspicions that terrorists have used both cities as safe havens.

Above: USS *Theodore Roosevelt* Tomcat Double D's (2005–06)

Left: A former Marine sergeant and A-6 pilot who transitioned to the Tomcat in the mid-1990s following the retirement of the Intruder, Cdr Rick "Twig" LaBranche led VF-31 during its OIF IV cruise. This photograph was taken on January 16, 2006 after he had achieved the milestone of 1,000 carrier landings, LaBranche being at the controls of "Felix 100" (BuNo 164342) at the time. He was subsequently made commander of CVW-17 in 2010. (US Navy)

Lt Justin Halligan found himself in the right place at the right time on November 8 when he received the call to provide live fire support for troops in contact with the enemy in Husaybah:

We were getting mid-cycle gas while on patrol just west of Baghdad when we received a call from "King Pin" [the overall mission controller for Iraq who had a direct line to the general in charge of air operations in the CAOC – dubbed the "voice of God" by aviators in-theater]. The controller told my section leader that the Marine Corps needed us to prosecute a TST. It took us just 15 minutes, flying at 0.95 mach [729mph/1,173km/h], to reach the target area in north-western Iraq.

We cleared airspace deconfliction issues with the various controlling agencies in-country such as the DASC as we headed west, and with this "admin" out of the way we were able to make direct contact with our JTAC on-scene as soon as we were overhead the target area. The Marine JTAC was very easy to work with, and he appeared far more at ease handling fast jets than the US Army controllers we were used to dealing with in the eastern half of Iraq.

Upon our arrival, my section leader and I rolled in and the JTAC simply gave us the target co-ordinates, which we copied in, and cleared us "hot" to drop our JDAM! Again, we "pickled" a single GBU-38 off each jet in the section, which was all we were carrying at the time as we had a

Showing clear signs of heavy use during the high operational tempo maintained by CVW-8 on deployment, VF-31's "Felix 100" takes flight after riding *Theodore Roosevelt*'s bow catapult two in early February 2006. This aircraft accrued 384.6 flying hours during the course of 131 sorties on cruise, and it also dropped a single GBU-38 in anger. VF-2's "Bullet 106" in OIF I (during which it dropped 51 bombs and also strafed), this aircraft was stricken on September 28, 2006 and loaned to the Wings Over Miami Air Museum in Florida. (Richard Cooper)

mixed load-out of one JDAM and one GBU-12 LGB per Tomcat. The first one hit the target, and the Marine JTAC immediately responded with "Good hit. Standby for the next co-ordinates." Once my bomb had gone, the JTAC asked us if we had any more JDAM, and when we said no, he sent us on our way, as they needed to get more JDAM droppers overhead ASAP.

The Marines had taken fire while in the process of clearing out buildings in Husaybah. They had watched their insurgent adversaries run into houses in an effort to take up new firing positions, so the Marines had made the call for us to take out two different buildings in TST attacks. The houses were surrounded by numerous other dwellings, so they needed the pinpoint accuracy of JDAM in order to avoid the collateral damage issues that occasionally arise with the employment of LGBs. The LANTIRN was cued to the target, and we watched the section leader's bomb score a direct hit on the building. The JTAC responded, "That's a shack. Standby for the next target." We then got our co-ordinates and bombed with the same level of accuracy.

I personally derived more satisfaction from my second bomb drop, as we were directly helping out troops in contact with the enemy. I viewed it as my chance to protect the guys on the ground, as the effects of our bombs in Husaybah were felt immediately by the Marines who were taking direct fire as we rolled in to drop our JDAM.

VF-213 had actually been the first unit in CVW-8 to drop ordnance in support of *Steel Curtain*, expending both GBU-12s and GBU-38s in TST strikes on houses in Husaybah and Al-Qa'im on November 6 and 7. Lt Cdr Robb Soderholm was involved in the first of these missions:

I got to drop my only bombs of the cruise when my pilot and I delivered a GBU-12 and a GBU-38 against targets in the vicinity of Al-Qa'im – we also guided our wingman's LGB too when his laser broke. CVW-8 had been kept out of the fight for the first 48 hours of the offensive, but on the 6th we were called in to help while conducting a day mission. My wingman also made two extremely effective strafing runs that resulted in confirmed kills just before the insurgents got the chance to fire RPGs at our ground forces.

We took out three targets in total during the 90 minutes that we spent on station working with the Marine JTACs. All three buildings that we attacked were some distance apart, and we helped out different mobile units on each occasion that we were cleared to bomb. The Marines were effectively flushing out insurgents on a block-by-block basis, surrounding an area and then putting the vices on it by slowly moving in. We had to receive a target talk-on every time we were asked to provide support, and I had to convince my JTAC that we did indeed have the target that he wanted hit in our sights. Such cross-checking prior to dropping ordnance was crucial, as our troops were often very close to the buildings that were being targeted for destruction.

The buildings that we hit were low-rise, single story structures located in heavily built-up suburbs. The Marines were in the process of clearing out the area street by street when they had taken sniper fire from the buildings that we were instructed to bomb. The JTAC controlling our section gave

This self-portrait of "nugget" NFO Lt(jg) Scott Timmester was taken during an OIF IV mission on October 13, 2005. One of a handful of first-cruise aircrew assigned to VF-213, his photographic efforts during the unit's OIF IV deployment produced some of the finest imagery even seen of the Tomcat on operations. (Lt(jg) Scott Timmester)

us an accurate talk-on to the target, and because they had sufficient equipment at hand to give us precise co-ordinates for the GBU-38s, we were cleared to drop these weapons. We also expended a GBU-12 about 40 minutes later, using our LANTIRN pod for accurate target designation.

We had remained on-station overhead Al-Qa'im after our first attack, talking to a JTAC, while our wingman went off to receive mid-cycle gas from a nearby tanker. The Tomcat crew had just checked back in when the JTAC radioed "I need a bomb now." I already had the target acquired following previous reports from our troops working in the area, so I instructed my wingman to join up and drop his bomb upon my command.

When budding lasing for my wingman on this mission, I told him to position himself on my left side and fly form as close as he could. I then told him that I would call for him to release his LGB when my targeting symbology told me to drop, should I have had a weapon on my jet. He was essentially a "truck" in this attack, simply transporting the bomb to the right spot so that my LANTIRN could then lase it once the weapon has been "pickled."

Surveillance

Only a fraction of the 1,163 combat missions flown by VF-31 and VF-213 in the NAG resulted in aircraft dropping bombs or strafing targets. For much of the 6,876 flight hours spent either over Iraq or the waters of the NAG, crews were using their onboard sensors to provide critical surveillance for troops on the ground. One of the key pieces of equipment employed by the Tomcat squadrons only reached them in early December, some three months into their final cruise. A press release issued by the US Navy at the time described the equipment upgrade:

> In keeping with its history of being adaptable to new challenges, the Tomcat soared to a new level during its last deployment when it became the first Navy aircraft to make use of the ROVER system. This equipment allows ground forces to view what the aircraft's sensors are seeing via real-time images transmitted down to laptop computers that are configured to receive this video feed. Troops now have the ability to view their surroundings from the aircraft's point of view in real time, ROVER ultimately providing better reconnaissance and target identification, which are essential to combat air support missions in Iraq.
>
> Previously, ROVER had been used by the Air Force, and with a few modifications from personnel at NAS Oceana and Naval Air Depot Jacksonville, it became one of the last great upgrades to the Tomcat.
>
> Before ROVER capability, ground controllers had to rely on "visual talk-ons" to hunt for IEDs, track insurgents or follow suspicious vehicles. The ground controller would have a map he used to guide the pilots where they needed to go. "The ground controllers are excited because it eliminates talk-ons," said VF-213 RIO Lt(jg) Will Parish. "It gives them a lot more confidence when making decisions such as dropping bombs, because they have the same real-time bird's eye view as the aircrew."

CVW-8 made the request for ROVER, which is usually found in UAVs in-theater, in early November 2005, and within six weeks the program had moved from paper to installing the data transfer systems in the 22 Tomcats aboard *Theodore Roosevelt* at the cost of just (US)$800.00 per jet. Such an expeditious turnaround allowed several days for aircrew to train with the technology before the all-important parliamentary election in Iraq on December 15. The first ROVER mission was performed by VF-213 on December 11, and squadron CO Cdr Dan Cave made extensive use of the system during the increased operational tempo that surrounded the government election:

> ROVER was pretty much a hands-off system for the crew, the RIO simply activating the transmitting switch for the equipment in the rear cockpit and then returning to his business. You didn't have to change your tactics to accommodate the system, as it transmitted whatever picture was being recorded on the LANTIRN. We relied on the guy on the ground, who was viewing our picture on

Venting fuel in order to lighten the jet prior to landing back onboard *Theodore Roosevelt* on December 13, 2005, the pilot of "Black Lion 213" (BuNo 164602) rolls into the low-holding pattern several miles astern of the carrier. This aircraft dropped a GBU-12 and a GBU-38 into buildings housing insurgents equipped with RPGs in Al-Qa'im during a *Steel Curtain* mission on November 6, 2005. Out of bombs, its crew then expended 500 rounds of 20mm ammunition in a series of strafing passes. "Black Lion 205" (BuNo 161163) also dropped an LGB and a JDAM during this mission. Twenty-four hours later, both aircraft attacked targets in Husaybah with single GBU-38s. (Lt(jg) Scott Timmester)

> his ROVER-capable laptop, to give us directions where to slew the LANTIRN in order to improve his vision of the target. We could also zoom in or out and change polarity with our targeting pod according to his instructions.
>
> We maintained a pretty steady pace to our operations following our arrival in the NAG, flying between 16–18 sorties per day. Things ramped up during the constitutional referendum in Iraq, which was held on October 15, followed by the government elections on December 15. VF-213 specifically flew missions over the polling stations in Baghdad, Mosul and Tikrit on these dates, while the air wing also covered several other major cities.
>
> We would keep our LANTIRN pod locked on to the polling station that we were guarding while flying a constant right-hand banking turn directly overhead at medium altitude. One two-jet

VF-213's "Black Lion 211" (BuNo 159629) circles overhead Mosul at 15,000ft on December 7, 2005, the jet being armed with a GBU-12 and a GBU-38, as well as a solitary AIM-9M and 500 rounds of 20mm ammunition for its M61A1 Vulcan cannon. Only a fraction of the 1,163 combat sorties flown by VF-31 and VF-213 in the NAG resulted in aircraft dropping bombs or strafing targets. For much of the 6,876 flying hours tallied by Tomcat crews either over Iraq or the waters of the NAG, they were using their onboard sensors to provide critical surveillance for troops on the ground. (Lt(jg) Scott Timmester)

section would cover several major polling stations during the course of a six-hour mission, keeping a weather eye out for anything that looked suspicious.

The Iraqi authorities greatly helped our cause on these dates by limiting the amount of vehicle traffic allowed into the cities. This meant that we could quickly pick up cars driving in the vicinity of the polling stations and have their occupants stopped and checked by troops patrolling the city centers. We performed this mission day-in, day-out following "TR's" arrival in the NAG in October, the air wing being instructed to look at the various areas in these big cities where ground forces expected problems to occur.

The arrival of ROVER in-theater also saw CVW-8 shift its section make-up from pairs of Tomcats patrolling Iraq to mixed formations of F-14Ds and F/A-18Cs. Lt Cdr Soderholm explained why this switch came about:

CVW-8 was keen to try and get as wide a selection of weapons as possible in the air over our troops at any one time. Therefore, a ROVER-equipped Tomcat, armed with a GBU-12 and a GBU-38, would be teamed up with an ATFLIR-equipped F/A-18C carrying a single 500lb [226kg] JDAM and an AGM-65E laser Maverick. This suite of weaponry could now be swiftly employed if needed thanks to the superior sensor mix of the LANTIRN/ROVER and ATFLIR.

VF-213/VF-31 Ho Ho Ho Baby! (2005)

ROVER helped us to better perform one of our primary missions while in the NAG – checking for suspicious activity along main roads and in urban areas. During a typical sortie in-country, we would be told to go and look for IED placements along key roads used by our convoys. We used our LANTIRN when conducting this mission, as its IR capability allowed us to pick up "hot spots" alongside the tarmac. The sensors were powerful enough to locate disturbances in the earth, which typically denoted the burial of IEDs – the insurgents' preferred way of placing explosives targeting our convoys and road patrols.

Once we had detected a hot spot, we relayed co-ordinates to the closest Stryker team on the ground, and they would go and check it out while we remained in an orbiting pattern overhead, feeding them live ROVER imagery. The latter allowed us to guide the troops directly on to the suspected IED location via our LANTIRN picture feed, and this greatly improved our mission effectiveness when it came to checking out possible roadside bombs. The troops were very impressed with our ROVER/LANTIRN combination, which led them to potential targets as small as trashcans in heavily urbanized areas.

The squadron also conducted convoy escorts, and during a number of these missions Tomcat crews detected hot spots some distance ahead and vehicles were diverted around them. The section on station would relay co-ordinates for the disturbed earth down to the ground, and when soldiers went back to these locations once the convoy had passed, they usually found IEDs. Such pick-ups meant that we had directly saved the lives of our troops, and these missions were among the most rewarding that we flew in the NAG.

If we had indeed detected an IED, we would be called on to overfly the immediate area while an Explosive Ordnance Disposal team was brought in to take the device apart. It was our job to look out for any insurgents who may have attempted to detonate the bomb while it was being defused, and to keep track of anybody seen fleeing the immediate area.

Guided by the hand signals of the yellow-shirted deck handler, the pilot of "Black Lion 211" carefully taxies his aircraft on to bow catapult two. The oldest aircraft assigned to VF-213 during its 2005–06 deployment, the jet had been delivered new to VF-24 in December 1975. The seventh F-14A to be remanufactured as a D-model, it subsequently saw combat in OIF I with VF-213 in 2003. The fighter was stricken immediately after returning to Oceana following the completion of the OIF IV cruise and put on display in Veteran's Park in Kenner, Louisiana. (Richard Cooper)

ROE

The ROE governing whether Tomcat crews could employ ordnance while in Iraq was strictly adhered to throughout the deployment. The possibility of inflicting collateral damage to military personnel, civilians and property was paramount in the minds of both the JTACs and CVW-8's Tomcat and Hornet crews. As a direct result of this concern, only 9,500lb (4,309kg) of ordnance was dropped by the air wing in four months of combat operations across Iraq. VF-31's Lt Dan Komar explained how the ROE in-theater worked:

I came close to dropping ordnance on two occasions on cruise, but both times my wingman and I were timed out on-station by our relief section while waiting for clearance to drop. There was a sequence of events that had to be strictly adhered to in order for a crew to be given approval for a bomb to be expended. Various agencies both inside Iraq and in the CAOC in Qatar had to be spoken to first, and we were relieved on station on both occasions as ground forces were trying to work through the proper channels to secure bomb release.

We had been responding to troops who had received mortar fire from insurgent positions, and my RIO and I were doing our level best to locate the weaponry through our LANTIRN while in a circling pattern immediately overhead our forces. The latter were able to pass us some grid co-ordinates for the mortar positions, and we had these dialed into both our LANTIRN and the single GBU-38s that my wingman and I were carrying on our jets. However, the clearances did not come through quickly enough for us to strike the sites.

Approval from the chain of command to hit pop-up threats could take some time to reach us after the first request went into the system, as everybody was concerned about inflicting "blue-on-blue" casualties. Being carrier-based, we were time-limited in-theater by the location of *Theodore Roosevelt* hundreds of miles south of our main operating areas of central and western Iraq.

Fighting 31 Last Show Baby!!! (2005)

Lt Komar's CO, Cdr Rick LaBranche, was philosophical about expending ordnance during the course of VF-31's myriad OIF IV missions over the beach:

Whether we dropped bombs on the insurgents or simply scared them away by our physical presence in the area, saving the lives of our troops meant we had achieved our stated aim while in-theater.

I derived great satisfaction from providing protection to our forces on the ground in Iraq. Flying such missions gave my squadron, and the air wing overall, a real purpose in life, and motivated us to maintain the increased op tempo that became the norm during the course of the deployment. You could hear in the tone of the JTACs' voices when we checked in over Iraq that they were pleased to see us back overhead.

On one occasion when we returned to the line after a port call, during our first mission back over the beach one of the JTACs came up on the radio and told us "we missed you"! Such a call was a first for me. We were visited by a group of JTACs from Mosul in early January 2006, and their comments were truly heart-warming for the squadron. They told us that they felt far safer when they had Tomcats overhead. Their remarks left me feeling very proud of my occupation, and thankful that I was flying such a capable jet in-theater.

Although I did not get to drop a bomb in anger during the cruise, I did not feel as though I had failed in my mission when I chopped out of the NAG. Far from it, in fact, as I derived plenty of mission satisfaction performing show-of-force flybys and simply being overhead the troops providing a physical presence that helped to keep the bad guys' heads down.

One of my cruise highlights was chasing down insurgents fleeing from an al-Qaeda cell meeting that had been interrupted by an Army Stryker patrol in the desert west of Baghdad. I spotted

The officers of VF-31 pose in front of "Felix 105" for an early cruise photograph in October 2005. The pilots and RIOs of the unit (augmented by the staff officers of CVW-8 and Rear Admiral Winnefeld) flew 1,595 sorties during the deployment, totaling 4,931.2 flying hours. Standing fifth from the right in the front row is squadron CO, Cdr Rick Labranche, and opposite him is his XO, Cdr Jim Howe. (VF-31)

an SUV [sports utility vehicle] fleeing from the scene at about 120mph [193km/h] along a nearby road, and it was up to my RIO and I to get the vehicle to turn around, as the Strykers could not keep up with it. I flew right past the SUV at low altitude and then banked across the road directly in front of it. The driver quickly made a 180-degree turn once he saw us looming above him!

On another mission, we received a report that one of our Strykers had hit an IED in the road and had been blown on to its back. My wingman and I subsequently provided high cover for the rescue effort that saw ambulances driven in to help take out the wounded.

VF-31 Last Tomcat Cruise (2005–06)

Very few of our sorties into Iraq were benign in nature. Virtually every time we flew over the beach we were told to go and provide surveillance in an area where insurgent activity had been reported, and we generally found something thanks to the Tomcat having the best sensors in-theater for this type of mission at that time. Sadly, we also regularly received calls asking us to provide aerial cover in the aftermath of an incident involving the insurgency, where our troops on the ground had been attacked.

The Tomcat also proved to be an excellent platform for show-of-force fly-bys, and I performed just such a pass in the aftermath of an IED explosion in the middle of a city. We dropped down to low altitude, pinned the wings back, opened the throttles and accelerated downtown. At nearly 63ft in length, a Tomcat with its wings folded back doing 600+ knots [1,111km/h] at less than 3,000ft tended to get people's attention very quickly. You had to be constantly on the lookout for small arms fire, shoulder-launched SAMs and RPGs when making these passes.

On one such flyby we were engaged by no fewer than seven RPGs that were fired simultaneously at us as we were circling over an urban area. They certainly got my attention, and proved once again that you cannot get complacent or too comfortable when patrolling over Iraq. As a matter of routine, I always pumped out chaff and flares on these show-of-force fly-bys, as I was not going to wait for my enemy to shoot an infrared heat-seeking missile before I reacted – I may not have seen it until it was too late.

The new year (2006) brought fewer opportunities for the F-14 squadrons to put "warheads on foreheads," although both units continued to fly thrice daily patrols over Iraq and myriad *Sea Dragon* sorties. With *Theodore Roosevelt*'s time on station drawing to a close at the end of the first week of February, a VF-31 Tomcat flown by Lts Justin Halligan and Bill Frank dropped a single GBU-38 in support of coalition troops in contact with insurgent forces near Balad on the 7th. RIO Frank recalled: "We were called on to drop, and that's what we did. It's special, and it's something I can say I did, but what's more important is the work of the sailors

Capt Bill Sizemore, CAG of CVW-8, peers through the HUD of "Black Lion 207" (BuNo 161166) as he carefully positions the jet behind a C-2A of VRC-40 Det 1 for the benefit of photographer Richard Cooper during a mission on February 2, 2006. Like Cdr "Twig" LaBranche, Capt Sizemore also joined the exclusive "Grand Club" when he made his 1,000th carrier landing in "Black Lion 213" on December 16, 2005. Flying almost 500 hours on deployment, BuNo 161166 fired 163 rounds during a strafing pass on November 17, 2005 – it did not employ any other ordnance, however. Originally delivered as an A-model in March 1981, this aircraft served with VF-11, VF-142 and VF-143 prior to being converted into an F-14D in 1993. A veteran of OIF I (with VF-213), it was retired to the Carolinas Aviation Museum in Charlotte, North Carolina, shortly after completing the OIF IV deployment. (Richard Cooper)

who made it possible. They have worked so hard during this cruise to make every Tomcat operational."

Fittingly, this solitary JDAM was dropped on what proved to be CVW-8's final mission over Iraq. The last F-14 to land back aboard the carrier was VF-213's "Black Lion 204," flown by the air wing's CAG, and veteran Naval Aviator, Capt William Sizemore and Lt(jg) Jim Cunningham. The aircraft trapped at 0035hrs on the morning of February 8, 2006, thus bringing to an end a famous chapter in Naval Aviation history.

Tomcat Tweaker

Despite the Tomcat being very much in the twilight of its career with the US Navy in 2005–06, the aircraft made a significant contribution to CVW-8's mission tally during the air wing's time in the NAG. VF-31 amassed 582 combat sorties and VF-213 581, these impressive totals indicating just how active US and coalition forces were on the ground at the time as Iraq fell prey to the growing insurgency.

"Our primary mission throughout this cruise has been to support our troops on the ground, and the bulk of our sorties have seen us doing just that," Capt Sizemore told the author in January 2006. "We have certainly flown appreciably more missions than I thought we would do, and this is primarily because we have proven to the CAOC that we are capable of handling everything they have asked us to do. We have been blessed with good weather and our aircraft have remained reliable thanks to the hard work of the 1,100+ maintainers within CVW-8."

Heading up VF-31's small army of Maintenance/Material Control personnel was Lt Tom Ober, whose 210 sailors were responsible for the health and well-being of 11 jets:

I was around Tomcats for more than two decades, completing eight deployments with as many squadrons in that time, supporting combat operations during four of them. Yet I have never seen an operational tempo sustained at that level for as long as we did in CVW-8 in 2005–06. The Tomcat was the ideal aircraft for this type of scenario, as it was a solid workhorse of a jet that could perform virtually anything asked of it as long as it was maintained properly.

I have never seen Tomcats with systems working as well as the jets that we had on the ship during the "TR" cruise. That was primarily because the aircraft was doing what it did best for much of the deployment – fly a lot. And we had the best maintainers in the navy keeping them in an airworthy state. They were so dedicated to their jobs that I often had to order them to go to bed. They would work well beyond their shifts in order to keep the aircraft in an "up" state on the flightdeck. This

"Felix 103" (BuNo 164350) has the antenna for its AN/APG-71 radar worked on in preparation for the jet's next mission. This aircraft was VF-31's high-time aircraft on deployment, logging 589.2 flying hours. It was second only to VF-213's "Black Lion 204" (BuNo 161159), which flew an astonishing 724 hours. For every flying hour that the F-14 completed over the beach in Iraq, squadron maintainers up on *Theodore Roosevelt*'s "roof" or in the vessel's cavernous hangar bay spent around 60 hours mending weeping hydraulic lines, troubleshooting temperamental avionics or just simply turning the jet around between missions. In total, the 420 maintenance personnel assigned to VF-31 and VF-213 notched up more than 720,000 man-hours fettling the F-14Ds charged to their care for the duration of the cruise. By comparison, the figure for the 20 F/A-18Cs that shared deck space with the Tomcats aboard the carrier was just over 200,000 maintenance hours for the same amount of flying time – like the two Tomcat units, VFA-15 and VFA-87 each completed around 1,000 flying hours per month while on cruise. (Gert Kromhout)

A sight to gladden the hearts of Tomcat proponents the world over. VF-31 and VF-213 get ready for a mass launch of 22 F-14Ds from *Theodore Roosevelt* off the Virginia coast to mark the end of the jet's final operational deployment on March 10, 2006. CVW-8's Tomcats racked up a total of 9,856.2 flying hours during the cruise. (US Navy)

was because they took pride in their work, they enjoyed what they were doing and they knew that they were among the best maintainers in the fleet. They were also motivated by the fact that they were taking part in the final operational cruise of the F-14.

I certainly wanted to be a part of the Tomcat legacy, and that was why I came to VF-31. Every cruise I had made in my 21 years with the navy up to 2005 had seen me working on Tomcats, so I made the decision some time ago that my final cruise would also be with the jet.

Operational to the End

Although all of VF-213's Tomcats had gone by mid-April 2006 as the unit transitioned to the F/A-18F Super Hornet, VF-31 maintained its operational capabilities with the F-14D due to CVW-8's "surge" readiness status. This meant aircrew had to retain their mission and carrier qualifications in case the air wing was tasked with a short notice deployment.

One of VF-31's highlights in its final months as the US Navy's last operator of the Tomcat was the June 7, 2006 ship-sinking exercise (known in the US Navy as a Sink-Ex) carried out 300 miles (483km) off the North Carolina coast. Conducted as part of CVW-8's "surge" status classification, VF-31 assigned five F-14Ds to the exercise, while VFA-87 committed four F/A-18Cs. In the first Sink-Ex held on the Atlantic coast in almost two years, the targets for the day were the 28-year-old 8,000-ton *Spruance* class destroyers *Stump* and *Comte de Grasse*, both of which had been decommissioned and stricken from the Atlantic Fleet some years earlier.

Four of the F-14Ds would drop ordnance as part of the Sink-Ex, with the crew of the fifth jet acting as safety observers for their squadron mates. Three Tomcats were loaded with two 1,000lb (454kg) Mk 83 GP bombs and the fourth jet carried a pair of 2,000lb (907kg) Mk 84 GP bombs. At approximately 0930hrs, the Tomcats and Hornets departed Oceana to engage the destroyers. On the way to the target area,

Right: VF-31 Tomcatters Last Cat Standing Baby! (2005–06)

Below: Four of the five VF-31 jets involved in the June 7, 2006 Sink-Ex, which took place 300 miles (483km) off the North Carolina coast, rendezvous with an Omega Aerial Refueling Services Boeing 707 for mid-mission fuel prior to attacking two decommissioned *Spruance* class destroyers. Each aircraft is armed with two Mk 84 GP bombs, these weapons being the final ordnance expended by US Navy F-14s prior to the aircraft's retirement. (Jon M Houghtaling)

the aircraft were joined by an S-3B Viking from VS-24, which was conducting a sea control mission as part of CVW-8's involvement in the Sink-Ex.

All ten jets then met up with an Omega Air Boeing 707 tanker and topped off their fuel, prior to attacking *Comte de Grasse*. *Stump* had already been sunk by then following accurate shelling by three destroyers and a cruiser from Norfolk-based Destroyer Squadron 28, as well as several bombing runs from USAF B-52Hs and B-1Bs.

Leading the VF-31 strike package during the Sink-Ex was VF-31 XO, Cdr Curt Seth, who recalled:

The ship was about 450 miles (724km) east of Oceana, and we each took a few thousand pounds of gas from the tanker to give us some cushion. Surface combatants shot at the ships first, and *Comte de Grasse* was taking on water when we arrived. As we rolled in on the target, the bow of the ship was sticking straight up, and in the process of sinking fast. The first two Tomcats were able to hit the target prior to the ship going under, and the last two jets bombed the water where the vessel had sunk, thus helping push the old destroyer down to her watery grave.

When conducting our bombing runs on the vessel, we rolled in from about 15,000ft at a 45-degree angle. Typically, when using GP bombs, the only accurate delivery method is visually through the HUD. The higher the delivery angle, the less the error introduced to bomb fall. When using the HUD, we employ the CCIP delivery method, which gives the crew an updated impact point on the HUD – a "death dot," if you will. This means that if the "pickle" button is depressed when the symbology is over the target, that is precisely where the bombs will hit. The Tomcat's ability to carry out CCIP deliveries made it a highly accurate visual bomber.

Once in our dive on the *Comte de Grasse*, the ordnance was pickled at 8,000ft so as to keep all attacking aircraft above the frag [detonation] altitude. This release height also gave us a large margin of safety for low pull-ups. With each aircraft only being able to make a single run at the target due to the rapidity at which it was sinking, the four jets were separated by just 15–20 seconds in their dives. There was no difference in the aim points for each F-14, since only the bow was sticking out of the water by the time we were called in to bomb the ship.

Our efforts to conduct a tactical attack were limited somewhat by the restricted run in lines and altitudes that we had to observe in order to deconflict with other aircraft and surface ships in the immediate vicinity of the target. However, we did simulate expendable usage [flares and chaff] as we descended into the high-threat area for infrared weapons.

Although we did not have a full ship to drop ordnance on, we did better than our F/A-18 brethren from VFA-87, as they were still on the tanker when the ship sunk. They ended up having to take their bombs back to Oceana!

I noted this mission in my logbook as two hours of boredom – transit to and from the target area, as well as tanking – and five minutes of sheer fun.

With every one of VF-31's weapons fusing on target and hitting their mark, the Tomcat had proven yet again when delivering the last live ordnance dropped by a US Navy F-14 that the jet was going out on top of its game. The unit conducted its final underway period between July 18–28, when CVW-8 and *Theodore Roosevelt* participated in a joint task force exercise

Tomcats rolled in from about 15,000ft at a 45-degree angle and some 15–20 seconds apart when attacking *Comte de Grasse*. Ordnance was "pickled" at 8,000ft, with the pilots placing their HUD "death dot" (impact point) on the upright bow of the ship. All four F-14s struck their target. (US Navy)

LAST CATFIGHT

Last Time Baby! (Tomcat Sunset) (2006)

Hornets By Mandate Tomcats By Choice (2006)

Four Tomcats from VF-31 perform a tight formation flypast with F/A-18Cs of VFA-34 and VFA-87 and F/A-18Fs from VFA-11 during the Oceana airshow on September 10, 2006. By month-end VF-31 had retired all bar one of its beloved F-14Ds and commenced conversion to the F/A-18E Super Hornet as VFA-31. (US Navy)

in the western Atlantic. At 1300hrs on the 28th Lts Chris Rattigan and Paul Dort made the final Tomcat trap (in BuNo 164346), and at 1642hrs Lt Blake Coleman and Lt Cdr Dave Lauderbaugh were the last crew to launch from a carrier (in BuNo 164341).

VF-31 stood down from "surge" status on September 13, and the unit sortied all ten of its Tomcats that day, prior to commencing the jets' retirement. The official final flight date was September 22, 2006, although the squadron continued to operate a dwindling number of aircraft from Oceana until October 4. On that date, at 1140hrs, Lt Cdrs Chris Richard and Robert Gentry took off in VF-31's last Tomcat and headed to Republic Airport, on Rhode Island, where the jet was eventually put on display at Northrop Grumman's Bethpage plant (the aircraft was subsequently moved to the Cradle of Aviation Museum in Garden City, New York, and fully restored in 2023).

Right: Return To Bethpage Back Home, Baby! (2006)

Below: Devoid of any VF-31 markings, "Felix 102" was supposed to perform the last official flight of an F-14 in US Navy service on September 22, 2006 at Oceana. However, its port engine "refused to light off," so the jet was taxied out of sight of the assembled crowd and the crew jumped into the strategically placed "Felix 107," which, of course, proudly featured VF-31's "Felix and the bomb" insignia on its twin fins! "Felix 102" was subsequently shipped to the Pacific Aviation Museum at Pearl Harbor on Ford Island, Honolulu, shortly after its retirement. (US Navy)

GLOSSARY

AAA	Anti-Aircraft Artillery
AARP	Advanced Attack Readiness Program
ACM	Air Combat Maneuvering
AGL	Above Ground Level
AMARC	Aerospace Maintenance and Regeneration Center
AMRAAM	Advanced Medium-Range Air-to-Air Missile
AOR	Area of Responsibility
ASPJ	Airborne Self-Protection Jammer
ATFLIR	(AN/ASQ-228) Advanced Targeting Forward-Looking Infrared
ATO	Air Tasking Order
AWACS	Airborne Warning and Control System
BHA	Bomb Hit Assessment
B/N	Bombardier/Navigator
CALCM	Conventional Air-Launched Cruise Missile
CAOC	Combined Air Operations Center
CAP	Combat Air Patrol
CAS	Close Air Support
CBU	Cluster Bomb Unit
CCIP	Constantly Computed Impact Point
CENTAF	Central Command Air Forces
CENTCOM	(US) Central Command
CEP	Circular Error of Probability
CINC	Commander-In-Chief
CJTFEX	Combined Joint Task Force Exercise
CNO	Chief of Naval Operations
CO	(Squadron) Commanding Officer
COMFITAEWWINGPAC	Commander, Fighter and Airborne Early Warning Wing, US Pacific Fleet
COMNAVAIRLANT	Commander, Naval Air Force, US Atlantic Fleet
COMNAVAIRPAC	Commander, Naval Air Forces, US Pacific Fleet
COMPTUEX	Composite Training Unit Exercise
CONOPS	Concept of Operations
CSAR	Combat Search and Rescue
CTF	Commander Task Force
CTGT	Computer Target
CVIC	Carrier Intelligence Centre
CVW	Carrier Air Wing
DASC	Direct Air Support Center
DCA	Defensive Counter Air
DFCS	Digital Flight Control System
DI	Digital Imaging
DMPI	Designated Mean Point of Impact
ECM	Electronic Counter Measures
ELINT	Electronic Intelligence
ESM	Electronic Support Measures
FAC(A)	Forward Air Control (Airborne)
FFARP	Fleet Fighter ACM Readiness Program
FITWINGLANT	Fighter Wing, US Atlantic Fleet
FITWINGONE	Fighter Wing One, US Atlantic Fleet
FITWINGPAC	Fighter Wing, US Pacific Fleet
FLEETEX	Fleet Exercise
FLIR	Forward-Looking Infrared
FOB	Forward Operating Base
FRS	Fleet Replacement Squadron
FTI	Fast Tactical Imagery
GBU	Guided Bomb Unit
GCU	Guidance Control Unit
GP	General Purpose (bomb)
GPS	Global Positioning System

HARM	(AGM-88) High-speed Anti-Radiation Missile	MCAS	Marine Corps Air Station
HEI	High-Explosive Incendiary	MEF	Marine Expeditionary Force
HUD	Head-Up Display	MEZ	Missile Exclusion Zone
		MFD	Multi-Function Display
IADS	Integrated Air Defense System	MIL STD	Military Standard
IED	Improvised Explosive Device	MMCAP	Multi-Mission Capability Avionics Program
IFF	Identification Friend of Foe	MMCPO	Maintenance Master Chief Petty Officer
IMU	Inertial Measurement Unit		
INS	Inertial Navigation System	NAF	Naval Air Facility
IP	Initial Point	NAG	Northern Arabian Gulf
IR	Infrared	NAS	Naval Air Station
IrAF	Iraqi Air Force	NATC	Naval Air Test Center
IRST	Infrared Search and Track	NATO	North Atlantic Treaty Organisation
ISR	Intelligence, Surveillance, Reconnaissance	NATOPS	Naval Air Training and Operating Procedures Standardization
JCAS	Joint Close Air Support	NAVAIR	Naval Air Systems Command
JDAM	Joint Direct Attack Munition	NAVSTA	Naval Station
JMEM	Joint Munitions Effectiveness Manual	NAWC	Naval Air Warfare Center
JO	Junior Officer	NAWS	Naval Air Weapons Station
Joint STARS	Joint Surveillance Target Attack Radar System	NORAD	North American Aerospace Defense Command
JTAC	Joint Terminal Attack Controller	NSAWC	Naval Strike and Air Warfare Center
JTACC	Joint Tactical Air Control Course	NSWC	Naval Strike Warfare Center
JTF-SWA	Joint Task Force-Southwest Asia	NVG	Night Vision Goggles
JTIDS	Joint Tactical Information Distribution System		
		OEF	Operation *Enduring Freedom*
LANTIRN	Low Altitude Navigation and Targeting Infrared for Night	OFP	Operational Flight Program
LAT	Low Altitude Training	OIF	Operation *Iraqi Freedom*
LDGP	Low Drag General Purpose	OPNAV	Office of the Chief of Naval Operations
LGB	Laser-Guided Bomb	OSD	Office of the Secretary of Defense
LGTR	Laser-Guided Training Round	ONW	Operation *Northern Watch*
LMAV	Laser Maverick (AGM-65E)	OSW	Operation *Southern Watch*
LST	Laser Spot Tracker		
LTS	LANTIRN Targeting System	PMTC	Pacific Missile Test Center
LVTRS	LANTIRN Video Transmitter Receiver System	PGM	Precision-Guided Munition
		Pk	Probability of kill
MANPADS	Man-Portable Air Defense System	PMA	Program Manager, Air
MAWTS	Marine Air Weapons and Tactics Squadron	PTID	Programmable Tactical Information Display

GLOSSARY

R&R	Rest and Recreation	TAC D&E	Tactical Development and Evaluation
RAF	Royal Air Force	TACP	Tactical Air Control Party
RAAF	Royal Australian Air Force	TACTS	Tactical Aircrew Combat Training System
RHAW	Radar Homing and Warning	TALD	Tactical Air-Launched Decoy
RIO	Radar Intercept Officer	TAMPS	Tactical Automated Mission Planning System
ROVER	Remotely Operated Video Enhanced Receiver	TARPS	Tactical Airborne Reconnaissance Pod System
RoE	Rules of Engagement	TARPS-CD	Tactical Airborne Reconnaissance Pod System – Completely Digital
RPG	Rocket-Propelled Grenade		
RWR	Radar Warning Receiver	TARPS-DI	Tactical Airborne Reconnaissance Pod System – Digital Imagery
		TASS	Tomcat Advanced Strike Syllabus
SA	Situational Awareness	TID	Tactical Information Display
SAFIRE	Surface-to-Air Fire	TLAM	Tomahawk Land Attack Missile
SAM	Surface-to-Air Missile	TOO	Target Of Opportunity
SAPHEI	Semi-Armor-Piercing High-Explosive Incendiary	ToT	Time-on-Target
SCAR	Strike Co-ordination And Reconnaissance	TRAM	Target Recognition and Attack Multi-Sensor
SEAD	Suppression of Enemy Air Defenses	TST	Time-Sensitive Target
SES	Self-Escort Strike	TYCOMS	Type Commanders
SFAM	Strike Familiarization		
SFARP	Strike Fighter Advanced Readiness Program	UAV	Unmanned Aerial Vehicle
SFTI	Strike Fighter Tactics Instructor	UN	United Nations
SFWSL	Strike Fighter Weapons School, US Atlantic Fleet		
SFWT	Strike Fighter Weapons and Tactics	VBIED	Vehicle-Borne Improvised Explosive Device
SIGINT	Signals Intelligence	VFX	Naval Fighter Experimental
SLAM	Standoff Land Attack Missile		
SLAM-ER	Standoff Land Attack Missile-Expanded Response	WARP	Wing Aerial Refueling Pod
SLATS	Strike Leader Attack Training Syllabus	WBB	Whitney, Bradley & Brown
SME	Subject Matter Expert	WESTPAC	Western Pacific Ocean
SOF	Special Operations Forces	WMD	Weapons of Mass Destruction
SPINS	Special Instructions	WIP	Weapon Impact Point
SUV	Sports Utility Vehicle	WTI	Weapons and Tactics Instructor
SWATSLANT	Strike Weapons and Tactics School, US Atlantic Fleet		
		XCAS	On-Call exercise Close Air Support
T3	Tomcat Tactical Targeting	XO	(Squadron) Executive Officer
TACAIR	Tactical Air		

BIBLIOGRAPHY

Books/PDFs
Baranek, Dave "Bio", *Tomcat RIO* (Skyhorse Publishing, 2020)

Brown, David F, *Tomcat Alley – A Photographic Roll Call of the Grumman F-14 Tomcat* (Schiffer Military, 1998)

Brown, David F, *Legends of Warfare Aviation – F-14 Tomcat* (Schiffer Military, 2019)

Coremans, Danny, *Uncovering the Grumman F-14A/B/D Tomcat* (DACO Publications, 2006)

Crutch, Mike, *CVW – US Navy Carrier Air Wing Aircraft 1975–2015 Volume One F-4 Phantom II, F-8 Crusader, F-14 Tomcat* (Michael Crutch/A9Aviation, 2022 (pdf sixth edition)

Crutch, Mike, *CVW – US Navy Carrier Air Wing Aircraft 1975–2015 Volume Two A-3 Skywarrior, A-4 Skyhawk, A-5 Vigilante, A-6 Intruder, EA-6B Prowler, A-7 Corsair II, AV-8 Harrier* (Michael Crutch/A9Aviation, 2022 [pdf first edition])

Elward, Brad, *TOPGUN – The Legacy* (Schiffer Military, 2021)

Gillcrist, Rear Admiral Paul T, *Tomcat! The Grumman F-14 Story* (Schiffer Military, 1994)

Hildebrandt, Erik, *Anytime, Baby! – Hail and Farewell to the US Navy F-14 Tomcat* (Cleared Hot Media, 2006)

Holmes, Tony, *Osprey Combat Aircraft 52 – US Navy F-14 Tomcat Units of Operation Iraqi Freedom* (Osprey Publishing, 2005)

Holmes, Tony, *Osprey Combat Aircraft 70 – US Navy F-14 Tomcat Units of Operation Enduring Freedom* (Osprey Publishing, 2008)

Kinzey, Bert, *Colors & Markings of US Navy F-14 Tomcats Part 1; Atlantic Coast Squadrons* (Detail & Scale Aviation Publications, 2015)

Kinzey, Bert, *Colors & Markings of US Navy F-14 Tomcats Part 2; Pacific Coast Squadrons* (Detail & Scale Aviation Publications, 2018)

Lake, Jon (editor), *Grumman F-14 Tomcat – Shipborne Superfighter* (Aerospace Publishing, 1998)

Morse, Stan (general editor), *Gulf Air War Debrief* (Aerospace Publishing, 1991)

Parsons, Dave, Hall, George and Lawson, Bob, *Grumman F-14 Tomcat – Bye, Bye, Baby... !* (Zenith Press, 2006)

Romano, Angelo (with contributions by Robert L Lawson), *US Navy Squadron Histories No. 301 – Black Knights Rule! (BKR) – A Pictorial History of VBF-718/VF-68A/VF-837/VF-154/VFA-154 1946-2013* (Ginter Books, 2014)

Romano, Angelo, *US Navy Squadron Histories No. 306HB – World Class Diamondbacks – A Pictorial History of Strike Fighter Squadron 102 (VFA-102)* (Ginter Books, 2020)

Wilcox, Robert K, *Black Aces High – The story of a modern fighter squadron at war* (Thomas Dunne Books/St. Martin's Press, 2002)

Periodicals
The Hook (various issues from 1990 through to 2006)

International Air Power Review Volume 3 – "Northrop Grumman F-14" by Jon Lake (AIRtime Publishing, 2002)

F-14 Bombcat – The US Navy's Ultimate Precision Bomber (Key Publishing, 2015)

F-14 Tomcat – The Real Star of TOP GUN (Key Publishing, 2020)

Official Publications
VF-2, VF-11 and VF-41 command histories (various years)

Commander, Fighter Wing One command histories 1990–2001

Commander, Fighter and Airborne Early Warning Wing, US Pacific Fleet command history 1994

OIF medal citations

Author Interviews/Combat Accounts
Aircrew, maintainers and intelligence personnel from VF-2, VF-11, VF-14, VF-21, VF-24, VF-31, VF-32, VF-41, VF-84, VF-101, VF-102, VF-103, VF-124, VF-143, VF-154, VF-211 and VF-213, CVW-2, CVW-5, CVW-8 and CVW-14, NSAWC, TOPGUN, SWATSLANT and SFWSL were interviewed or provided combat accounts for inclusion in this book. Thank you one and all.